COMPUTER SIMULATION OF SHAPED CHARGE PROBLEMS

COMPUTER SIMULATION OF SHAPED CHARGE PROBLEMS

WEN HO LEE

formerly Los Alamos National Laboratory, USA

World Scientific

NEW JERSEY • LONDON • SINGAPORE • BEIJING • SHANGHAI • HONG KONG • TAIPEI • CHENNAI

Published by

World Scientific Publishing Co. Pte. Ltd.

5 Toh Tuck Link, Singapore 596224

USA office: 27 Warren Street, Suite 401-402, Hackensack, NJ 07601

UK office: 57 Shelton Street, Covent Garden, London WC2H 9HE

British Library Cataloguing-in-Publication Data
A catalogue record for this book is available from the British Library.

COMPUTER SIMULATION OF SHAPED CHARGE PROBLEMS

ISBN-13 978-981-256-623-2
ISBN-10 981-256-623-6

Typeset by Stallion Press
Email: enquiries@stallionpress.com

In memory of my mother Ko Shen and my brother Lee Wun Yee

CONTENTS

PREFACE

Computer simulation of material flow with strong shocks became possible after von Neumann and Richtmyer introduced the artificial viscosity for calculating the hydrodynamic shocks in 1950. This book describes the numerical methods for solving the shaped charge problems which involve the high pressure compression of solid material by high explosives, the location of the artificial voscosity applied to the Lagrangian coordinate zones, the Particle-in-Cell method for defining the material interfaces in the Eulerian coordinate system, and the equation of state and constitutive relationships pertinent to the material properties.

In 2D Lagrangian code the treatment of the sliding line between different materials is crucial, while in 2D Eulerian code the material interface tracking and handling of the flux terms are most important. In this book, these problems are solved by some accurate methods. The 3D Eulerian code is used for computing the theater missile defense (TMD) problems.

For shaped charges and explosive formed projectiles, one needs good equation of state and material constitutive models to obtain the optimal design. Usually, one starts with a device (e.g. shaped charge) which has been well-tuned with experimental data; then, a small variation (i.e. less than 5%) is applied to the device configurations to produce a new device. Meanwhile, all parameters associated with the EOS and constitutive models are fixed. This is how the devices are designed in this book.

During the collapsing process of the shaped charge liner, the liner particle may be ejected into the material which is attached to the forward side of the liner, assuming the high explosive is located on the backward side of the liner. A good two-phase flow model is necessary to calculate the mixture of the liner with the material located in the forward section.

Eleven shaped charges are presented in Chapter 12 with their important characteristics pertinent to the optimal design for oil well perforation applications.

I should like to express my special gratitude to F. H. Harlow and P. P. Whalen for their advice and suggestions during my 23 years at Los Alamos

National Laboratory. E. J. Caramana's 2D Lagrangian code was used by G. N. Lee to finish part of the work reported in Chapter 5. B. A. Kashiwa helped me to finish the work described in Chapter 6. R. W. Lyczkowski's contribution to Chapter 7 is appreciated. The 2D PIC code documented in Chapter 3 was developed with the help of J. W. Painter. The 2D Eulerian code used in Chapter 12 was originally developed by D. Youngs and further developed by S. P. Clancy.

Wen Ho Lee
Albuquerque, New Mexico
September 2005

CHAPTER 1

INTRODUCTION

Today, most scientific researchers are challenged with the pace of the dynamic software and hardware industries. For example, computer scientists are using the C^{++} programming language for object-oriented programming while the large scientific codes are still maintained in Fortran 77 programming language. It is very difficult for a good physicist to write a complex code that solves the pertinent partial differential equations using a modern language like C^{++}. However, the gap between the computer scientist and the physicist can be narrowed down by introducing a better numerical method. This is the hope and the purpose of this book.

For multi-material problems, Lagrangian method is the most accurate tool for tracking the material interfaces, assuming that the material deformation is minimal. As soon as the deformation becomes large, the computational mesh starts to form a long thin zone or slender zone that will terminate the calculation due to an unaffordable small time step.

For the last 35 years, the scientists working on Lagrangian codes spend most of their time trying to fix these mesh tangling problems.

Due to its fixed grid mesh, Eulerian method can handle large deformation. However, the accuracy of calculating the mass, momentum, and energy flux across the zone boundary can be challenging. The other difficulty associated with the Eulerian method is the tracking of the material interface that tends to be smeared and fuzzy. Therefore, most of the code developers using Eulerian equations are battling with the flux or the advection and the material interfacial problems.

The main idea of Arbitrary Lagrangian and Eulerian (ALE) method is to create a code that preserves the advantages of both Lagrangian and Eulerian methods. At the same time, the ALE method aims to avoid the weaknesses of the Lagrangian and Eulerian methods. However, no matter how clever you are, mother nature seems to always work against you. Frequently, the ALE method will end up taking the weaknesses of both the Lagrangian and Eulerian methods. Using the ALE method without falling into the pitfalls is a delicate art.

Free Lagrangian method was first brought into attention at Lawrence Livermore National Laboratory in early 1970. A few years later, two-dimensional free Lagrangian codes were developed and used by many researchers throughout the world. This method becomes very popular during 1985–1990. After 1990, the research work on the free Lagrangian method decreased drastically due to the difficulty in calculating the particles when they are moving too closely. Today, most of the researchers are working on "Smooth Particle Hydrodynamic" that is a subset of the free Lagrangian method.

Free Lagrangian method uses nodes but no grid to represent the fluid and solid material. The advantage is its fast computation that consumes less CPU. The problem of this method arises when the material is squeezed spherically from the outside towards the center. Since too many nodes are confined in a small volume, the calculation starts to lose its accuracy. Also at this moment, the nearest neighborhood searching becomes very confusing.

The Riemann solver, or its family member, such as the Godunov or the Total Variation Diminishing (TVD) methods, is the best numerical scheme for solving one dimensional shock problems. The essence of this method lies in its dividing the original shock problem into two regions using the shock front as the connecting boundary for these two regions which have no discontinuity inside its won domain. The Riemann solver has been used in both Lagrangian and Eulerian coordinates and the results from these two approaches are all excellent. Many researchers are trying to extend this method into multi-dimensional shock wave problems with the real material equation of state.

For shaped charge problems, most of the wave codes use programmed burn for computing the chemical released energy from high explosive. The programmed burn model, although not as accurate as the reactive burn model, is simple and robust. If one would modify the detonation speed while considering the burning front curvature's effect, then the programmed burn model will become more accurate for the insensitive explosive. In principle, the reactive burn model is more desirable due to its accuracy. However, in the large code calculation, it becomes impractical for the reactive burn approach because it requires an enormous large core memory and CPU.

The performance of the shaped charge jet is mostly dependent on the grade of the liner material and the energy of the high explosive. The pressure wave interactions between the burned explosive gas and the neighboring material, i.e. liner or casing, are so complicated that it is very difficult for the laboratory test to identify these material characteristics. By using the two- or three-dimensional Eulerian hydro codes, it is possible to design a good shaped charge while specifying the thickness and geometrical shape of the liner and casing quantitatively. The grain size and the uniform density in the high explosive are also important in the formation of the shaped charge jet.

Explosive formed projectile (EFP) is made up of casing, detonator, high explosive and liner. When the projectile hits the target, the EFP speed usually exceeds 2 km/s and the mechanical power can be huge due to its speed and mass. During 1970s, the EFP technology advanced dramatically through the use of 2D and 3D Lagrangian hydro codes. Today, designers use Lagrangian or finite element code exclusively to obtain flared EFP, finned EFP, and long-rod EFP by varying the thickness and the geometrical shape of the liner. A thicker casing may increase the explosive impulse and increase the liner speed as a result. However, there exists an optimal thickness of the casing for any special EFP.

Penetration always involves a penetrator and a target. For example, a tungsten rod penetrates an aluminum plate. Usually, the penetrator will have a relatively high speed as compared with the target. In the first stage of the penetration process, the kinetic energy of the penetrator becomes heat energy which is shared by the penetrator and the target at the contact interface. As the temperature rises, the material of both the penetrator and the target at the interface start to melt and, eventually, a crater forms inside the target.

In Appendix A, the Lagrangian method described in Chapter 2 is extended to include the radiation transport which is solved by using the Variable Eddington (VE) approximation. The VE approximation uses both the Rosseland and Planck mean opacities and is capable of describing the radiation transport in optically thin and thick systems. Energy exchange between the material and the radiation field due to Compton scattering and the attendant spectrum changes are also taken into account.

The numerical method for calculating the thermonuclear burn of deuterium-tritium sphere is described in Appendix B. In this appendix, the couplings between the photon and the electron, and between the electron and the ion, are discussed in a Lagrangian coordinate system which is described in Chapter 2. The logical procedure and the calculation steps for solving the non-homogeneous and non-equilibrium transport equations are also discussed.

CHAPTER 2

LAGRANGIAN METHOD

Notations

c	speed of light (3×10^{10} cm/s)
E	radiation energy density (jerks/cm^3)
E^m	material internal energy per unit volume (jerks/cm^3)
$\dot{e}^{ij}$	strain rate deviator tensor (1/shake)
$\vec{F}$	radiation flux vector (jerks/cm^2/shake)
G	shear modulus of elasticity (jerks/cm^3)
$I(\vec{r}, v, \vec{\Omega}, t)$	specific intensity of the radiation field defined as the rate of energy flow per unit frequency and solid angle across a unit area oriented normal to the direction of propagation at point $\vec{r}$, frequency ν, in the direction $\vec{\Omega}$, at time t (jerks/cm^2)
J	volume Jacobian (cm^3)
j	Jacobian of the transformation between (R, Z) and (k, l) (cm^2)
K	Lagrangian coordinate used in figures and difference equation
k	Lagrangian coordinate
L	Lagrangian coordinate used in figures and difference equation
l	Lagrangian coordinate
M	mass constant (g)
P^m	material pressure (jerks/cm^3)
$(P^r)^{ij}$	radiation energy tensor (jerks/cm^3)
Q	artificial viscosity (jerks/cm^3)
Q_A	artificial viscosity, either ${}_1Q_A$ or ${}_3Q_A$ as defined by Eqs. (2.69) and (2.70)
Q_B	artificial viscosity, either ${}_2Q_B$ or ${}_4Q_B$ as defined by Eqs. (2.71) and (2.72)
$\tilde{R}$	normal vector to $\vec{r}$, i.e. $\tilde{R} = (Z, -R)$
$\hat{R}$	$\hat{R} = R$ for cylindrical coordinate and $\hat{R} = 1$ for Cartesian coordinate

$\vec{r}$	position vector (R, Z)
S^{ij}	stress deviator tensor (jerks/cm^3)
$\dot{S}^{ij}$	stress rate deviator tensor (jerks/cm^3/shake)
$S^{RR}, S^{ZZ}, S^{RZ}, S^{\theta\theta}$	stress deviator components (jerks/cm^3)
t	time (shake)
u	absolute value of the velocity vector, i.e. $u = \lvert\vec{u}\rvert$ (cm/shake)
$\vec{u}$	velocity vector (U, V) (cm/shake)
W	energy source (jerks/g $\cdot$ shake)
$\dot{W}$	rate of energy source (jerks/g $\cdot$ shake^2)
Y^0	yield stress in simple tension (jerks/cm^3)

Greek Letters

γ	ratio of the specific heat, i.e. $\gamma = C_P/C_V$
δ^{RR}, δ^{ZZ}	correction terms for the rigid body rotation (jerks/cm^3/shake)
ε	specific internal energy (jerks)
η	weighting function defined in Ref. [2.1]
ξ	weighting function defined in Ref. [2.1]
ρ	density (g/cm^3)
τ	specific volume (cm^3/g)
ϕ	azimuthal angle (radian)
$\vec{\Omega}$	unit vector in the direction of the photon transport

Subscripts

0	initial value
R	derivative with respect to R coordinate
t	derivative with respect to time
Z	derivative with respect to Z coordinate

Superscripts

n	time at n time-step, i.e. $t^n = t_0 + n \cdot \Delta t$

2.1 Introduction

Lagrangian method is most suitable for small deformation problems. However, it can solve large deformation problems if the code has the rezone and the sliding line capabilities. The governing equations of the mass, the momentum, and the energy may be discretized and solved by finite difference or finite element methods. In this book, only finite difference Lagrangian method will be discussed.

Two-dimensional Lagrangian methods are still used extensively in solving multi-material problems with strong shock. In this chapter, we use Schulz's two-dimensional radiation hydrodynamic code as a framework [2.1]. Three new features are implemented into the basic code, namely, material strength model, artificial viscosity in the direction of local acceleration, and sliding interfaces. Finite difference equations for deviatoric stresses and strains, shear modulus and yield strengths are written compatibly with Schulz's special formulations. Material properties (i.e. shear modulus and yield strength) are dependent on pressure, compression and equivalent plastic strain. Modifications to the artificial viscosity are made so that the acceleration from the artificial viscosity will project onto the unit vector in the direction of local acceleration. This projection eliminates 'false heating' and helps maintain stable computational grids. Also, it will obtain a very sharp shock front location through this method.

In many two-dimensional Lagrangian radiation hydrodynamic calculations, shear along material boundaries is a serious problem. Since the mesh is tied to the material, this shear will cause distortions in the mesh that make the calculation very difficult. What is needed to handle this situation is a method of allowing material flow along interfaces. A number of methods have been implemented to take care of this problem. One of the most successful of these methods for hydrodynamic calculations has been to turn the interface into two lines in the mesh that move semi-independently. However, this method makes it necessary to use some approximations at the interface for computing special physics; for example, the radiation hydrodynamic coupling problems. The approach described in this chapter adopts a semi-Eulerian calculation that allows material flow through a Lagrangian mesh along an interface. It is not necessary to have any special mesh, and all calculations, other than the hydrodynamics, are unchanged. The basic idea is to consider each point along the slip line temporarily as a double point. One would then calculate the motion of each point separately and then pull the points back together using the automatic rezone method [2.2]. This would allow the materials to shear along the interface.

2.2 Definition of Variable and Notation

In this section, we describe the definitions of variable and notation which will be used for the governing equations such as the conservations of mass, momentum, and energy.

A typical zone of the present computation is shown in Fig. 2.1. On this mesh, we then have two types of variables, zone variables, defined at zone centers and point variables, defined at mesh points. The zone variables are defined so that the mass of zone k, ℓ appears as $M_{k-\frac{1}{2},\ell-\frac{1}{2}}$ and similarly for the other zone-centered quantities. We will occasionally, for brevity, use k, ℓ instead of $k - \frac{1}{2}, \ell - \frac{1}{2}$, and $k - 1, \ell$ instead of $k - \frac{3}{2}, \ell$, etc, in Fig. 2.1, the following definitions hold:

$R(k, \ell, t) =$ Eulerian coordinate (Cartesian or cylindrical) in cm.
$Z(k, \ell, t) =$ Eulerian coordinate (always Cartesian) in cm.
$\vec{R} =$ Vector (R, Z).
$k =$ Lagrangian coordinate.
$\ell =$ Lagrangian coordinate.
$j =$ Jacobian of the transformation between (R, Z) and (k, ℓ) in cm^2.

Letting $R_k = \partial R/\partial k$, $R_\ell = \partial R/\partial \ell$, etc. then

$$j = R_k Z_\ell - R_\ell Z_k \tag{2.1}$$

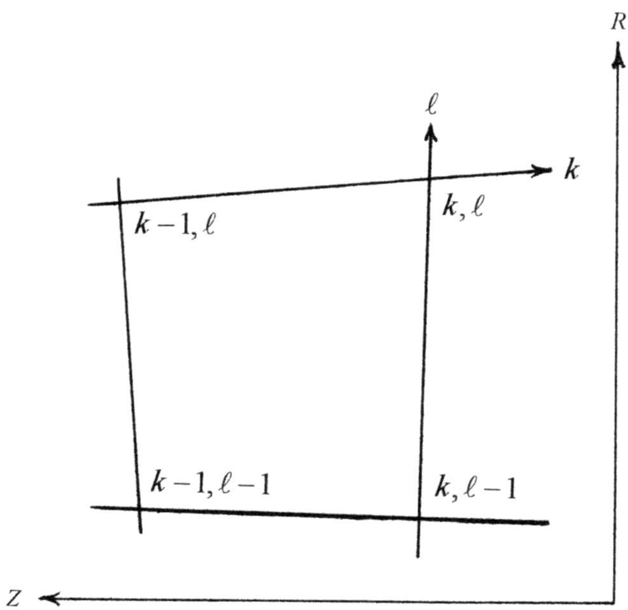

Figure 2.1. A typical computational zone for cell (k, ℓ).

Define $\hat{R} = R$ for cylindrical coordinates and $\hat{R} = 1$ for Cartesian coordinate, then a volume Jacobian may be defined as

$$J = \hat{R}j\,. \tag{2.2}$$

We now want to obtain the relations which take us from Eulerian space derivatives to their corresponding Lagrangian counterparts:

$$\frac{\partial}{\partial R} = \frac{\partial k}{\partial R}\frac{\partial}{\partial k} + \frac{\partial \ell}{\partial R}\frac{\partial}{\partial \ell} \tag{2.3}$$

$$\frac{\partial}{\partial Z} = \frac{\partial k}{\partial Z}\frac{\partial}{\partial k} + \frac{\partial \ell}{\partial Z}\frac{\partial}{\partial \ell} \tag{2.4}$$

Expressions are required which relate $\partial k/\partial R \ldots \partial \ell/\partial Z$ to $R_k \ldots Z_\ell$. For arbitrary g, we have

$$\frac{\partial g}{\partial k} = R_k\frac{\partial g}{\partial R} + Z_k\frac{\partial g}{\partial Z}\,, \tag{2.5}$$

$$\frac{\partial g}{\partial \ell} = R_\ell\frac{\partial g}{\partial R} + Z_\ell\frac{\partial g}{\partial Z}\,. \tag{2.6}$$

Letting $g = k$, we can solve for $\partial k/\partial R$ and $\partial k/\partial Z$, and letting $g = \ell$ provides $\partial \ell/\partial R$ and $\partial \ell/\partial Z$. The result is

$$\frac{\partial k}{\partial R} = \frac{Z_\ell}{j}\,, \tag{2.7}$$

$$\frac{\partial \ell}{\partial R} = \frac{Z_k}{j}\,, \tag{2.8}$$

$$\frac{\partial k}{\partial Z} = \frac{R_\ell}{j}\,, \tag{2.9}$$

and

$$\frac{\partial \ell}{\partial Z} = \frac{R_k}{j}\,, \tag{2.10}$$

which gives

$$\frac{\partial}{\partial R} = \frac{Z_\ell}{j}\frac{\partial}{\partial k} - \frac{Z_k}{j}\frac{\partial}{\partial \ell}\,, \tag{2.11}$$

and

$$\frac{\partial}{\partial Z} = -\frac{R_\ell}{j}\frac{\partial}{\partial k} + \frac{R_k}{j}\frac{\partial}{\partial \ell}\,. \tag{2.12}$$

Now define a vector $\overleftrightarrow{R}$ which lags $\vec{R}$ by 90° as the normal vector to $\vec{R}$, thus

$$\overleftrightarrow{R} = (Z, -R)\,. \tag{2.13}$$

We now define the gradient operator in Lagrange space as $\vec{\nabla} \to \vec{D}$ where

$$\vec{D} = \frac{1}{j}\left[\overleftrightarrow{R}_\ell \frac{\partial}{\partial k} - \overleftrightarrow{R}_k \frac{\partial}{\partial \ell}\right] \tag{2.14}$$

$$= \frac{1}{j}\left[\frac{\partial}{\partial k}\left(\overleftrightarrow{R}_\ell \ldots\right) + \frac{\partial}{\partial \ell}\left(\overleftrightarrow{R}_k \ldots\right)\right]. \tag{2.15}$$

Hence, for arbitrary function f and vector $\vec{f}$, we have

$$\vec{\nabla} f = \vec{D} f \tag{2.16}$$

and

$$\vec{\nabla} \cdot \vec{f} = \frac{1}{R}\vec{D} \cdot (\hat{R}\vec{f})\,. \tag{2.17}$$

Lagrange time derivatives, i.e. partial derivatives with respect to time with k and ℓ fixed, are written as

$$u(k, \ell, t) = \frac{\partial R}{\partial t} = \dot{R} = R_t = R \text{ velocity in cm/shake}\,,$$

and

$$v(k, \ell, t) = \frac{\partial Z}{\partial t} = \dot{Z} = Z_t = Z \text{ velocity in cm/shake}\,,$$

with $\vec{u}$ = the vector (u, v) and 1 shake $= 10^{-8}$ seconds.

In addition to the variables already defined, i.e. R, Z, u, v, which are point variables and j, a zone variable, we have the following definitions where (Z) implies a zone variable, (P) a point variable, and (I) an interface variable:

M = mass (g) (Z)

$\tau = 1/\rho$ = specific volume (cm^3/g) (Z)

P^m = material pressure (jerks/cm^3) (Z)

where 1 jerk = 10^{16} ergs

E^m = material energy per unit volume (jerks/cm^3) (Z)

ε^m = material specific internal energy (jerks/cm^3) (Z)

c_v = material specific heat at constant specific volume (jerks/g/keV) (Z)

c = speed of light = 300 (cm/shake) (Z)

$I(\vec{r},\nu,\vec{\Omega},t)$ = specific intensity of the radiation field defined as the rate of energy flow per unit frequency and solid angle across a unit area oriented normal to the direction of propagation at point $\vec{r}$, frequency ν, in direction $\vec{\Omega}$, at time t. (jerks/cm^2) (Z)

$$E = \frac{1}{c}\int_0^\infty d\nu \int_{4\pi} I(\nu,\vec{\Omega})d\vec{\Omega}$$
$$= \text{ radiation energy density (jerks/cm}^3\text{) } (Z) \tag{2.18}$$

$$\vec{F} = \int_0^\infty d\nu \int_{4\pi} \vec{\Omega} I(\nu,\vec{\Omega})d\vec{\Omega}$$
$$= \text{ radiative flux (jerks/cm}^2\text{/shake) } (I) \tag{2.19}$$

$$P^r = \frac{1}{c}\int_0^\infty d\nu \int_{4\pi} \vec{\Omega}\vec{\Omega} I(\nu,\vec{\Omega})d\vec{\Omega}$$
$$= \text{radiation energy tensor (jerks/cm}^3\text{) } (Z) \tag{2.20}$$

$$\mu_a(\nu) = \text{ absorption coefficient at frequency } \nu \text{ (cm}^{-1}\text{) } (Z)$$
$$\mu_a'(\nu) = \mu_a(\nu)(1-e^{-h\nu/kT}) = \text{ absorption coefficient corrected for induced emission (cm}^{-1}\text{) } (Z) \tag{2.21}$$

$\bar{\mu}^P$ = Planck absorption coefficient (cm^{-1}) (Z)

$K^P = \bar{\mu}^P\tau$ = Planck opacity (cm^2/g) (Z)

$\bar{\mu}^R$ = Rosseland absorption coefficient (cm^{-1}) (Z, I)

$K^R = \bar{\mu}^R\tau$ = Rosseland opacity (cm^2/g) (Z, I)

T = material temperature (keV) (Z)

$\varphi = aT^4$ = radiative source function (jerks/cm^3) (Z)

a = radiation constant (0.0137 jerks/cm^3/keV^4) (Z)

Difference over the k variable will be represented by Δ, i.e.

$$\Delta = \frac{\partial}{\partial k} \quad \text{and}$$
$$\Delta \vec{R}^n_{k+\frac{1}{2},\ell} = \vec{R}^n_{k+1,\ell} - \vec{R}^n_{k,\ell}\,. \tag{2.22}$$

Similarly, we use δ for $\frac{\partial}{\partial \ell}$ and

$$\delta \vec{R}^n_{k,\ell+\frac{1}{2}} = \vec{R}^n_{k,\ell+1} - \vec{R}^n_{k,\ell}\,. \tag{2.23}$$

A typical zone in k, ℓ space is shown in Fig. 2.2 with pressure defined at cell center. There are four artificial viscosities ${}_1q, {}_2q, {}_3q$ and ${}_4q$ defined along the four sides. We further break up each zone into four triangles as shown in Fig. 2.3. Each of these triangles has an associated material internal energy, E_1, E_2, E_3, E_4, and likewise for the pressure and artificial viscosity.

There are two weighted functions which must be defined for use in the momentum equation. They are obtained as follows. Define

$$\omega_{k+\frac{1}{2},\ell} = \left[\frac{1}{4}\left(\Delta \vec{R}^{+}_{k+\frac{1}{2},\ell+\frac{1}{2}} + \Delta \vec{R}_{k+\frac{1}{2},\ell-\frac{1}{2}}\right)^2\right]^{\frac{1}{2}}, \tag{2.24}$$

$$\omega_{k,\ell+\frac{1}{2}} = \left[\frac{1}{4}\left(\delta \vec{R}^n_{k+\frac{1}{2},\ell+\frac{1}{2}} + \delta \vec{R}^n_{k-\frac{1}{2},\ell+\frac{1}{2}}\right)^2\right]^{\frac{1}{2}}, \tag{2.25}$$

$$\xi'_{k,\ell} = \frac{2\omega_{k-\frac{1}{2},\ell}}{\omega_{k+\frac{1}{2},\ell} + \omega_{k-\frac{1}{2},\ell}}, \tag{2.26}$$

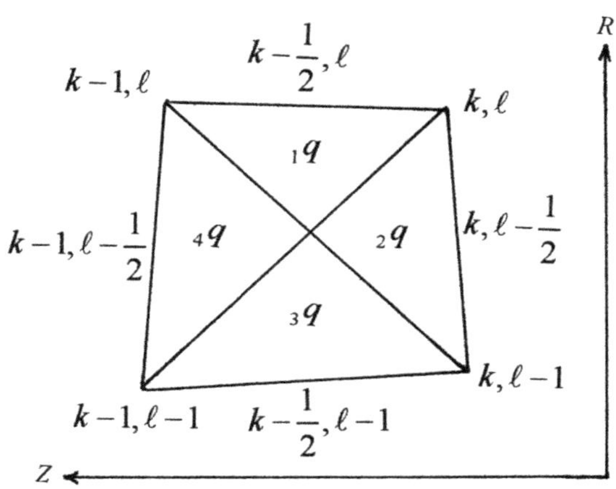

Figure 2.2. A typical quadrilateral zone in (k, ℓ) space.

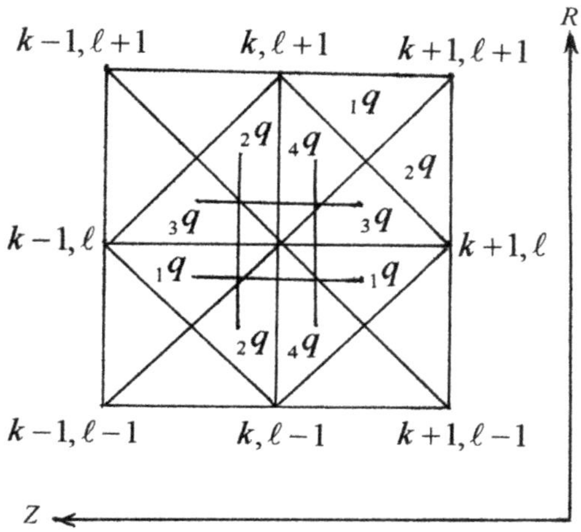

Figure 2.3. Centering of q, p, E with triangular subzone, only ${}_1q, {}_2q, {}_3q$ and ${}_4q$ are shown.

and

$$\eta'_{k,\ell} = \frac{2\omega_{k,\ell-\frac{1}{2}}}{\omega_{k,\ell+\frac{1}{2}} + \omega_{k,\ell-\frac{1}{2}}} . \tag{2.27}$$

The weighted functions are then given by

$$\xi_{k,\ell} = \max\left\{0.6, \min\left[\xi'_{k,\ell}, 1.4\right]\right\} , \tag{2.28}$$

and

$$\eta_{k,\ell} = \max\left\{0.6, \min\left[\eta'_{k,\ell}, 1.4\right]\right\} . \tag{2.29}$$

Essentially these weighted functions are used to weigh the various pressure difference and artificial viscosity difference terms in the momentum equation based on the proximity of the point of their formation to the point k, ℓ. Actually, it's more of an inverse weight.

2.3 The Governing Equation

The basic Lagrangian equation for the radiation hydrodynamics are given by Pomraning [2.3]. With the addition of the stress deviatoric tensors, we have the position equation

$$\frac{\partial \vec{r}}{\partial t} = \vec{u}\,, \tag{2.30}$$

the continuity equation

$$\rho j = M\,, \tag{2.31}$$

the momentum equation

$$\rho \frac{D}{Dt}\left(\vec{u} + \frac{\vec{F}}{\rho c^2}\right) + \nabla(P^m + Q) + \nabla \cdot \left[(P^r)^{ij} - \frac{\vec{u}\vec{F}}{c^2}\right] - S^{ij}_{,j} = 0\,, \tag{2.32}$$

and the energy equation

$$\frac{D}{Dt}\left[\frac{u^2}{2} + \frac{1}{\rho}(E^m + E)\right] + \nabla \cdot [\vec{F} + (P^m + Q - E)\vec{u}] - (S^{ij}\vec{u})_{,j} = W\,. \tag{2.33}$$

Note that $(P^r)^{ij}$ in Eq. (2.32) and S^{ij} in both Eqs. (2.32) and (2.33) are tensors with $,j$ representing the derivative.

The governing Eqs. (2.30)–(2.33) are split into two major blocks and solved by finite difference method. The first block solves the hydrodynamics and the elastic-plastic flow to obtain new values of pressure, energy, density, and velocity as described in Section 2.5. The radiation transport equation with the energy source is solved by radiation diffusion method in the second block that is described in [2.4]. At the end of the radiation calculations, we update the pressure and the energy due to their changes in the radiation process. Therefore, in the first block, the conservation equations (2.30)–(2.33) in cylindrical coordinates without the external work become

$$\frac{\partial \vec{r}}{\partial t} = \vec{u}\,, \tag{2.34}$$

$$\rho j = M\,, \tag{2.35}$$

$$\rho \frac{D\vec{u}}{Dt} + \nabla(P^m + Q) - S^{ij}_{,j} = 0\,, \tag{2.36}$$

and

$$\frac{D}{Dt}\left(\frac{u^2}{2} + \frac{E^m}{\rho}\right) + \nabla \cdot (P^m + Q)\vec{u} - (S^{ij}\vec{u})_{,j} = 0\,. \tag{2.37}$$

2.4 Equation of State

A stress-supporting medium deforming under a wide range of stress exhibits a variety of different physical characteristics. Depending on its retention of elastic character, the flow may be elastic or plastic. When melting, it behaves like a fluid; therefore, in each regime of flow, we need appropriate models or equations of state and criteria defining the transition from one regime to another; for example, when material flows elastically, Hook's law for the deviator stresses can be written as

$$S_t^{RR} = 2G\left(U_R + \frac{\rho_t}{3\rho}\right) + \delta^{RR}, \tag{2.38}$$

$$S_t^{ZZ} = 2G\left(V_Z + \frac{\rho_t}{3\rho}\right) + \delta^{ZZ}, \tag{2.39}$$

$$S_t^{RZ} = G(U_Z + V_R), \tag{2.40}$$

and

$$S_t^{\theta\theta} + S_t^{RR} + S_t^{ZZ} = 0. \tag{2.41}$$

For plastic flows, Prandtl and Reuss considered both plastic and elastic strain simultaneously and arrived at the following flow equation

$$\dot{S}^{ij} = 2G\dot{e}^{ij} - \frac{G\dot{W}}{\frac{1}{3}(Y^0)^2}S^{ij}, \tag{2.42}$$

where $\dot{W}\left(\Sigma_{ij}S^{ij}\dot{e}^{ij}\right)$ is the plastic work/unit volume, G and Y are obtained from Steinberg-Guinan model which is described in Chapter 8 and Ref. [8.4].

2.5 Calculation Procedures and Finite Differences

Before we express Eqs. (2.34)–(2.37) in finite difference form, it is useful for the reader to get familiar with the notations and some derivatives as given in Section 2.2. The scalar form of the conservation equations (2.34)–(2.37) can be written as the position equations

$$\frac{\partial R}{\partial t} = U, \tag{2.43}$$

$$\frac{\partial Z}{\partial t} = V, \tag{2.44}$$

the continuity equation

$$\rho\left(\frac{\partial R}{\partial k}\frac{\partial Z}{\partial \ell}-\frac{\partial R}{\partial \ell}\frac{\partial Z}{\partial k}\right)=\rho_0\,, \tag{2.45}$$

the momentum equations

$$\begin{aligned}\frac{\partial U}{\partial t}=&-\frac{1}{\rho j}\left(\frac{\partial Z}{\partial \ell}\frac{\partial P}{\partial k}-\frac{\partial Z}{\partial k}\frac{\partial P}{\partial \ell}+\frac{\partial Z}{\partial k}\frac{\partial S^{RR}}{\partial \ell}-\frac{\partial Z}{\partial \ell}\frac{\partial S^{RR}}{\partial k}\right)\\&+\frac{1}{\rho j}\left(\frac{\partial R}{\partial k}\frac{\partial S^{RZ}}{\partial \ell}-\frac{\partial R}{\partial \ell}\frac{\partial S^{RZ}}{\partial k}\right)\\&+\frac{1}{\rho R}(S^{RR}-S^{\theta\theta})-\frac{1}{M}\frac{\partial}{\partial k}\left[RQ_A\frac{\frac{\partial Z}{\partial \ell}\frac{\partial U}{\partial k}-\frac{\partial R}{\partial \ell}\frac{\partial V}{\partial k}}{\left(\frac{\partial U}{\partial k}\right)^2+\left(\frac{\partial V}{\partial k}\right)^2}\left(\frac{\partial U}{\partial k}\right)\right]\\&+\frac{1}{M}\frac{\partial}{\partial \ell}\left[RQ_B\frac{\frac{\partial Z}{\partial k}\frac{\partial U}{\partial \ell}-\frac{\partial R}{\partial k}\frac{\partial V}{\partial \ell}}{\left(\frac{\partial U}{\partial \ell}\right)^2+\left(\frac{\partial V}{\partial \ell}\right)^2}\left(\frac{\partial U}{\partial \ell}\right)\right],\end{aligned} \tag{2.46}$$

$$\begin{aligned}\frac{\partial V}{\partial t}=&-\frac{1}{\rho j}\left(\frac{\partial R}{\partial \ell}\frac{\partial P}{\partial k}-\frac{\partial R}{\partial k}\frac{\partial P}{\partial \ell}+\frac{\partial R}{\partial k}\frac{\partial S^{ZZ}}{\partial \ell}-\frac{\partial R}{\partial \ell}\frac{\partial S^{ZZ}}{\partial k}\right)\\&+\frac{1}{\rho j}\left(\frac{\partial Z}{\partial \ell}\frac{\partial S^{RZ}}{\partial k}-\frac{\partial Z}{\partial k}\frac{\partial S^{RZ}}{\partial \ell}\right)+\frac{S^{RZ}}{\rho R}\\&+\frac{1}{M}\frac{\partial}{\partial k}\left[RQ_A\frac{\frac{\partial Z}{\partial \ell}\frac{\partial U}{\partial k}-\frac{\partial R}{\partial \ell}\frac{\partial V}{\partial k}}{\left(\frac{\partial U}{\partial k}\right)^2+\left(\frac{\partial V}{\partial k}\right)^2}\left(\frac{\partial V}{\partial k}\right)\right]\\&-\frac{1}{M}\frac{\partial}{\partial \ell}\left[RQ_B\frac{\frac{\partial Z}{\partial k}\frac{\partial U}{\partial \ell}-\frac{\partial R}{\partial k}\frac{\partial V}{\partial \ell}}{\left(\frac{\partial U}{\partial \ell}\right)^2+\left(\frac{\partial V}{\partial \ell}\right)^2}\left(\frac{\partial V}{\partial \ell}\right)\right],\end{aligned} \tag{2.47}$$

and the energy equation

$$\begin{aligned}\rho\frac{\partial \varepsilon}{\partial t}=&-P\nabla\cdot\vec{u}-\frac{Q_A}{j}\left(\frac{\partial Z}{\partial \ell}\frac{\partial U}{\partial k}-\frac{\partial R}{\partial \ell}\frac{\partial V}{\partial k}\right)-\frac{Q_B}{j}\left(-\frac{\partial Z}{\partial k}\frac{\partial U}{\partial \ell}+\frac{\partial R}{\partial k}\frac{\partial V}{\partial \ell}\right)\\&+\frac{S^{RR}}{j}\left(\frac{\partial Z}{\partial \ell}\frac{\partial U}{\partial K}-\frac{\partial Z}{\partial k}\frac{\partial U}{\partial \ell}\right)+S^{\theta\theta}\left(\frac{U}{R}\right)+\frac{S^{ZZ}}{j}\left(\frac{\partial R}{\partial k}\frac{\partial V}{\partial \ell}-\frac{\partial R}{\partial \ell}\frac{\partial V}{\partial k}\right)\\&+\frac{S^{RZ}}{j}\left(\frac{\partial R}{\partial k}\frac{\partial U}{\partial \ell}-\frac{\partial R}{\partial \ell}\frac{\partial U}{\partial k}+\frac{\partial Z}{\partial \ell}\frac{\partial V}{\partial k}-\frac{\partial Z}{\partial k}\frac{\partial V}{\partial \ell}\right).\end{aligned} \tag{2.48}$$

Since all of the dependent variables appearing in Eqs. (2.38)–(2.48) are functions of time and space, we discretize them using different space locations as well as time level. In Fig. 2.4, the logical meshes (i.e. K and L) are

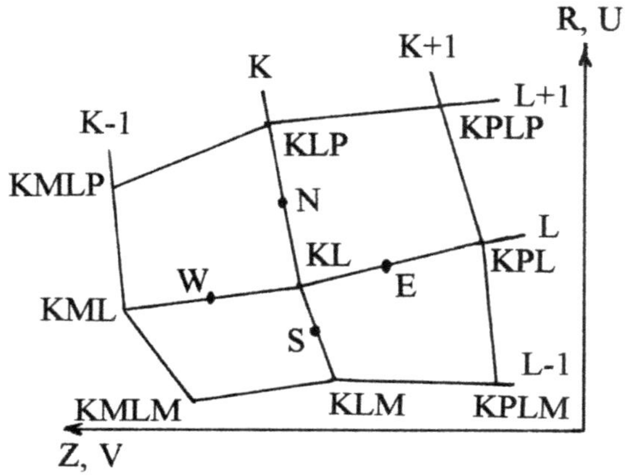

Figure 2.4. The logical meshes and index as plotted on the physical plane (i.e. R, Z coordinates).

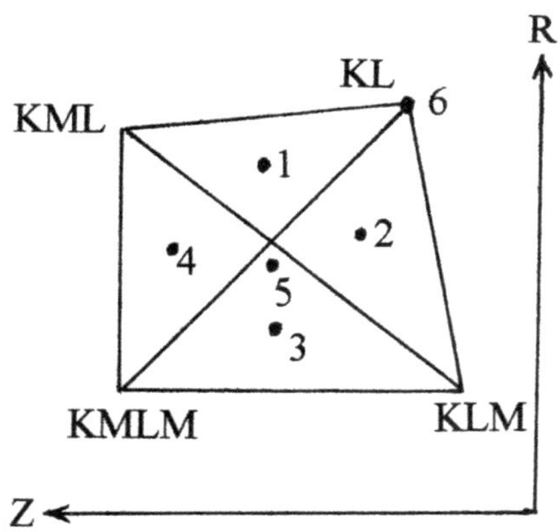

Figure 2.5. M, j, ρ, ε and S^{ij} are defined at cell center point 5, while U, V, R, Z are defined at vertex point 6, pressure and artificial viscosity are at points 1, 2, 3 and 4.

plotted on the physical plane (i.e. R and Z coordinates). The cell centered quantities, i.e. point 5, are M, j, ρ, ε and S^{ij} with the vertex quantities, i.e. point 6, U, V, R and Z as shown in Fig. 2.5. Each quadrilateral zone is further divided into four triangular subzones with pressure P and artificial viscosity Q defined at the centroid of each triangle, i.e. point 1, 2, 3 and 4. Therefore, for each quadrilateral cell, we have four P's and Q's.

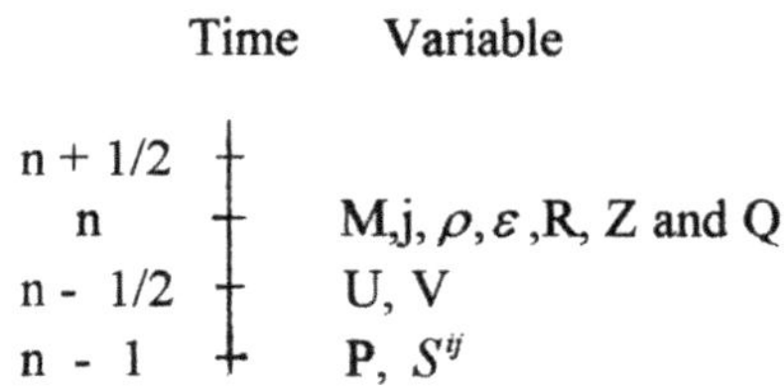

Figure 2.6. Variables are defined at different time level.

For time level n, we define $M, j, \rho, \varepsilon, R, Z$ and Q at t^n and U and V at $t^{n-1/2}$. But pressure P and deviator stress S^{ij} are defined at t^{n-1} as shown in Fig. 2.6. In summary, we have:
cell-centered quantities

$$M_{k-\frac{1}{2},\ell-\frac{1}{2}}, j^n_{k-\frac{1}{2},\ell-\frac{1}{2}}, \rho^n_{k-\frac{1}{2},\ell-\frac{1}{2}}, (S^{ij})^{n-1}_{k-\frac{1}{2},\ell-\frac{1}{2}}, G^{n-1}_{k-\frac{1}{2},\ell-\frac{1}{2}}, Y^{n-1}_{k-\frac{1}{2},\ell-\frac{1}{2}}$$
$$\text{and} \quad \varepsilon^{n-1}_{k-\frac{1}{2},\ell-\frac{1}{2}},$$

cell vertex quantities

$$U^{n-1/2}_{k,\ell}, V^{n-1/2}_{k,\ell}, R^n_{k,\ell}, \quad \text{and} \quad Z^n_{k,\ell},$$

and triangular subzone quantities

$${}_1P^{n-1}_{k,\ell}, {}_2P^{n-1}_{k,\ell}, {}_3P^{n-1}_{k,\ell}, {}_4P^{n-1}_{k,\ell}, {}_1Q^n_{k,\ell}, {}_2Q^n_{k,\ell}, {}_3Q^n_{k,\ell} \quad \text{and} \quad {}_4Q^n_{k,\ell}.$$

Based on the known dependent variables at time $t = t^n$, the following steps are taken logically:

Step 1. Calculate U_R, U_Z, V_R and V_Z.
Step 2. If the sliding interface treatment is required, then update $U^{n+1/2}$, $V^{n+1/2}$, $R^{n+1/2}$ and $Z^{n+1/2}$ due to the effects of sliding.
Step 3. Calculate ρ^{n+1} and ρ_t, from R^n, Z^n and M.
Step 4. Calculate G and Y from Chapter 8.
Step 5. Calculate $(S^{ij})^n$ from Eqs. (2.38) through (2.41) for pure elastic deformation.
Step 6. Check von Mises yield condition.
Step 7. Calculate $(S^{ij})^n$ with Prandtl-Reuss plastic flow term, i.e. Eq. (2.42).
Step 8. Calculate equivalent plastic strain.
Step 9. Add the rigid body rotation correction terms to the deviatoric stresses.
Step 10. Calculate the principle stresses to check the fracture conditions.

Step 11. Calculate velocities $U^{n+1/2}$ and $V^{n+1/2}$ from Eqs. (2.46) and (2.47).

Step 12. Calculate the artificial viscosity ${}_1Q^n_{k,\ell}$, ${}_2Q^n_{k,\ell}$, ${}_3Q^n_{k,\ell}$ and ${}_4Q^n_{k,\ell}$.

Step 13. Calculate energy ε^{n+1} from Eq. (2.48).

Step 14. Calculate pressure P^n from equation of state, i.e. $P = f(\rho, \varepsilon)$. In our case, it is a table lookup.

Steps 1–14 complete the one time step calculations, and we go back to Step 1 for the next time increment computations.

For finite difference, we can solve Eq. (2.38) (i.e. Step 5) by

$$(\dot{S}^{RR})_{k-\frac{1}{2},\ell-\frac{1}{2}} = 2G_{k-\frac{1}{2},\ell-\frac{1}{2}} \left[\frac{1}{j} \left(Z_\ell \frac{\partial U}{\partial k} - Z_k \frac{\partial U}{\partial \ell} \right) + \frac{\rho_t}{3\rho} \right]^n_{k-\frac{1}{2},\ell-\frac{1}{2}} \tag{2.49}$$

to obtain the deviator stress for pure elastic deformation.

Similar formula are used for computing $(\dot{S}^{ZZ})_{k-\frac{1}{2},\ell-\frac{1}{2}}$ and $(\dot{S}^{RZ})_{k-\frac{1}{2},\ell-\frac{1}{2}}$ from Eqs. (2.39) and (2.40). Once the elastic regime is finished, we can add the Prandtl-Reuss plastic flow and the correction due to rigid body rotations.

The finite difference formula for the momentum equations is more complicated. For the sake of simplicity, we like to rewrite the momentum equation in R component, i.e. Eq. (2.46) as

$$U_t = {}_1U_t + {}_2U_t \tag{2.50}$$

where

$$\begin{aligned} {}_1U_t = &-\frac{1}{M}\frac{\Delta}{\Delta k}\left(RQ_A \frac{Z_\ell U_k - R_\ell V_k}{U_k^2 + V_k^2} U_k \right) \\ &+\frac{1}{M}\frac{\Delta}{\Delta \ell}\left(RQ_B \frac{Z_k U_\ell - R_k V_\ell}{U_\ell^2 + V_\ell^2} U_\ell \right), \end{aligned} \tag{2.51}$$

and

$$\begin{aligned} {}_2U_t = &-\frac{1}{\rho j}\left(Z_\ell \frac{\Delta P}{\Delta k} - Z_k \frac{\Delta P}{\Delta \ell} + Z_k \frac{\Delta S^{RR}}{\Delta \ell} - Z_\ell \frac{\Delta S^{RR}}{\Delta k} \right) \\ &+\frac{1}{\rho j}\left(R_k \frac{\Delta S^{RZ}}{\Delta \ell} - R_\ell \frac{\Delta S^{RZ}}{\Delta k} \right) + \frac{1}{\rho R}(S^{RR} - S^{\theta\theta}). \end{aligned} \tag{2.52}$$

Since there are four Q's in one cell; namely, ${}_1Q^n_{k,\ell}$, ${}_2Q^n_{k,\ell}$, ${}_3Q^n_{k,\ell}$ and ${}_4Q^n_{k,\ell}$ as calculated in Eqs. (4.27)–(4.30) (see Fig. 2.5), we divide the right-hand side of Eq. (2.51) into four parts. The first two parts are associated

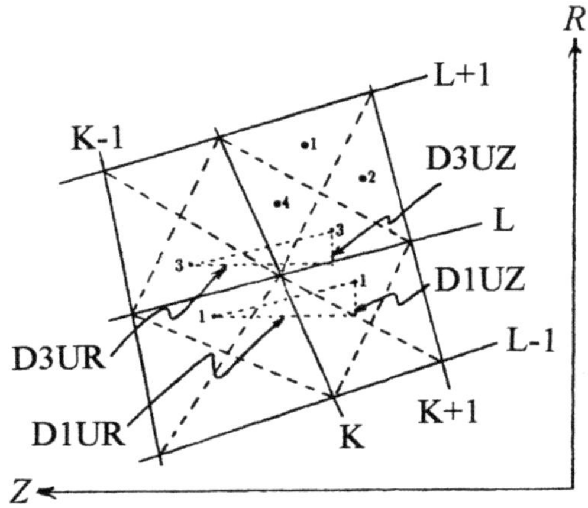

Figure 2.7. Each cell has four triangular subzones where pressure and artificial viscosity are defined. The momentum components D1UZ, D1UR, D3UZ and D3UR are also shown.

with Q_A, i.e. $Q_A = f({}_1Q, {}_3Q)$ and the last two parts are with Q_B, i.e. $Q_B = f({}_2Q, {}_4Q)$. Figure 2.7 shows four cells; each has four triangular subzones. Within each subzone, we assign one pressure and one artificial viscosity at the centroid of the triangle. When we solve the momentum equations for the vertex node (K, L), we connect point 1–1 and point 3–3 as shown in Fig. 2.7. The vertical component of line 1–1 is called *D1 UZ*, while the horizontal component is *D1 UR*. The vertical and horizontal components for line 3–3 are *D3 UZ* and *D3 UR*. The same idea is applied to point 2 and point 4 as shown in Fig. 2.8, which gives *D2 UZ, D2 UR, D4 UZ* and *D4 UR*. The finite difference expressions using the notations shown in Fig. 2.4 are

$$\begin{aligned} D1\,UZ = \frac{1}{M_S}\Bigg\{&[(Z_{KL} - Z_{KLM})(U_{KPL} - U_{KL})(R_{KL} - R_{KLM}) \\ &\times (V_{KPL} - V_{KL})] \cdot \frac{(R_1Q)_{KPL}(U_{KPL} - U_{KL})}{(U_{KPL} - U_{KL})^2 + (V_{KPL} - V_{KL})^2} \\ &+ [(R_{KL} - R_{KLM})(V_{KL} - V_{KML})(Z_{KL} - Z_{KLM})(U_{KL} - U_{KML})] \\ &\times \frac{(R_3Q)_{KLP}(U_{KL} - U_{KML})}{(U_{KL} - U_{KML})^2 + (V_{KL} - V_{KML})^2}\Bigg\}, \end{aligned} \tag{2.53}$$

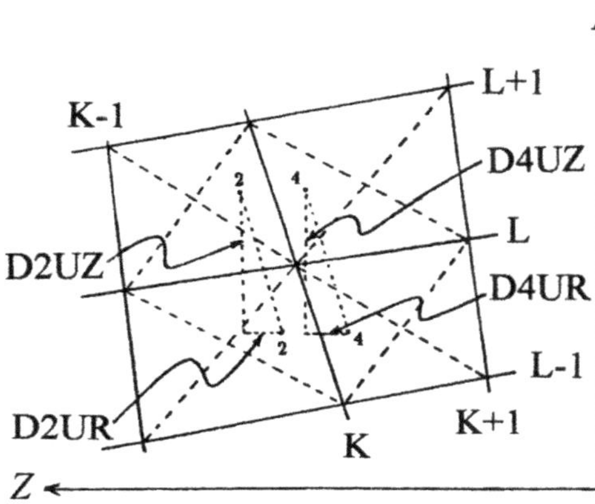

Figure 2.8. The momentum components D2UZ, D2UR, D4UZ and D4UR are shown.

$$D3UZ = \frac{1}{M_N}\Bigg\{[(Z_{KLP} - Z_{KL})(U_{KPL} - U_{KL}) - (R_{KLP} - R_{KL})$$

$$\times (V_{KPL} - V_{KL})] \cdot \frac{(R_3Q)_{KPLP}(U_{KPL} - U_{KL})}{(U_{KPL} - U_{KL})^2 + (V_{KPL} - V_{KL})^2}$$

$$+ [(R_{KLP} - R_{KL})(V_{KL} - V_{KML}) - (Z_{KLP} - Z_{KL})$$

$$\times (U_{KL} - U_{KML})] \cdot \frac{(R_3Q)_{KLP}(U_{KL} - U_{KML})}{(U_{KL} - U_{KML})^2 + (V_{KL} - V_{KML})^2}\Bigg\}, \quad (2.54)$$

$$D2UZ = \frac{1}{M_W}\Bigg\{[(Z_{KL} - Z_{KML})(U_{KLP} - U_{KL}) - (R_{KL} - R_{KML})$$

$$\times (V_{KPL} - V_{KL})] \cdot \frac{(R_2Q)_{KLP}(U_{KLP} - U_{KL})}{(U_{KLP} - U_{KL})^2 + (V_{KLP} - V_{KL})^2}$$

$$+ [(R_{KL} - R_{KML})(V_{KL} - V_{KLM}) - (Z_{KL} - Z_{KML})$$

$$\times (U_{KL} - U_{KLM})] \cdot \frac{(R_2Q)_{KL}(U_{KL} - U_{KLM})}{(U_{KL} - U_{KLM})^2 + (V_{KL} - V_{KLM})^2}\Bigg\} \quad (2.55)$$

$$D4UZ = \frac{1}{M_E}\Bigg\{[(Z_{KPL} - Z_{KL})(U_{KLP} - U_{KL}) - (R_{KPL} - R_{KL})$$

$$\times (V_{KLP} - V_{KL})] \cdot \frac{(R_4Q)_{KPLP}(U_{KLP} - U_{KL})}{(U_{KLP} - U_{KL})^2 + (V_{KLP} - V_{KL})^2}$$

$$+ [(R_{KPL} - R_{KPLM})(V_{KL} - V_{KLM}) - (Z_{KPL} - Z_{KL}) \times (U_{KL} - U_{KLM})] \cdot \frac{(R_4Q)_{KPL}(U_{KL} - U_{KLM})}{(U_{KL} - U_{KLM})^2 + (V_{KL} - V_{KLM})^2}\Bigg\}, \quad (2.56)$$

$$D1\,UR = \frac{-1}{M_S}\Bigg\{[(Z_{KL} - Z_{KLM})(U_{KPL} - U_{KL}) - (R_{KL} - R_{KLM}) \times (V_{KPL} - V_{KL})] \cdot \frac{(R_1Q)_{KPL}(V_{KPL} - V_{KL})}{(U_{KPL} - U_{KL})^2 + (V_{KPL} - V_{KL})^2} + [(R_{KL} - R_{KLM})(V_{KL} - V_{KML}) - (Z_{KL} - Z_{KLM}) \times (U_{KL} - U_{KML})] \cdot \frac{(R_1Q)_{KL}(V_{KL} - V_{KML})}{(U_{KL} - U_{KML})^2 + (V_{KL} - V_{KML})^2}\Bigg\}, \quad (2.57)$$

$$D3\,UR = \frac{-1}{M_N}\Bigg\{[(Z_{KLP} - Z_{KL})(U_{KPL} - U_{KL}) - (R_{KLP} - R_{KL}) \times (V_{KPL} - V_{KL})] \cdot \frac{(R_3Q)_{KPLP}(V_{KPL} - V_{KL})}{(U_{KPL} - U_{KL})^2 + (V_{KPL} - V_{KL})^2} + [(R_{KLP} - R_{KL})(V_{KL} - V_{KML}) - (Z_{KLP} - Z_{KL}) \times (U_{KL} - U_{KML})] \cdot \frac{(R_3Q)_{KLP}(V_{KL} - V_{KML})}{(U_{KL} - U_{KML})^2 + (V_{KL} - V_{KML})^2}\Bigg\}, \quad (2.58)$$

$$D2\,UR = \frac{-1}{M_W}\Bigg\{[(Z_{KL} - Z_{KML})(U_{KLP} - U_{KL}) - (R_{KL} - R_{KML}) \times (V_{KPL} - V_{KL})] \cdot \frac{(R_2Q)_{KLP}(V_{KLP} - V_{KL})}{(U_{KLP} - U_{KL})^2 + (V_{KLP} - V_{KL})^2} + [(R_{KL} - R_{KML})(V_{KL} - V_{KLM}) - (Z_{KL} - Z_{KML}) \times (U_{KL} - U_{KLM})] \cdot \frac{(R_2Q)_{KL}(V_{KL} - V_{KLM})}{(U_{KL} - U_{KLM})^2 + (V_{KL} - V_{KLM})^2}\Bigg\}, \quad (2.59)$$

and

$$D4\,UR = \frac{-1}{M_E}\Bigg\{[(Z_{KPL} - Z_{KL})(U_{KLP} - U_{KL}) - (R_{KPL} - R_{KL}) \times (V_{KLP} - V_{KL})] \cdot \frac{(R_4Q)_{KPLP}(V_{KLP} - V_{KL})}{(U_{KLP} - U_{KL})^2 + (V_{KLP} - V_{KL})^2} + [(R_{KPL} - R_{KPLM})(V_{KL} - V_{KLM}) - (Z_{KPL} - Z_{KL}) \times (U_{KL} - U_{KLM})] \cdot \frac{(R_4Q)_{KPL}(V_{KL} - V_{KLM})}{(U_{KL} - U_{KLM})^2 + (V_{KL} - V_{KLM})^2}\Bigg\}. \quad (2.60)$$

In Eqs. (2.53)–(2.60), we use $M_S = (\rho J)_{k,\ell-\frac{1}{2}}$, $M_N = (\rho J)_{k,\ell+\frac{1}{2}}$, $M_W = (\rho J)_{k-\frac{1}{2},\ell}$, and $M_E = (\rho J)_{k+\frac{1}{2},\ell}$. The R in the parentheses (R_1Q), $(R_2Q), \ldots$ etc. means the average of R coordinates, e.g. $(R_1Q)_{KPL} = 0.25(R_{KPL} + R_{KL} + R_{KLM} + R_{KPLM}) \cdot ({}_1Q)_{KPL}$.

Since there are four pressures in one cell, the finite difference of $\partial P/\partial k$ and $\partial P/\partial \ell$ will have similar formulations as in $(\partial Q/\partial k)$ and $(\partial Q/\partial \ell)$. For the reason of simplicity, we just write Eq. (2.52) as

$$({}_2U_t)_{K,L} = \left\{ \left(-\frac{1}{\rho j}\right) \left[Z_\ell \frac{\partial P}{\partial k} - Z_k \frac{\partial P}{\partial \ell} + Z_k S_\ell^{RR} - Z_\ell S_k^{RR} \right] \right.$$
$$\left. + \left(\frac{1}{\rho j}\right)(R_k S_\ell^{RZ} - R_\ell S_k^{RZ}) + \left(\frac{1}{\rho R}\right)(S^{RR} - S^{\theta\theta}) \right\}_{K,L}^{n} . \quad (2.61)$$

Using the Eqs. (2.53) through (2.61), we get

$$U_{K,L}^{n+1} = U_{K,L}^{n} + \Delta t\{D1\,UZ - D2\,UZ + D3\,UZ - D4\mathrm{UZ} + ({}_2U_t)_{K,L}\}\,, \quad (2.62)$$

for the new time velocity in the R direction, and

$$V_{K,L}^{n+1} = V_{K,L}^{n} + \Delta t\{D1\,UR - D2\,UR + D3\,UR - D4\,UR + ({}_2V_t)_{K,L}\}\,, \quad (2.63)$$

for that in Z direction where $({}_2V_t)_{K,L}$ is similar to Eq. (2.52).

The internal energy equation can be rewritten as

$$\frac{\partial \varepsilon}{\partial t} = -P\frac{\partial}{\partial t}\left(\frac{1}{\rho}\right) - \frac{Q_A(Z_\ell U_k - R_\ell V_k)}{\rho j} - \frac{Q_B(-Z_k U_\ell + R_k V_\ell)}{\rho j}$$
$$+ \frac{S^{RR}}{\rho j}\left(Z_\ell \frac{\Delta U}{\Delta k} - Z_k \frac{\Delta U}{\Delta \ell}\right) + \frac{S^{\theta\theta}}{\rho}\left(\frac{U}{R}\right)$$
$$+ \frac{S^{ZZ}}{\rho j}\left(R_k \frac{\Delta V}{\Delta \ell} - R_\ell \frac{\Delta V}{\Delta k}\right)$$
$$+ \frac{S^{RZ}}{\rho j}\left(R_k \frac{\Delta U}{\Delta \ell} - R_\ell \frac{\Delta U}{\Delta k} + Z_\ell \frac{\Delta V}{\Delta k} - Z_k \frac{\Delta V}{\Delta \ell}\right). \quad (2.64)$$

Let the energy source due to stresses be W, then

$$W = \left[\frac{S^{RR}}{\rho j}\left(Z_\ell \frac{\Delta U}{\Delta k} - Z_k \frac{\Delta U}{\Delta \ell}\right) + \frac{S^{\theta\theta}}{\rho}\left(\frac{U}{R}\right) + \frac{S^{ZZ}}{\rho j}\left(R_k \frac{\Delta V}{\Delta \ell} - R_\ell \frac{\Delta V}{\Delta k}\right) \right.$$
$$\left. + \frac{S^{RZ}}{\rho j}\left(R_k \frac{\Delta U}{\Delta \ell} - R_\ell \frac{\Delta U}{\Delta k} + Z_\ell \frac{\Delta V}{\Delta k} - Z_k \frac{\Delta V}{\Delta \ell}\right) \right]_{K,L}^{n} . \quad (2.65)$$

Therefore, the new time energy is

$$\varepsilon^{n+1} = \varepsilon^n - P_{K,L}^{n-1}\left(\frac{1}{\rho^{n+1}} - \frac{1}{\rho^n}\right)_{K,L} - (\Delta t)\left(\frac{1}{\rho j}\right)_{K,L}^{n+1} \times [Q_A(Z_\ell U_k - R_\ell V_k) + Q_B(-Z_k U_\ell - R_k V_\ell)]_{K,L}^n + (\Delta t)W\,. \quad (2.66)$$

With new energy ε^{n+1} and density ρ^{n+1}, one can obtain new pressure P^{n+1} by Equation of State, i.e. $P^{n+1} = f(\rho^{n+1}, \varepsilon^{n+1})$.

2.6 Sliding Interface Treatments

First, we consider the slip interface as having two additional mass points: one associated with the two zone masses above the interface and the other associated with the two zone masses below the interface. We calculate the velocity of these two assumed points separately by using a modified form of the Schulz [2.4] hydrodynamic equations.

If we refer to Schulz's Report UCRL-6776 [2.4], pp 28–29 where he gives his acceleration equations, we see that at each node he has four terms (see Fig. 2.9), each weighted by a η or ξ which are derived from the spacing. The differenced form of the momentum equation without the radiation and material strength now becomes

$$\vec{u}_{k,\ell}^{\,n+1/2} = \vec{u}_{k,\ell}^{\,n-1/2} + D\,\vec{u}_{k,\ell}^{\,n} \quad (2.67)$$

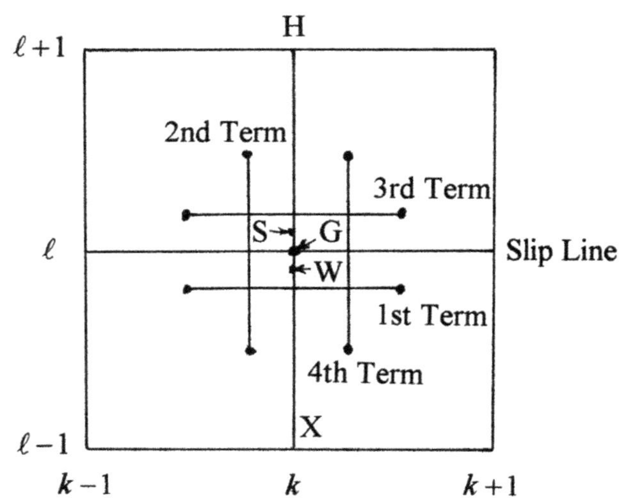

Figure 2.9. The Lagrangian zones (or logical zones) with slip line along $\ell = \ell$ line. The 1st, 2nd, 3rd and 4th terms represent the terms on the right hand side of Eq. (2.68).

where

$$D\,\vec{u}^{\,n}_{k,\ell} = Dt^n \left\{ \eta_{k,\ell} \left[\frac{\partial\vec{R}/\partial\ell}{2\rho j}\frac{\partial_1 P}{\partial k} + \frac{1}{2M}\frac{\partial}{\partial k}\left(\hat{R}\frac{\partial\vec{R}}{\partial\ell}{}_1Q_A\right)\right]^n_{k,\ell-\frac{1}{2}} \right.$$

$$- \xi_{k,\ell}\left[\frac{\partial\vec{R}/\partial k}{2\rho j}\frac{\partial_2 P}{\partial\ell} + \frac{1}{2M}\frac{\partial}{\partial\ell}\left(\hat{R}\frac{\partial\vec{R}}{\partial k}{}_2Q_B\right)\right]^n_{k-\frac{1}{2},\ell}$$

$$+ (2-\eta_{k,\ell})\left[\frac{\partial\vec{R}/\partial\ell}{2\rho j}\frac{\partial_3 P}{\partial k} + \frac{1}{2M}\frac{\partial}{\partial k}\left(\hat{R}\frac{\partial\vec{R}}{\partial\ell}{}_3Q_A\right)\right]^n_{k,\ell+\frac{1}{2}}$$

$$\left. - (2-\xi_{k,\ell})\left[\frac{\partial\vec{R}/\partial k}{2\rho j}\frac{\partial_4 P}{\partial\ell} + \frac{1}{2M}\frac{\partial}{\partial\ell}\left(\hat{R}\frac{\partial\vec{R}}{\partial k}{}_4Q_B\right)\right]^n_{k+\frac{1}{2},\ell} \right\}. \tag{2.68}$$

The artificial viscosities, ${}_1Q^n_A, {}_3Q^n_A, {}_2Q^n_B$ and ${}_4Q^n_B$, appeared in Eq. (2.68) are computed from

$$\begin{aligned} {}_1Q^n_A = \frac{1}{2}\left[\frac{3}{4}\left({}_1Q^{n-1/2}_{k-\frac{1}{2},\ell-\frac{1}{2}}\right) + \frac{1}{4}\left({}_3Q^{n-1/2}_{k-\frac{1}{2},\ell-\frac{1}{2}}\right)\right. \\ \left. + \frac{3}{4}\left({}_1Q^{n+1/2}_{k-\frac{1}{2},\ell-\frac{1}{2}}\right) + \frac{1}{4}\left({}_3Q^{n+1/2}_{k-\frac{1}{2},\ell-\frac{1}{2}}\right)\right], \end{aligned} \tag{2.69}$$

$$\begin{aligned} {}_3Q^n_A = \frac{1}{2}\left[\frac{3}{4}\left({}_3Q^{n-1/2}_{k-\frac{1}{2},\ell-\frac{1}{2}}\right) + \frac{1}{4}\left({}_1Q^{n-1/2}_{k-\frac{1}{2},\ell-\frac{1}{2}}\right)\right. \\ \left. + \frac{3}{4}\left({}_3Q^{n+1/2}_{k-\frac{1}{2},\ell-\frac{1}{2}}\right) + \frac{1}{4}\left({}_1Q^{n+1/2}_{k-\frac{1}{2},\ell-\frac{1}{2}}\right)\right], \end{aligned} \tag{2.70}$$

$$\begin{aligned} {}_2Q^n_B = \frac{1}{2}\left[\frac{3}{4}\left({}_2Q^{n-1/2}_{k-\frac{1}{2},\ell-\frac{1}{2}}\right) + \frac{1}{4}\left({}_4Q^{n-1/2}_{k-\frac{1}{2},\ell-\frac{1}{2}}\right)\right. \\ \left. + \frac{3}{4}\left({}_2Q^{n+1/2}_{k-\frac{1}{2},\ell-\frac{1}{2}}\right) + \frac{1}{4}\left({}_4Q^{n+1/2}_{k-\frac{1}{2},\ell-\frac{1}{2}}\right)\right], \end{aligned} \tag{2.71}$$

and

$$\begin{aligned} {}_4Q^n_B = \frac{1}{2}\left[\frac{3}{4}\left({}_4Q^{n-1/2}_{k-\frac{1}{2},\ell-\frac{1}{2}}\right) + \frac{1}{4}\left({}_2Q^{n-1/2}_{k-\frac{1}{2},\ell-\frac{1}{2}}\right)\right. \\ \left. + \frac{3}{4}\left({}_4Q^{n+1/2}_{k-\frac{1}{2},\ell-\frac{1}{2}}\right) + \frac{1}{4}\left({}_2Q^{n+1/2}_{k-\frac{1}{2},\ell-\frac{1}{2}}\right)\right]. \end{aligned} \tag{2.72}$$

On the right-hand sides of Eqs. (2.69)–(2.72), the artificial viscosities, ${}_1Q^n_A, {}_3Q^n_A, {}_2Q^n_B$ and ${}_4Q^n_B$, are located at the center of the triangles as indicated by 1, 2, 3 and 4 shown in Fig. 2.7. The pressure ${}_1P^n, {}_2P^n, {}_3P^n$ and ${}_4P^n$ appeared in the right-hand side of Eq. (2.68), are obtained from the similar formula of ${}_1Q^n_A, {}_3Q^n_A, {}_2Q^n_B$ and ${}_4Q^n_B$ as given by Eqs. (2.69)–(2.72).

In Fig. 2.9, let us take the line $\ell = \ell$ (for $k = k - 1$, k and $k + 1$) as the slip line. We temporarily assume that point $G(k, \ell)$ is two points, i.e. S and W (above and below ℓ). We calculate an acceleration, velocity and new positions of S and W. Then we pull the two positions back together with the rezone technique to form a single point. The mesh will be moved through the material allowing the materials on both sides of the line to move with respect to each other. We will leave the new above ℓ and below ℓ velocities as they are. After the rezone, an appropriate velocity for the single point will be formed for use in the normal code computation.

For the above ℓ calculation, we drop out the first term on the right-hand side of Eq. (2.68) altogether (see Fig. 2.10), i.e. $\eta_{k,\ell}= 0$. Since $\eta_{k,\ell}= 0$, we have the coefficient of the third term equals to 2. The coefficients of 2nd and 4th terms are unchanged. Therefore, Eq. (2.68) becomes

$$
\begin{aligned}
D\,\vec{u}^{\,n}_{k,\ell} = Dt^n \Bigg\{ & -\xi_{k,\ell}\left[\frac{\partial\vec{R}/\partial k}{2\rho j}\frac{\partial_2 P}{\partial \ell} + \frac{1}{2M}\frac{\partial}{\partial \ell}\left(\hat{R}\frac{\partial\vec{R}}{\partial k}{}_2Q_B\right)\right]^n_{k-\frac{1}{2},\ell} \\
& + 2\left[\frac{\partial\vec{R}/\partial \ell}{2\rho j}\frac{\partial_3 P}{\partial \ell} + \frac{1}{2M}\frac{\partial}{\partial k}\left(\hat{R}\frac{\partial\vec{R}}{\partial \ell}{}_3Q_A\right)\right]^n_{k,\ell+\frac{1}{2}} \\
& - (2-\xi_{k,\ell})\left[\frac{\partial\vec{R}/\partial k}{2\rho j}\frac{\partial_4 P}{\partial \ell} + \frac{1}{2M}\frac{\partial}{\partial \ell}\left(\hat{R}\frac{\partial\vec{R}}{\partial k}{}_4Q_B\right)\right]^n_{k+\frac{1}{2},\ell}\Bigg\}. \quad (2.73)
\end{aligned}
$$

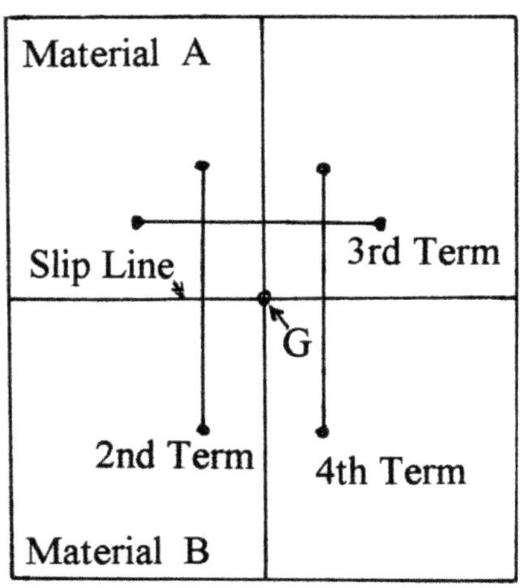

Figure 2.10. Only 2nd, 3rd, and 4th terms are used for computing the momentum equation, Eq. (2.73), for the above ℓ calculation.

When the momentum equation, i.e. Eq. (2.32), is solved by finite difference, it is split into three steps; the first step solves the hydrodynamics, the stresses, and the corrections to the artificial viscosity, i.e. Eq. (2.46); the second step solves the sliding interface, i.e. Eq. (2.68); and the third step solves the radiation diffusion.

For the below ℓ calculation, we drop out the third term on the right-hand side of Eq. (2.68) altogether, i.e. make $2-\eta_{k,\ell}=0$ or $\eta_{k,\ell}=2$; again the coefficients of 2nd and 4th terms are unchanged. Therefore, Eq. (2.68) becomes

$$\begin{aligned} D\,\vec{u}^{\,n}_{k,\ell} = Dt^n \Bigg\{ & 2\left[\frac{\partial\vec{R}/\partial\ell}{2\rho j}\frac{\partial_1 P}{\partial k} + \frac{1}{2M}\frac{\partial}{\partial k}\left(\hat{R}\frac{\partial\vec{R}}{\partial\ell}{}_1Q_A\right)\right]^n_{k,\ell-\frac{1}{2}} \\ & - \xi_{k,\ell}\left[\frac{\partial\vec{R}/\partial k}{2\rho j}\frac{\partial_2 P}{\partial\ell} + \frac{1}{2M}\frac{\partial}{\partial\ell}\left(\hat{R}\frac{\partial\vec{R}}{\partial k}{}_2Q_B\right)\right]^n_{k-\frac{1}{2},\ell} \\ & - (2-\xi_{k,\ell})\left[\frac{\partial\vec{R}/\partial k}{2\rho j}\frac{\partial_4 P}{\partial\ell} + \frac{1}{2M}\frac{\partial}{\partial\ell}\left(\hat{R}\frac{\partial\vec{R}}{\partial k}{}_4Q_B\right)\right]^n_{k+\frac{1}{2},\ell}\Bigg\}. \end{aligned} \tag{2.74}$$

The results are shown in Fig. 2.11. In calculating the momentum equations, we use term 3 as it would be used in non-slip calculations. This use

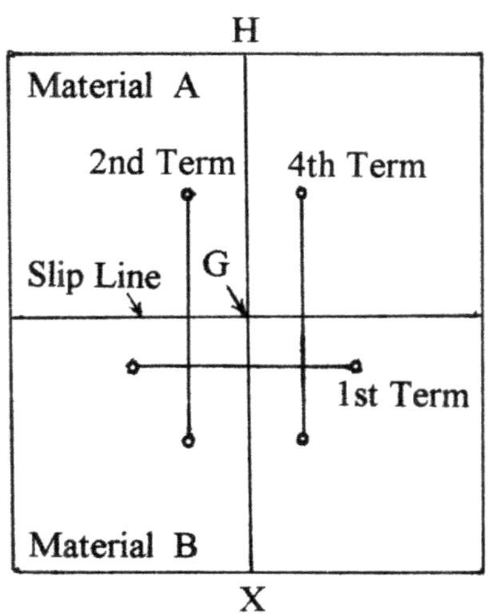

Figure 2.11. Only 1st, 2nd, and 4th terms are used for computing the momentum equation, Eq. (2.74), for the below ℓ calculation.

is important because it means that the pressure in material B affects the velocities in material A and vice versa. Therefore, the separate accelerations are not completely independent. However, we are allowing the two sets of mass to move differently, if they are so inclined, because different accelerations applied to two points with the same position will give different velocities and hence different new positions. Using the results of Eqs. (2.67) and (2.73), we can get the new velocity vector at time $n+1/2$. This new velocity vector can be written as

$$\vec{u}_{k,\ell}^{\,n+1/2} = \left(U_{k,\ell}^{n+1/2}, V_{k,\ell}^{n+1/2}\right). \tag{2.75}$$

The new location S for the above ℓ zones originally may move to a position below the ℓ-line (see Fig. 2.12). The new position of S is computed from

$$\vec{r}_{k,\ell}^{\,n+1/2} = \left(R_{k,\ell}^{n} + Dt^{n+1/2}U_{k,\ell}^{n+1/2}, Z_{k,\ell}^{n} + Dt^{n+1/2}V_{k,\ell}^{n+1/2}\right), \tag{2.76}$$

and the new location W for the below ℓ zones is obtained by using Eqs. (2.67), (2.74), (2.75) and (2.76). Figure 2.12 shows the zones along the interface after the two separate motions have been calculated. These zones may overlap, separate, etc. In the present example, as shown in Fig. 2.13, points S and W have penetrated into the other material. Now we have the two points in different positions. We now pick a point where they should coincide (such as the average position). There are a number of possible ways to pick this position.

Let M_5 be the mass of the zone $\overline{DHSE}$ and M_6 of the zone $\overline{HFPS}$, as shown in Fig. 2.13. Similarly, M_7 is the mass of the left zone below ℓ and

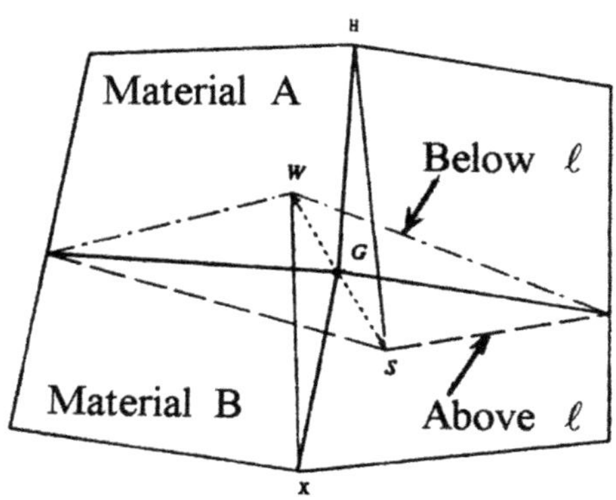

Figure 2.12. Point S is the new location of G for the above ℓ calculation via Eq. (2.73). Similarly, the new location W is from Eq. (2.74) for the below ℓ calculation.

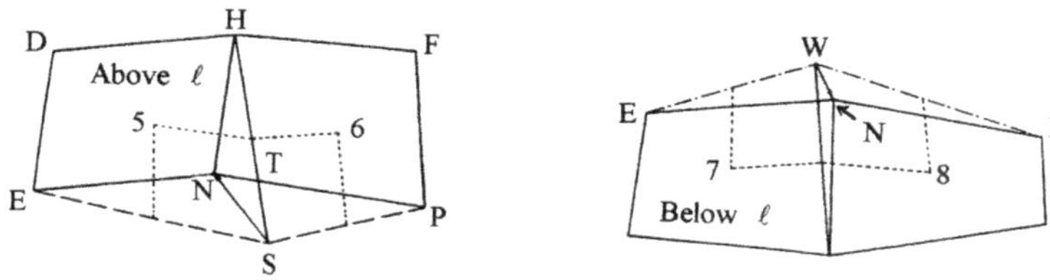

Figure 2.13. The left figure is the above ℓ zones. After a mass weighting of velocities, the coincident point N is defined. Points 5 and 6 indicate the zone center of $\overline{DHSE}$ and $\overline{HFPS}$. Similar notation are used for the below ℓ zones as described in the right figure.

M_8 is that of the right zone. To get the position of the coincident point N we use the mass-weighted average of the two slip velocities. For example, the coordinates of N are

$$R_N^{n+1} = R_{k,\ell}^n + Dt^{n+1/2}\big(U_{k,\ell}^{n+1/2}\big)_{average}, \tag{2.77}$$

and

$$Z_N^{n+1} = Z_{k,\ell}^n + Dt^{n+1/2}\big(V_{k,\ell}^{n+1/2}\big)_{average}, \tag{2.78}$$

where

$$\big(U_{k,\ell}^{n+1/2}\big)_{average} = \frac{(M_5+M_6)\big(U_{k,\ell}^{n+1/2}\big)_a + (M_7+M_8)\big(U_{k,\ell}^{n+1/2}\big)_b}{M_5+M_6+M_7+M_8}, \tag{2.79}$$

and $\big(U_{k,\ell}^{n+1/2}\big)_a$ is obtained from Eqs. (2.67), (2.73), and (2.75), while $\big(V_{k,\ell}^{n+1/2}\big)_{average}$ is from Eqs. (2.67), (2.74), and (2.75). $\big(V_{k,\ell}^{n+1/2}\big)_{average}$ is calculated similar to $\big(U_{k,\ell}^{n+1/2}\big)_{average}$.

Picking a coincident point N, for the two points, S and W, defines a displacement for both of the points which is used in rezoning. We now use the rezone [2.2] package to move the points together with the material outside the zones treated as a vacuum. Material may transfer between zone $\overline{DHSE}$ and zone $\overline{HFPS}$ which have the same kind of material, but material in M_5 cannot move to M_7. In Fig. 2.13, the material in ΔHTN will be removed from M_5 and added to M_6. After the rezone calculations, the kinetic energy or the internal energy may be transferred across the material interface. The mesh line inside the zones, e.g. $\overline{HT}$, is moved through the material, leaving material in its new position, thus allowing slip along the interface. In the resultant mesh, the temporary separation of the point is eliminated. The two velocities on the temporary points are preserved for use on the next cycle. This process is repeated for each point along the slip line. The slip along a k line is handled in a similar way.

References

2.1 Schulz, WD, Two-dimensional Lagrangian hydrodynamic difference-equation, Method Comput. Physics, 3, pp 1–45 (1964).
2.2 Wallick, KB, REZONE: A method for automatic rezoning in two-dimensional Lagrangian hydrodynamic problems, Los Alamos National Laboratory report LA-10829-MS (1987).
2.3 Wilson, HL, and Pomraning, GC, Theoretical and numerical radiation hydrodynamics, SAI-73-525-LJ, AD-530919 (1974).
2.4 Schulz, WD, Two-dimensional Lagrangian hydrodynamic difference equation, Lawrence Radiation Laboratory report UCRL-6776 (1963).

CHAPTER 3

EULERIAN METHOD

Notations

E	specific internal energy per unit volume (Mbar – cm^3/cm^3)
e	equivalent plastic strain (no unit)
G	shear modulus of elasticity (Mbar)
I	specific internal energy per unit mass (Mbar – cm^3/g)
M	cell mass (g)
P	pressure (Mbar)
R, r	radial coordinate (cm)
S^{ij}	stress deviator tensor (Mbar)
$S^{rr}, S^{zz}, S^{rz}, S^{\theta\theta}$	stress deviator components (Mbar)
t	time (μs)
Δt	time step (μs)
$\Delta t_{\max}$	maximum allowed time step (μs)
u	velocity in r direction (cm/μs)
$\bar{U}$	average velocity defined by Eq. (3.56) (cm/μs)
$\lvert U\rvert$	in Eq. (3.57), may be $\lvert U\rvert$ or $\lvert V\rvert$ dependent on the direction of calculation (cm/μs)
v	velocity in z direction (cm/μs)
$\bar{V}$	average velocity similar to Eq. (3.56) (cm/μs)
$\dot{W}$	rate of energy source due to work hardening (Mbar – $cm^3/cm^3/\mu s$)
Y	flow stress of elasticity (Mbar)
Z, z	axial coordinate (cm)

Greek Letters

δ^{ij}	Kronnecker delta
ε	specific internal energy (Mbar – cm^3)
γ	ratio of the specific heat, i.e. $\gamma = C_P/C_V$ (no unit)
η	normalized density ($= \rho/\rho_0$) (no unit)
θ	angular coordinate

λ	half of the grid size in R direction, i.e. $\lambda = \Delta R/2$ (cm)
μ	normalized density minus one $(= \eta - 1 = \rho/\rho_0 - 1)$ (no unit)
ρ	density (g/cm^3)
σ^{ij}	stress tensor (Mbar)
$\sigma^{rr}, \sigma^{zz}, \sigma^{rz}, \sigma^{\theta\theta}$	stress deviator components (Mbar)
τ	specific volume (cm^3/g)

Subscripts

0	initial value
e	elastic regime
p	plastic regime
r	derivative with respect to r coordinate
t	derivative with respect to time
z	derivative with respect to z coordinate

Superscripts

n	time at n time-step, i.e. $t^n = t_0 + n \cdot \Delta t$

3.1 Introduction

There is considerable interest in solving multi-material compressible flow problems with material interface. A Lagrangian approach would be quite a natural choice — for example HEMP [3.1], TOODY [3.2], and MAGEE [3.3 and 3.4]. However, for a large material distortion, the Lagrangian calculations can no longer be continued, so an Eulerian or a combined Lagrangian/Eulerian-type scheme has been applied.

Although many schemes have been invented in the various Eulerian code developments, materials are treated in only two ways, namely the particle-in-cell (PIC) method [3.5], where materials are represented by discrete mass points called Particle, and the continuous Eulerian methods, as in SOIL [3.6], HELP [3.7], and CSQ [3.8]. Computing economy is gained in the continuous method; however, the Lagrangian-type capability of the PIC method is lost and subsequently is replaced by various interface treatment.

In developing of the present code, the goal is to retain PIC capability while improving the accuracy and computing economy. Recently, Clark [3.9] proved the accuracy of the standard PIC method to be a first-order scheme in hydrodynamic computation of multi-material problems and proposed a second-order scheme. Here a similar second-order scheme is used to compute a multi-material elastic-plastic flow, including phase transition and spall. The truncation error, stability analysis, and other details of computing are presented here.

3.2 General Description of Physical Formulation

3.2.1 The Conservation Equation for a Stress-Supporting Medium

The conservation equation in a two-dimensional cylindrical or plane Eulerian coordinate can be written as
mass

$$\frac{D\rho}{Dt} + \rho\left(\frac{\partial U}{\partial R} + \alpha\frac{U}{R} + \frac{\partial V}{\partial Z}\right) = 0\,, \tag{3.1}$$

momentum

$$\rho\frac{DU}{Dt} = \frac{\partial \sigma^{RR}}{\partial R} + \frac{\partial \sigma^{RZ}}{\partial Z} + \frac{\alpha}{R}(\sigma^{RR} - \sigma^{\theta\theta})\,, \tag{3.2a}$$

$$\rho\frac{DV}{Dt} = \frac{\partial \sigma^{ZZ}}{\partial Z} + \frac{\partial \sigma^{RZ}}{\partial R} + \frac{\alpha}{R}\sigma^{RZ}\,, \tag{3.2b}$$

and
energy

$$\rho\frac{D\varepsilon}{Dt} = \sigma^{RR}\frac{\partial U}{\partial R} + \alpha\sigma^{\theta\theta}\frac{U}{R} + \sigma^{ZZ}\frac{\partial V}{\partial Z} + \sigma^{RZ}\left(\frac{\partial U}{\partial Z} + \frac{\partial V}{\partial R}\right), \tag{3.3}$$

where $\alpha = 0$ for plane geometry and $\alpha = 1$ for cylindrical geometry with the following definitions

$$\frac{D}{Dt} = \frac{\partial}{\partial t} + U\frac{\partial}{\partial R} + V\frac{\partial}{\partial Z}\,, \tag{3.3a}$$

$$\sigma^{ij} = S^{ij} - P\delta_{ij}\,, \tag{3.3b}$$

$$P = -\frac{1}{3}\sigma^{ij} = \frac{1}{3}(\sigma^{RR} + \sigma^{ZZ} + \sigma^{\theta\theta})\,, \tag{3.3c}$$

$$S^{ij} = \text{stress deviator}\,, \tag{3.3d}$$

and

$$\varepsilon = \text{specific internal energy}. \tag{3.3e}$$

3.2.2 Equation of State

A stress-supporting medium flowing under a wide range of stresses exhibits a variety of different physical characteristics. Depending on its retention of elastic character, the flow may be elastic or plastic. Here a straightforward approach is taken, yet there does not seem to exist any better model.

3.2.2.1 Stress in Elastic Regime

When a material flows elastically, Hooke's law in current form can be written as

$$\frac{DS^{RR}}{Dt} = 2G\left(\frac{\partial U}{\partial R} + \frac{1}{3\rho}\frac{D\rho}{Dt}\right) + \delta^{RR}, \tag{3.4a}$$

$$\frac{DS^{ZZ}}{Dt} = 2G\left(\frac{\partial V}{\partial Z} + \frac{1}{3\rho}\frac{D\rho}{Dt}\right) + \delta^{ZZ}, \tag{3.4b}$$

$$\frac{DS^{RZ}}{Dt} = G\left(\frac{\partial U}{\partial Z} + \frac{\partial V}{\partial R}\right) + \delta^{RZ}, \tag{3.4c}$$

$$\frac{DS^{\theta\theta}}{Dt} = 2G\left(\alpha\frac{U}{R} + \frac{1}{3\rho}\frac{D\rho}{Dt}\right) = \frac{D}{Dt}(-S^{RR} - S^{ZZ}). \tag{3.4d}$$

and

$$S^{RR} + S^{ZZ} + S^{\theta\theta} = 0, \tag{3.4e}$$

where

$$G = \text{modulus of elasticity in shear}$$
$$= f(\rho, \varepsilon, \text{material}),$$

and

$$\delta^{ij} = \text{correction for rigid body rotation}.$$

In tensor form

$$\frac{DS^{ij}}{Dt} = 2G\dot{e}^{ij} + \delta^{ij}, \tag{3.5}$$

where

$$\dot{e}^{ij} = \text{strain rate deviator}$$

$$= \begin{bmatrix} \dfrac{\partial U}{\partial R} + \dfrac{1}{3\rho}\dfrac{D\rho}{Dt} & \dfrac{1}{2}\left(\dfrac{\partial U}{\partial Z} + \dfrac{\partial V}{\partial R}\right) & 0 \\ \dfrac{1}{2}\left(\dfrac{\partial U}{\partial Z} + \dfrac{\partial V}{\partial R}\right) & \dfrac{\partial V}{\partial Z} + \dfrac{1}{3\rho}\dfrac{D\rho}{Dt} & 0 \\ 0 & 0 & \alpha\dfrac{U}{R} + \dfrac{1}{3\rho}\dfrac{D\rho}{Dt} \end{bmatrix}. \quad (3.6)$$

3.2.2.2 Stresses in Plastic Regime

For plastic flows, Prandtl and Reuss consider both plastic and elastic strain simultaneously and arrived at the following flow equation.

$$\dot{S}^{ij} = 2G\dot{e}^{ij} - \frac{G\dot{W}}{\frac{1}{3}(Y^0)^2}S^{ij}, \quad (3.7)$$

where $\dot{W} = \sum_{ij} S^{ij}\dot{e}^{ij}$ is the plastic work/unit volume and Y^0 = yield stress in simple tension.

Expanding Eq. (3.7), we get

$$\frac{DS^{RR}}{Dt} = 2G\left(\frac{\partial U}{\partial R} + \frac{1}{3\rho}\frac{D\rho}{Dt}\right) - \frac{G\dot{W}}{\frac{1}{3}(Y^0)^2}S^{RR} + \delta^{RR}, \quad (3.8a)$$

$$\frac{DS^{ZZ}}{Dt} = 2G\left(\frac{\partial V}{\partial Z} + \frac{1}{3\rho}\frac{D\rho}{Dt}\right) - \frac{G\dot{W}}{\frac{1}{3}(Y^0)^2}S^{ZZ} + \delta^{ZZ}, \quad (3.8b)$$

$$\frac{DS^{RZ}}{Dt} = G\left(\frac{\partial U}{\partial Z} + \frac{\partial V}{\partial R}\right) - \frac{G\dot{W}}{\frac{1}{3}(Y^0)^2}S^{RZ} + \delta^{RZ}, \quad (3.8c)$$

$$S^{\theta\theta} = -(S^{RR} + S^{ZZ}), \quad (3.8d)$$

and

$$\begin{aligned}\dot{W} &= \left(\frac{\partial U}{\partial R} + \frac{1}{3\rho}\frac{D\rho}{Dt}\right)S^{RR} + \left(\frac{\partial U}{\partial Z} + \frac{\partial V}{\partial R}\right)S^{RZ} + \left(\frac{\partial V}{\partial Z} + \frac{1}{3\rho}\frac{D\rho}{Dt}\right)S^{ZZ} \\ &\quad + \left(\alpha\frac{U}{R} + \frac{1}{3\rho}\frac{D\rho}{Dt}\right)S^{\theta\theta} \\ &= \left(\frac{\partial U}{\partial R}\right)S^{RR} + \left(\frac{\partial U}{\partial Z} + \frac{\partial V}{\partial R}\right)S^{RZ} + \left(\frac{\partial V}{\partial Z}\right)S^{ZZ} + \left(\alpha\frac{U}{R}\right)S^{\theta\theta}. \quad (3.9)\end{aligned}$$

Physically $\dot{W}$ is the work done per unit volume in changing the shape of the medium. Eqs. (3.8a)–(3.8d) hold at the yield limit and for $\dot{W} \geq 0$. When $\dot{W} < 0$, the material is unloading elastically from a plastic state and Eqs. (3.4a)–(3.4d) are to be used. Therefore, Eqs. (3.8a)–(3.8d) and (3.9) have to be solved for a plastic regime of flow. The appearance of $\dot{S}^{ij}$ and S^{ij} greatly complicates the equation mathematically. However, we can solve these equations quite nicely without iteration in the present numerical scheme. The details will be shown later.

3.2.2.3 Hydrostatic Pressure

Various forms of equation of state for pressure are available; we will simply write this as below.

$$P = f(\rho, \varepsilon, \text{Material})\,. \tag{3.10}$$

3.2.2.4 Yield Criterion

To check whether a material is in elastic or plastic state, von Mises criterion is employed here, denoting three principal directions by subscripts 1, 2, and 3,

$$S_1 = \frac{1}{2}\left(S^{RR} + S^{ZZ}\right) + \frac{1}{2}\left[\left(S^{RR} - S^{ZZ}\right)^2 + \left(2S^{RZ}\right)^2\right]^{1/2}, \tag{3.11a}$$

$$S_2 = \frac{1}{2}\left(S^{RR} + S^{ZZ}\right) - \frac{1}{2}\left[\left(S^{RR} - S^{ZZ}\right)^2 + \left(2S^{RZ}\right)^2\right]^{1/2}, \tag{3.11b}$$

and

$$S_3 = S^{\theta\theta}\,. \tag{3.11c}$$

The second invariant of S^{ij} tensor is then defined by

$$\begin{aligned} \phi &= S_1^2 + S_2^2 + S_3^2 \\ &= \left(S^{RR}\right)^2 + \left(S^{ZZ}\right)^2 + \left(S^{\theta\theta}\right)^2 + 2\left(S^{RZ}\right)^2. \end{aligned} \tag{3.12}$$

Then, von Mises yield criterion says that a material is in elastic state when $\phi \leq \frac{2}{3}(Y^0)^2$ or plastic state when $\phi > \frac{2}{3}(Y^0)^2$. Now, let's suppose that a small time step, Δt, is chosen such that the Prandtl-Reuss relation is valid. During the time increment Δt, a material can be in an elastic period Δt_1, then in a plastic period Δt_2 — that is, $\Delta t = \Delta t_1 + \Delta t_2$. Then for a rigorous computation of S^{ij}, Δt_1 has to be obtained such that

$$\phi_{\Delta t_1} = \frac{2}{3}(Y^0)^2\,, \tag{3.13}$$

is satisfied. An iterative procedure might be considered. However, considering the accuracy of the yield criterion itself and plastic flow model, a simple method seems practical. An experience with a Lagrangian computation shows that the following scheme proposed by Wilkins [3.1] is adequate in computing stress deviators on yield surface, $\left(S^{ij}\right)^*$, such that

$$\left(S^{ij}\right)^* = \left(S^{ij}\right)\frac{\sqrt{\frac{2}{3}}Y^0}{\phi}\,. \tag{3.14}$$

Therefore, if the flow becomes plastic during Δt period, S^{ij} is first calculated based on elastic assumptions. Then, the new stress deviators on yield surface $(S^{ij})^*$ are approximated by Eq. (3.14).

3.2.2.5 Stress Correction for Rigid Body Rotation

The magnitude of stresses does not change during a rigid body rotation, and new stresses, after rotation as well as after deformation, have to be expressed in terms of the original coordinate system. The corrections for rigid body rotation before and after deformation have to be expressed in terms of the original coordinate system. The correction for rigid body rotation is relatively straightforward. However, there seems to be no unique way to correct stresses for deformation. One possibility is to include this effect into governing differential equation (e.g. see Swegle [3.2]). Even though technically interesting, the inclusion of deformation correction does not appear to be very significant for practical purposes.

Suppose S^{ij} is rotated by ω during a time increment, Δt. Then, the new stress deviator in the original coordinate system, S_0^{ij}, can be written as

$$S_0^{RR} = S^{RR}\cos^2\omega + S^{ZZ}\sin^2\omega + 2S^{RZ}\sin\omega\cos\omega\,, \tag{3.15a}$$
$$S_0^{ZZ} = S^{RR}\sin^2\omega + S^{ZZ}\cos^2\omega - 2S^{RZ}\sin\omega\cos\omega\,, \tag{3.15b}$$

and

$$S_0^{RZ} = S^{RZ}(\cos^2\omega - \sin^2\omega) - (S^{RR} - S^{ZZ})\sin\omega\cos\omega\,, \tag{3.15c}$$

where $\sin\omega = \frac{\Delta t}{2}\left(\frac{\partial U}{\partial Z} - \frac{\partial V}{\partial R}\right)$.

Instead of including δ^{ij} to differential equation as in Eqs. (3.4) and (3.8), S^{ij} correction can be done to the resulting stresses at the end of each computational time step. Thus, it is convenient to define

$$\delta_0^{ij} = \Delta t \delta^{ij} = S_0^{ij} - S^{ij} \,, \tag{3.16a}$$

or

$$S_0^{ij} = S^{ij} + \delta_0^{ij} \,. \tag{3.16b}$$

Then

$$\delta_0^{RR} = \frac{1}{2}(S^{RR} - S^{ZZ})(\cos 2\omega - 1) + S^{RZ} \sin 2\omega \,, \tag{3.17a}$$

$$\delta_0^{ZZ} = -\delta_0^{RR} \,, \tag{3.17b}$$

and

$$\delta_0^{RZ} = S^{RZ}(\cos 2\omega - 1) - \frac{1}{2}(S^{RR} - S^{ZZ}) \sin 2\omega \,. \tag{3.17c}$$

For a small ω, we may further simplify Eqs. (3.17a)–(3.17c) by

$$\sin 2\omega = 2 \sin \omega = \Delta t \left(\frac{\partial U}{\partial Z} - \frac{\partial V}{\partial R} \right) . \tag{3.18}$$

3.2.3 Spall

Material failure can be considered under a number of different conditions. However, under intense, short-duration tension loading, a particular type of fracture called "spall" is frequently produced. For our current problems, the material failure is likely to occur in this form. In spalling, small, independent cracks or voids are produced and usually an extensive crack propogation does not take place. The shape of void thus produced may depend on material structure. However, effects of crystallographic orientation on gross damage as well as the stress distribution relative to the shape of the void seem to be insignificant.

Considering these characteristics of spall and considering that our problems do not require much computation once a material is significantly fractured, a simplistic model seems to be adequate. Therefore, for the present code, the following criteria are used to check material failure and phase transition.

3.2.3.1 Melting

When $\varepsilon >$ melt energy, the material is melted and subsequently is treated as a pure hydrodynamic fluid. Melt energy will be described in detail later.

3.2.3.2 Spall

When pressure gets down to a certain specified spall pressure ($P \leq P_{spall}$), the material is assumed to be spalled, as shown in Fig. 3.1. Once spalled, the material cannot support any stress, and P and S^{ij} are set to be zero until recombined. For recombining, if $\tau \leq \tau_{test}$, the material is regarded as recombined.

3.2.3.3 Fracture

Fracture criterion is based on tensile strength. If principal stresses, $\sigma_I >$ tensile strength, the material is assume to be fractured. A fractured cell is treated as spalled. For a more complete check, the compressive strength also has to be compared with principal stresses. However, for the class of problems we are solving, a material is probably melting before it reaches the compressive limit. Therefore, the fracture criteria based on compressive stress are not included. Computational detail for a spalled cell is given in Sections 3.3.5.2–3.3.6.3.

3.3 Computational Scheme

As briefly mentioned earlier, the second-order PIC, developed by Clark [3.9] in conjunction with hydrodynamic computation is extended to elastic-plastic flow here.

3.3.1 General Discussion

One of the first questions in computing an elastic-plastic flow is where to define stresses. Wherever defined in a continuum, stresses are physically

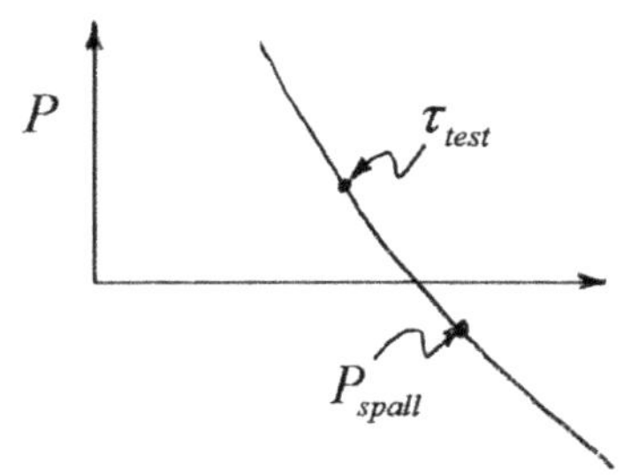

Figure 3.1. The pressure-specific volume curve for spall and τ_{test}.

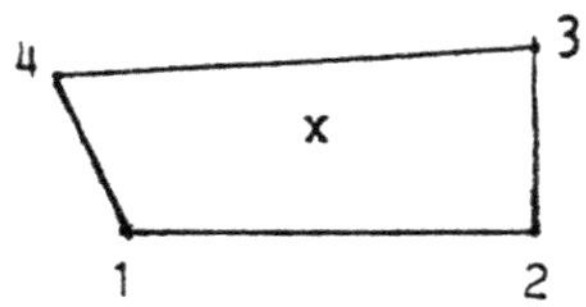

Figure 3.2. A typical Lagrangian mesh with velocities defined at vertices 1, 2, 3, and 4.

important. Depending on the choice of a differencing scheme, a particular definition can be more convenient than others. For example, in a typical Lagrangian grid (Fig. 3.2), the velocity at point 1, 2, 3, and 4 are known. Then, it seems natural to assign S^{ij} at the cell center to solve Eqs. (3.5) or (3.7). In PIC method, however, it is debatable whether a cell-centered stress is more advantageous than other possibilities, such as defining S^{RZ} along cell boundaries. For bookkeeping purposes, a cell-centered S^{ij} would be a very convenient choice. As will be shown, in remapping S^{ij} to original coordinate, a volume weighting scheme is second-order accurate, and a cell-centered S^{ij} is satisfactory.

A volume weighting of S^{ij} is essentially an interpolating scheme. Physically, it may be more sensible to balance forces on a small material element by defining S^{ij} on cell boundaries. However, since the force balance would require an interpolation scheme, conceptually it is not much different from the volume weighting.

3.3.2 Summary of Calculation Procedure

An operator splitting method is applied here. The governing set of cylindrical equation is split in radial and axial directions. Then, a separate calculation is performed in each direction. The order of this calculation is alternated for each time advancement to maintain the accuracy of a one-dimensional procedure. More detail description is given in the following sections. A sequence of computational procedure is presented briefly below.

3.3.2.1 Phase I: Lagrangian Phase

Based on known quantities at nth time step (denoted by a superscript n), Lagrangian quantities at $(n+1)$th time step are computed here (denoted by a wiggle). The following steps are taken logically.

Step 1: Calculate P and S^{ij} at $\left(n+\frac{1}{2}\right)$.
Step 2: Calculate $\tilde{U}_i, \tilde{\varepsilon}_i$ using $P_i^{n+1/2}$ and $(S^{ij})_i^{n+1/2}$. Then, calculate $(\tilde{S}^{ij})$.

Step 3: Temporarily assign the momentum (M_iU_i) and the internal energy $(M_i\varepsilon_i)$ to cell i.

3.3.2.2 Phase II: Particle Transport and Remapping

In this phase, particle are transported with proper velocities, which will be described in Sections 3.3.5.1 and 3.3.5.2. Along with particle, portions of momentum, internal energy and stresses are transported. These transported quantities are then summed up for each cell and new quantities at $(n+1)$th time step are calculated.

Step 1: Move particles.
Step 2: Remap; the new cell mass, momentum, internal energy, and stresses are calculated; subsequently, the velocities and specific internal energy are determined.

3.3.3 Lagrangian Phase

In this section, Lagrangian phase calculation is elaborated by equations in the radial direction after operator splitting. Similar equations can easily be obtained in the axial direction or in Cartesian coordinates by dropping radial terms.

3.3.3.1 Equations to be Solved in the Radial Direction

Mass

$$\rho_t = -\bar{U}\rho_R - \rho\left(U_R + \alpha\frac{U}{R}\right). \tag{3.19}$$

Momentum

$$U_t = -\bar{U}U_R + \frac{1}{\rho}\left[\sigma_R^{RR} + \frac{\alpha}{R}(\sigma^{RR} - \sigma^{\theta\theta})\right], \tag{3.20a}$$

and

$$V_t = -\bar{U}V_R + \frac{1}{\rho}\left[\sigma_R^{RZ} + \frac{\alpha}{R}(\sigma^{RZ})\right]. \tag{3.20b}$$

Energy

$$\varepsilon_t = -\bar{U}\varepsilon_R + \frac{1}{\rho}\left[\sigma^{RR}U_R + \frac{\alpha U}{R}(\sigma^{\theta\theta}) + \sigma^{RZ}V_R\right]. \tag{3.21}$$

Equation of State

Pressure : $$P = f(\rho, \varepsilon). \tag{3.22}$$

Elastic Regime:

$$S_t^{RR} = -\bar{U}S_R^{RR} + 2G\left[U_R + \frac{1}{3\rho}(\rho_t + U\rho_R)\right]$$

$$= -\bar{U}S_R^{RR} + 2G\left[\frac{1}{3}\left(2U_R - \frac{\alpha U}{R}\right)\right], \tag{3.23a}$$

$$S_t^{ZZ} = -\bar{U}S_R^{ZZ} + 2G\left[\frac{1}{3\rho}(\rho_t + U\rho_R)\right]$$

$$= -\bar{U}S_R^{ZZ} + 2G\left[-\frac{1}{3}\left(U_R + \frac{\alpha U}{R}\right)\right], \tag{3.23b}$$

$$S_t^{RZ} = -\bar{U}S_R^{RZ} + GV_R\,, \tag{3.23c}$$

and

$$S_t^{\theta\theta} = -(S^{RR} + S^{ZZ})\,. \tag{3.23d}$$

Plastic Regime:

$$S_t^{RR} = -\bar{U}S_R^{RR} + 2G\left[\frac{1}{3}\left(2U_R - \frac{\alpha U}{R}\right)\right] - \frac{G\dot{W}}{\frac{1}{3}(Y^0)^2}S^{RR}\,, \tag{3.24a}$$

$$S_t^{ZZ} = -\bar{U}S_R^{ZZ} + 2G\left[-\frac{1}{3}\left(U_R + \frac{\alpha U}{R}\right)\right] - \frac{G\dot{W}}{\frac{1}{3}(Y^0)^2}S^{ZZ}\,, \tag{3.24b}$$

$$S_t^{RZ} = -\bar{U}S_R^{RZ} + GV_R - \frac{G\dot{W}}{\frac{1}{3}(Y^0)^2}S^{RZ}\,, \tag{3.24c}$$

$$S^{\theta\theta} = -(S^{RR} + S^{ZZ})\,, \tag{3.24d}$$

and

$$\dot{W} = U_R S^{RR} + \frac{\alpha U}{R}S^{\theta\theta} + V_R S^{RZ}$$

$$= \left(U_R - \frac{\alpha U}{R}\right)S^{RR} - \frac{\alpha U}{R}S^{ZZ} + V_R S^{RZ}\,. \tag{3.24e}$$

3.3.3.2 Lagrangian Phase Calculation

Step 1: Calculate $P^{n+1/2}$ and $(S^{ij})^{n+1/2}$ as below

$$f^{n+1/2} = f^n + \frac{\delta t}{2}\frac{Df}{Dt} = f^n + \frac{\delta t}{2}(f_t + Uf_R)\,, \tag{3.25a}$$

and

$$P^{n+1/2} = g\left(\rho^{n+1/2}, \varepsilon^{n+1/2}\right). \tag{3.25b}$$

Step 2: Calculate $\tilde{U}, \tilde{\varepsilon}$, and $\tilde{S}^{ij}$. To visualize this Lagrangian phase calculation, let us consider the following sketch (Fig. 3.3) of a cylindrical section

σ^{RR} contribution to $\tilde{U}$

$$\begin{aligned}
&= \int_0^{\delta t} \frac{\left[\left(\sigma^{RR}\right)^+\left(R + \frac{\Delta R}{2}\right) - \left(\sigma^{RR}\right)^-\left(R - \frac{\Delta R}{2}\right)\right]\Delta\theta\Delta Z}{(R\Delta R\Delta\theta\Delta Z)\rho}dt \\
&= \int_0^{\delta t} \left\{\frac{\left[\left(\sigma^{RR}\right)^+ - \left(\sigma^{RR}\right)^-\right]}{\rho\Delta R} + \frac{\left[\left(\sigma^{RR}\right)^+ + \left(\sigma^{RR}\right)^-\right]}{2\rho R}\right\}dt \\
&= \delta t\left\{\left[\frac{\left(\sigma^{RR}\right)_R}{\rho}\right]^{n+1/2} + \left[\frac{\sigma^{RR}}{\rho R}\right]^{n+1/2}\right\}.
\end{aligned} \tag{3.26a}$$

$\sigma^{\theta\theta}$ contribution to $\tilde{U}$

$$\begin{aligned}
&= -\int_0^{\delta t} \frac{\left[\left(\sigma^{\theta\theta}\right)^+ + \left(\sigma^{\theta\theta}\right)^-\right]\Delta\theta\Delta R\Delta Z}{2(R\Delta R\Delta\theta\Delta Z)\rho}dt \\
&= -\delta t\left[\frac{\sigma^{\theta\theta}}{\rho R}\right]^{n+1/2}.
\end{aligned} \tag{3.26b}$$

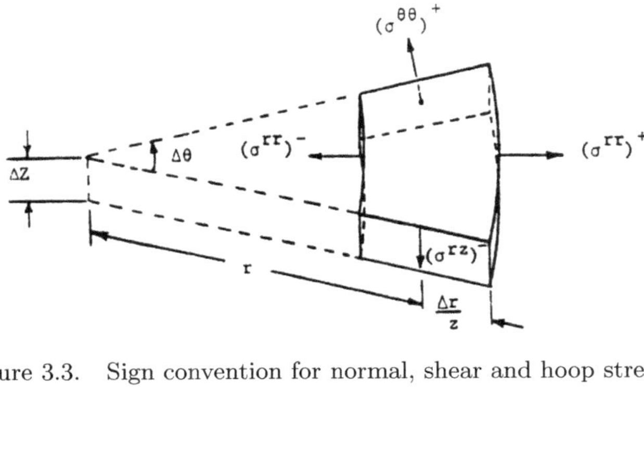

Figure 3.3. Sign convention for normal, shear and hoop stresses.

Adding Eqs. (3.26a) and (3.26b), we get

$$\tilde{U} = U^n + \delta t \left\{ \left[\frac{(\sigma^{RR})_R}{\rho} \right]^{n+1/2} + \alpha \left[\frac{\sigma^{RR} - \sigma^{\theta\theta}}{\rho R} \right]^{n+1/2} \right\}. \tag{3.26c}$$

Now, let us consider $\tilde{\varepsilon}$.

$\sigma^{RR} U$ contribution to energy

$$= \int_0^{\delta t} \frac{\left[(\sigma^{RR})^+ (R + \frac{\Delta R}{2}) U^+ - (\sigma^{RR})^- (R - \frac{\Delta R}{2}) U^- \right] \Delta\theta \Delta Z}{(R \Delta R \Delta\theta \Delta Z)\rho} dt$$

$$= \int_0^{\delta t} \left\{ \frac{\left[(\sigma^{RR} U)^+ - (\sigma^{RR} U)^- \right]}{\rho \Delta R} + \frac{\left[(\sigma^{RR} U)^+ + (\sigma^{RR} U)^- \right]}{2 \rho R} \right\} dt$$

$$= \delta t \left\{ \left[\frac{(\sigma^{RR} U)_R}{\rho} \right]^{n+1/2} + \left[\frac{\sigma^{RR} U}{\rho R} \right]^{n+1/2} \right\}. \tag{3.27a}$$

Then

$$\tilde{\varepsilon} + \frac{\bar{U}^2 + \bar{V}^2}{2} = \varepsilon^n + \frac{(U^n)^2 + (V^n)^2}{2} + \delta t \left\{ \left[\frac{(\sigma^{RR} U)_R}{\rho} \right]^{n+1/2} + \left[\frac{\sigma^{RR} U}{\rho R} \right]^{n+1/2} \right\} + [\sigma^{RZ} V \text{ contribution}], \tag{3.27b}$$

or

$$\tilde{\varepsilon} = \varepsilon^n + \delta t \left\{ \left[\frac{(\sigma^{RR} U)_R}{\rho} \right]^{n+1/2} + \left[\frac{\sigma^{RR} U}{\rho R} \right]^{n+1/2} \right\} - \delta t \left\{ \left[\frac{(\sigma^{RR})_R}{\rho} \right]^{n+1/2} + \alpha \left[\frac{\sigma^{RR} - \sigma^{\theta\theta}}{\rho R} \right]^{n+1/2} \right\} \bar{U} + [\sigma^{RZ} V \text{ terms}], \tag{3.27c}$$

$$= \varepsilon^n + \delta t \left\{ \left[\frac{\sigma^{RR} U_R}{\rho} \right]^{n+1/2} + \left[\frac{\sigma^{RR} U}{\rho R} \right]^{n+1/2} \right\} + [\sigma^{RZ} V \text{ terms}] + O(\delta t^2). \tag{3.27d}$$

Other quantities can be obtained similarly.

The Lagrangian phase calculation in the radial direction is summarized below

$$\tilde{U} = U^n + \delta t\left\{\left[\frac{(\sigma^{RR})_R}{\rho}\right]^{n+1/2} + \alpha\left[\frac{\sigma^{RR} - \sigma^{\theta\theta}}{\rho R}\right]^{n+1/2}\right\}, \quad (3.28a)$$

$$\tilde{V} = V^n + \delta t\left\{\left[\frac{(\sigma^{RZ})_R}{\rho}\right]^{n+1/2} + \alpha\left[\frac{\sigma^{RZ}}{\rho R}\right]^{n+1/2}\right\}, \quad (3.28b)$$

and

$$\tilde{\varepsilon} = \varepsilon^n + \delta t\left\{\left[\frac{\sigma^{RR} U_R}{\rho}\right]^{n+1/2} + \alpha\left[\frac{\sigma^{\theta\theta} U}{\rho R}\right]^{n+1/2} + \left[\frac{\sigma^{RZ} V_R}{\rho}\right]^{n+1/2}\right\}. \quad (3.28c)$$

In the elastic regime

$$\tilde{S}^{RR} = S^{RR} + \delta t\left\{2G\left[\frac{1}{3}\left(2U_R - \alpha\frac{U}{R}\right)\right]\right\}^{n+1/2}, \quad (3.29a)$$

$$\tilde{S}^{ZZ} = S^{ZZ} + \delta t\left\{2G\left[-\frac{1}{3}\left(U_R + \alpha\frac{U}{R}\right)\right]\right\}^{n+1/2}, \quad (3.29b)$$

and

$$\tilde{S}^{RZ} = S^{RZ} + \delta t(GV_R)^{n+1/2}. \quad (3.29c)$$

In the plastic regime

$$(\tilde{S}^{ij})_p = (\tilde{S}^{ij})_e - \delta t\left[\frac{3G\dot{W}}{(Y^0)^2}S^{ij}\right]^{n+1/2}, \quad (3.30)$$

where $(\tilde{S}^{ij})_e$ = elastic portion given by Eqs. (3.29a)–(3.29c). Eq. (3.30) is not straightforward; more details will be given later. An alternative form of Eqs. (3.28) and (3.29) is convenient for programming. Consider the following algebraic terms

$$\begin{aligned}
&\left(\frac{\sigma_R}{\rho}\right)^{n+1/2} - \frac{\sigma_R^{n+1/2}}{\rho^n} \\
&= \frac{\delta t}{2}\left[\frac{D}{Dt}\left(\frac{\sigma_R}{\rho}\right) - \frac{1}{\rho}\left(\frac{D\sigma}{Dt}\right)_R\right] \\
&= \frac{\delta t}{2}\left[\left(\frac{\sigma_R}{\rho}\right) + U\left(\frac{\sigma_R}{\rho}\right)_R - \frac{1}{\rho}\left(\sigma_t + U\sigma_R\right)_R\right] \\
&= \frac{\delta t}{2}\left\{\frac{\sigma_R}{\rho^2}\left[U\rho_R + \rho\left(U_R + \alpha\frac{U}{R}\right)\right] + \frac{\sigma_{tR}}{\rho} - \rho_R U\frac{\sigma_R}{\rho^2} + U\frac{\sigma_{RR}}{\rho^2}\right. \\
&= \left. -\frac{1}{\rho}\left[\sigma_{tR} + U_R\sigma_R + U\sigma_{RR}\right]\right\} \\
&= \alpha\frac{\delta t}{2}\left(\frac{\sigma_R}{\rho}\frac{U}{R}\right). && (3.31a)
\end{aligned}$$

Let $\Delta\sigma = \sigma^{RR} - \sigma^{\theta\theta}$, then

$$\begin{aligned}
&\alpha\left[\left(\frac{\Delta\sigma}{R\rho}\right)^{n+1/2} - \frac{(\Delta\sigma)^{n+1/2}}{R\rho^n}\right] \\
&= \frac{\alpha\delta t}{2}\left\{\left[\frac{1}{R}\left(\frac{\Delta\sigma}{\rho}\right)_t + U\left(\frac{\Delta\sigma}{\rho R}\right)_R\right] - \frac{1}{\rho R}[(\Delta\sigma)_t + U(\Delta\sigma)_R]\right\} \\
&= \frac{\alpha\delta t}{2}\left\{\frac{\Delta\sigma}{\rho^2 R}\left[U\rho_R + \rho\left(U_R + \frac{U}{R}\right)\right] + \frac{\rho_R U\Delta\sigma}{\rho^2 R} - \frac{U\Delta\sigma}{\rho R^2}\right\} \\
&= \frac{\alpha\delta t}{2}\frac{\Delta\sigma}{\rho R}U_R\,. && (3.31b)
\end{aligned}$$

Using $\bar{U} = \frac{1}{2}(U^n + \tilde{U}) = U^{n+1/2}$, one gets

$$\begin{aligned}
&\left(\frac{\sigma U_R}{\rho}\right)^{n+1/2} - \frac{\sigma^{n+1/2}\bar{U}_R}{\rho^n} \\
&= \sigma^{n+1/2}\left\{\left(\frac{U_R}{\rho}\right)^n + \frac{\delta t}{2}\left[\left(\frac{U_R}{\rho}\right)_t + U\left(\frac{U_R}{\rho}\right)_R\right]\right. \\
&\quad \left. - \left(\frac{U_R}{\rho}\right)^n - \frac{1}{\rho}\frac{\delta t}{2}(U_t + UU_R)\right\}
\end{aligned}$$

$$= \sigma^{n+1/2}\frac{\delta t}{2}\left[\frac{U_R}{\rho^2}\left(U\rho_R + \rho U_R + \frac{\alpha\rho U}{R}\right) - \rho_R\frac{UU_R}{\rho^2} - \frac{U_R^2}{\rho}\right]$$

$$= \sigma^{n+1/2}\frac{\delta t}{2}\frac{\alpha U U_R}{\rho R}, \qquad (3.31c)$$

$$\left(\frac{\sigma U}{\rho R}\right)^{n+1/2} - \frac{\sigma^{n+1/2}\bar{U}}{\rho R}$$

$$= \sigma^{n+1/2}U^{n+1/2}\left\{\left(\frac{1}{\rho R}\right)^n + \frac{\delta t}{2}\left[\left(\frac{1}{\rho R}\right)_t + U\left(\frac{1}{\rho R}\right)_R\right] - \left(\frac{1}{\rho R}\right)^n\right\}$$

$$= (\sigma U)^{n+1/2}\frac{\delta t}{2}\left[\frac{1}{\rho^2 R}\left(U\rho_R + \rho U_R + \frac{\alpha\rho U}{R}\right) - \frac{U}{\rho R^2} - \frac{U\rho_R}{R\rho^2}\right]$$

$$= \frac{\delta t}{2}\frac{\sigma U U_R}{\rho R}, \qquad (3.31d)$$

$$\left(\frac{SV_R}{\rho}\right)^{n+1/2} - \frac{S^{n+1/2}\bar{V}_R}{\rho}$$

$$= (S)^{n+1/2}\frac{\delta t}{2}\left[\left(\frac{V_R}{\rho}\right)_t + U\left(\frac{V_R}{\rho}\right)_R - \left(\frac{\bar{V}_R}{\rho}\right)\right]$$

$$= (S)^{n+1/2}\frac{\delta t}{2}\left[\frac{V_R}{\rho^2}\left(U\rho_R + \rho U_R + \alpha\frac{\rho U}{R}\right) + \frac{(V_R)_t}{\rho} + \frac{UV_{RR}}{\rho} - \frac{U\rho_R V_R}{\rho^2} - \frac{(V_t + UV_R)_R}{\rho}\right]$$

$$= \alpha\frac{\delta t}{2}\frac{SUV_R}{\rho R}. \qquad (3.31e)$$

$$(GU_R)^{n+1/2} - \bar{G}\bar{U}_R$$

$$= (GU_R)^n + \frac{\delta t}{2}[(GU_R)_t + U(GU_R)_R] - \left[G^n + \frac{\delta t}{2}(G_t + UG_R)\right]\left[U^n + \frac{\delta t}{2}(U_t + UU_R)\right]_R$$

$$= \frac{\delta t}{2}[(GU_R)_t + U(GU_R)_R - G(U_t + UU_R)_R - U_R(G_t + UG_R)] + O(\Delta^2)$$

$$= -\frac{\delta t}{2}G(U_R)^2. \qquad (3.31f)$$

$$\begin{aligned}
&\left(\frac{GU}{R}\right)^{n+1/2} - \frac{\bar{G}\bar{U}}{R} \\
&= \frac{GU}{R} + \frac{\delta t}{2}\left[\left(\frac{GU}{R}\right)_t + U\left(\frac{GU}{R}\right)_R\right] \\
&\quad - \frac{1}{R}\left[G + \frac{\delta t}{2}(G_t + UG_R)\right]\left[U + \frac{\delta t}{2}(U_t + UU_R)\right] \\
&= \frac{\delta t}{2}\left[\left(\frac{GU}{R}\right)_t + U\left(\frac{G_R U}{R} + \frac{GU_R}{R} - \frac{GU}{R^2}\right)_R\right] \\
&\quad - \frac{1}{R}\left[\frac{\delta t}{2}(UG_t + U^2 G_R + GU_t + GUU_R)\right] \\
&= -\delta t \frac{GU^2}{R^2}\,. \qquad (3.31g)
\end{aligned}$$

Therefore, Eqs. (3.28) and (3.29) can be rewritten as

$$\begin{aligned}
\tilde{U} &= U^n + \frac{\delta t}{\rho^n}\left[(\sigma^{RR})_R^{n+1/2} + \frac{\alpha}{R}(\Delta\sigma)^{n+1/2}\right] \\
&\quad + \frac{\delta t}{\rho^n}\left[\frac{\alpha\delta t}{2}\left(\sigma_R^{RR}\frac{U}{R} + \frac{\Delta\sigma U_R}{R}\right)\right], \qquad (3.32a) \\
\tilde{V} &= V^n + \frac{\delta t}{\rho^n}\left[(\sigma^{RZ})_R^{n+1/2} + \frac{\alpha}{R}(\sigma^{RZ})^{n+1/2}\right] \\
&\quad + \frac{\delta t}{\rho^n}\left[\frac{\alpha\delta t}{2}\left(\sigma_R^{RZ}\frac{U}{R} + \frac{\sigma^{RZ} U_R}{R}\right)\right], \qquad (3.32b) \\
\tilde{\varepsilon} &= \varepsilon^n + \frac{\delta t}{\rho^n}\left[(\sigma^{RR})^{n+1/2}\bar{U}_R + \frac{\alpha}{R}(\sigma^{\theta\theta})^{n+1/2}\bar{U} + (\sigma^{RZ})^{n+1/2}\bar{V}_R\right] \\
&\quad + \frac{\delta t}{\rho^n}\left[\frac{\alpha\delta t}{2}\left(\sigma^{RR}U_R\frac{U}{R} + \frac{\sigma^{\theta\theta}U_R U}{R} + \frac{U\sigma^{RZ}V_R}{R}\right)\right], \\
&= \varepsilon^n + \frac{\delta t}{\rho^n}\left[(\sigma^{RR})^{n+1/2}\frac{\partial(R\bar{U})}{R\partial R} - \frac{\bar{U}}{R}(\sigma^{RR} - \sigma^{\theta\theta})^{n+1/2} + \bar{V}_R(\sigma^{RZ})^{n+1/2}\right] \\
&\quad + \frac{\delta t}{\rho^n}\left[\frac{\alpha\delta t}{2}\left(\sigma^{RR}U_R\frac{U}{R} + \frac{\sigma^{\theta\theta}U_R U}{R} + \frac{U\sigma^{RZ}V_R}{R}\right)\right], \qquad (3.32c)
\end{aligned}$$

$$\tilde{S}^{RR} = (S^{RR})^n + \delta t\left\{\frac{2\bar{G}}{3}\left[2\bar{U}_R\left(1 - \frac{U_R\delta t}{2}\right) - \frac{\alpha\bar{U}}{R}\left(1 - \frac{U\delta t}{2R}\right)\right]\right\}, \qquad (3.33a)$$

$$\tilde{S}^{ZZ} = \left(S^{ZZ}\right)^n + \delta t \left\{ -\frac{2\bar{G}}{3} \left[\bar{U}_R \left(1 - \frac{U_R \delta t}{2}\right) + \frac{\alpha \bar{U}}{R} \left(1 - \frac{U \delta t}{2R}\right) \right] \right\}, \tag{3.33b}$$

and

$$\tilde{S}^{RZ} = \left(S^{RZ}\right)^n + \bar{G}\bar{V}_R \left(1 - \frac{U_R \delta t}{2}\right) \delta t \,. \tag{3.33c}$$

3.3.3.3 Summary of Equations in the Axial Direction

Equations to be solved:

Mass

$$\rho_t = -V \rho_Z - \rho V_Z \,. \tag{3.34}$$

Momentum

$$U_t = -V U_Z + \frac{\sigma_Z^{RZ}}{\rho} \,, \tag{3.35a}$$

$$V_t = -V V_Z + \frac{\sigma_Z^{ZZ}}{\rho} \,. \tag{3.35b}$$

Energy

$$\varepsilon_t = -U \varepsilon_Z + \frac{1}{\rho}\left(\sigma^{ZZ} V_Z + \sigma^{RZ} U_Z\right). \tag{3.36}$$

Equation of State:

Pressure

$$P = g(\rho, \varepsilon) \,.$$

Elastic Regime

$$\begin{aligned} S_t^{RR} &= -V S_Z^{RR} + 2G \left[\frac{1}{3\rho} (\rho_t + V \rho_Z) \right] \\ &= -V S_Z^{RR} - \frac{2 G V_Z}{3} \,, \end{aligned} \tag{3.37a}$$

$$\begin{aligned} S_t^{ZZ} &= -V S_Z^{ZZ} + 2G \left[V_Z + \frac{1}{3\rho} (\rho_t + V \rho_Z) \right] \\ &= -V S_Z^{ZZ} + \frac{4 G V_Z}{3} \,, \end{aligned} \tag{3.37b}$$

$$S_t^{RZ} = -V S_Z^{RZ} + G U_Z \,, \tag{3.37c}$$

and

$$S^{\theta\theta} = -(S^{RR} + S^{ZZ})\,. \tag{3.37d}$$

Plastic Regime

$$S_t^{ij} = \left(S_t^{ij}\right)_e - \left[\frac{3G\dot{W}S^{ij}}{(Y^0)^2}\right], \tag{3.38}$$

where $(S^{ij})_e$ = elastic component.

Lagrangian Phase:

$$\tilde{U} = U^n + \delta t \left(\frac{\sigma_Z^{RZ}}{\rho}\right)^{n+1/2} = U^n + \frac{\delta t}{\rho^n}\left(\sigma^{RZ}\right)_Z^{n+1/2}, \tag{3.39a}$$

$$\tilde{V} = V^n + \delta t \left(\frac{\sigma_Z^{ZZ}}{\rho}\right)^{n+1/2} = V^n + \frac{\delta t}{\rho^n}\left(\sigma^{ZZ}\right)_Z^{n+1/2}, \tag{3.39b}$$

$$\tilde{\varepsilon} = \varepsilon^n + \delta t \left(\frac{\sigma^{ZZ}V_Z}{\rho} + \frac{\sigma^{RZ}U_Z}{\rho}\right)^{n+1/2}$$

$$= \varepsilon^n + \frac{\delta t}{\rho^n}\left[\bar{V}_Z\left(\sigma^{ZZ}\right)^{n+1/2} + \bar{U}_Z\left(\sigma^{RZ}\right)^{n+1/2}\right], \tag{3.40a}$$

and

$$P^{n+1/2} = g\left(\rho^{n+1/2}, \varepsilon^{n+1/2}\right). \tag{3.40b}$$

In the elastic regime,

$$\tilde{S}^{RR} = \left(S^{RR}\right)^n - \frac{2\delta t}{3}(GV_Z)^{n+1/2} = \left(S^{RR}\right)^n - \frac{2\delta t}{3}(\bar{G}\bar{V}_Z)\left(1 - \frac{V_Z\delta t}{2}\right) \tag{3.41a}$$

$$\tilde{S}^{ZZ} = \left(S^{ZZ}\right)^n + \frac{4\delta t}{3}(GV_Z)^{n+1/2} = \left(S^{ZZ}\right)^n + \frac{4\delta t}{3}(\bar{G}\bar{V}_Z)\left(1 - \frac{V_Z\delta t}{2}\right), \tag{3.41b}$$

and

$$\tilde{S}^{RZ} = (S^{RZ})^n + \delta t(GU_Z)^{n+1/2} = (S^{RZ})^n + \delta t(\bar{G}\bar{U}_Z)\left(1 - \frac{V_Z\delta t}{2}\right). \tag{3.41c}$$

In the plastic regime,

$$(\bar{S}^{ij})_p = (\bar{S}^{ij})_e - \delta t \left[\frac{3G\dot{W}S^{ij}}{(Y^0)^2} \right]^{n+1/2}, \tag{3.42a}$$

where $(\bar{S}^{ij})_e$ = elastic portion given by Eqs. (3.37a)–(3.37d).

Note that

$$\left(\frac{\sigma_Z}{\rho} \right)^{n+1/2} = \frac{(\sigma_Z)^{n+1/2}}{\rho^n}, \tag{3.42b}$$

and

$$\left(\frac{\sigma V_Z}{\rho} \right)^{n+1/2} = \frac{(\sigma V_Z)^{n+1/2}}{\rho^n}. \tag{3.42c}$$

Now

$$\begin{aligned} &(GV_Z)^{n+1/2} - \bar{G}\bar{V}_Z \\ &= (GU_Z)^n + \frac{\delta t}{2}\left[(GU_Z)_t + V\,(GU_Z)_Z\right] \\ &\quad - \left[G^n + \frac{\delta t(G_t + VG_Z)}{2}\right]\left[U^n + \frac{\delta t(U_t + VU_Z)}{2}\right]_Z \\ &= -\frac{\delta t}{2}\,(GU_Z V_Z)\,. \end{aligned} \tag{3.42d}$$

Also

$$(GV_Z)^{n+1/2} - \bar{G}\bar{V}_Z = -\frac{\delta t}{2}\,(GV_Z V_Z)\,. \tag{3.42e}$$

3.3.4 Stress Calculation in the Plastic Regime of Flow

Since the Prandtl-Reuss equation of plastic flow is implicit in stresses, the solution procedure is not straightforward. An explicit method, consistent with the present second-order PIC, is described here in some detail for Lagrangian phase computation.

3.3.4.1 Radial Direction

Rewriting Eq. (3.24),

$$S_t^{ij} = -US_R^{ij} + 2G(\dot{e})^{ij} - A\dot{W}S^{ij}\,, \tag{3.43a}$$

where $\dot{W} = S^{ij}(\dot{e})^{ij}$,

$$(\dot{e})^{ij} = \begin{bmatrix} U_R + \frac{1}{3\rho}(\rho_t + U\rho_R) & \frac{V_R}{2} & 0 \\ \frac{V_R}{2} & \frac{1}{3\rho}(\rho_t + U\rho_R) & 0 \\ 0 & 0 & \frac{\alpha U}{R} + \frac{1}{3\rho}(\rho_t + U\rho_R) \end{bmatrix},$$

$$= \begin{bmatrix} \frac{1}{3}\left(2U_R - \frac{\alpha U}{R}\right) & \frac{V_R}{2} & 0 \\ \frac{V_R}{2} & -\frac{1}{3}\left(U_R - \frac{\alpha U}{R}\right) & 0 \\ 0 & 0 & -\frac{1}{3}\left(U_R - \frac{2\alpha U}{R}\right) \end{bmatrix}, \tag{3.43b}$$

and

$$A = \frac{3G}{(Y^0)^2}. \tag{3.43c}$$

When the above equation is encountered, $(S^{ij})^n$ and $(\dot{e}^{ij})^{n+1/2}$ are known. Therefore, for Lagrangian phase calculation, the following set of equations are to be solved for the stress deviator, $\tilde{S}^{ij}$.

$$\tilde{S}^{ij} = \left(S^{ij}\right)^n + \delta t\left(2G\dot{e}^{ij} - A\dot{W}S^{ij}\right)^{n+1/2}, \tag{3.44}$$

$$\left(\dot{W}\right)^{n+1/2} = \left(\dot{e}^{ij}S^{ij}\right)^{n+1/2}, \tag{3.45}$$

and

$$\left(S^{ij}\right)^{n+1/2} = \frac{1}{2}\left[\tilde{S}^{ij} + (S^{ij})^n\right]. \tag{3.46}$$

This looks messy but can be solved without iteration as shown below. From Eqs. (3.44) and (3.46),

$$\begin{aligned} \left(S^{ij}\right)^{n+1/2} &= \left(S^{ij}\right)^n + \frac{\delta t}{2}\left(2G\dot{e}^{ij} - A\dot{W}S^{ij}\right)^{n+1/2} \\ &= \left[\left(S^{ij}\right)^n + \delta t\left(2G\dot{e}^{ij}\right)^{n+1/2}\right] \Big/ \left[1 + \frac{\delta t}{2}(A\dot{W})^{n+1/2}\right] \\ &= \left[\left(S^{ij}\right)^n + \delta t\left(2G\dot{e}^{ij}\right)^{n+1/2}\right]\left[1 - \frac{\delta t}{2}(A\dot{W})^{n+1/2}\right] + O(\Delta^2) \\ &= \left(S^{ij}\right)^n + \delta t\left[\left(G\dot{e}^{ij}\right)^{n+1/2} - \frac{1}{2}(A\dot{W})^{n+1/2}(S^{ij})^n\right] + O(\Delta^2). \end{aligned} \tag{3.47a}$$

From Eqs. (3.45) and (3.47a),

$$\begin{aligned}(\dot{W})^{n+1/2} &= (\dot{e}^{ij})^{n+1/2}(S^{ij})^{n+1/2} \\ &= (\dot{e}^{ij})^{n+1/2}\left[(S^{ij})^n + \delta t\left(G\dot{e}^{ij}\right)^{n+1/2}\right] \Big/ \left[1 + \frac{\delta t}{2}(A\dot{W})^{n+1/2}\right] \\ &= (\dot{e}^{ij})^{n+1/2}\left\{(S^{ij})^n + \delta t\left[\left(G\dot{e}^{ij}\right)^{n+1/2} - \frac{1}{2}(A\dot{W})^{n+1/2}(S^{ij})^n\right]\right\} \\ &\quad + O(\Delta^2)\,. \end{aligned} \tag{3.47b}$$

Therefore,

$$\begin{aligned}&\left[1 + \frac{\delta t}{2}(A\dot{e}^{ij})^{n+1/2}(S^{ij})^n\right](\dot{W})^{n+1/2} \\ &\quad = (\dot{e}^{ij})^{n+1/2}(S^{ij})^n + \delta t(G\dot{e}^{ij}\dot{e}^{ij})^{n+1/2} + O(\Delta^2)\,,\end{aligned} \tag{3.47c}$$

and

$$\begin{aligned}(\dot{W})^{n+1/2} = (\dot{e}^{ij})^{n+1/2}(S^{ij})^n + \delta t\Big[&(G\dot{e}^{ij}\dot{e}^{ij})^{n+1/2} \\ &- \frac{1}{2}(A\dot{e}^{ij}\dot{e}^{ij})^{n+1/2}(S^{ij}S^{ij})^n\Big] + O(\Delta^2)\,.\end{aligned} \tag{3.48}$$

Since $G^{n+1/2}$, $A^{n+1/2}$, and $(\dot{e}^{ij})^{n+1/2}$ are known, $(\dot{W})^{n+1/2}$ can be calculated by Eq. (3.48). Subtituting the result into Eq. (3.47a), then from Eq. (3.44), $\tilde{S}^{ij}$ can be obtained. For programming convenience, we use

$$\begin{aligned}\dot{e}^{ij}\dot{e}^{ij} &= \begin{bmatrix} \frac{1}{3}\left(2U_R - \frac{\alpha U}{R}\right) & \frac{V_R}{2} & 0 \\ \frac{V_R}{2} & -\frac{1}{3}\left(U_R + \frac{\alpha U}{R}\right) & 0 \\ 0 & 0 & \frac{1}{3}\left(-U_R + \frac{2\alpha U}{R}\right) \end{bmatrix} \\ &\quad \times \begin{bmatrix} \frac{1}{3}\left(2U_R - \frac{\alpha U}{R}\right) & \frac{V_R}{2} & 0 \\ \frac{V_R}{2} & -\frac{1}{3}\left(U_R + \frac{\alpha U}{R}\right) & 0 \\ 0 & 0 & \frac{1}{3}\left(-U_R + \frac{2\alpha U}{R}\right) \end{bmatrix} \\ &= \frac{2}{3}\left(U_R^2 - \frac{\alpha U U_R}{R} + \frac{\alpha U^2}{R^2}\right) + \frac{V_R^2}{2}\,.\end{aligned} \tag{3.49}$$

$$\begin{aligned}\dot{e}^{ij}S^{ij} &= \frac{1}{3}\left(2U_R - \frac{\alpha U}{R}\right)S^{RR} + V_R S^{RZ} - \frac{1}{3}\left(U_R + \frac{\alpha U}{R}\right)S^{ZZ} \\ &\quad + \frac{1}{3}\left(-U_R + \frac{2\alpha U}{R}\right)S^{\theta\theta} \\ &= \left(U_R - \frac{\alpha U}{R}\right)S^{RR} + V_R S^{RZ} - \frac{\alpha U}{R}S^{ZZ}\,, \end{aligned} \tag{3.50}$$

$$\begin{aligned}\left(\frac{U}{R}\right)^{n+1/2} &= \left(\frac{U}{R}\right)^{n} + \frac{\delta t}{2}\left[\left(\frac{U}{R}\right)_t + U\left(\frac{U}{R}\right)_R\right] \\ &= \left(\frac{\bar{U}}{R}\right)\left(1 - \frac{\delta t}{2}\frac{U}{R}\right) + O(\Delta^2)\,. \end{aligned} \tag{3.51}$$

3.3.4.2 Axial Direction

The calculation in the axial direction can be done similarly. From Eq. (3.38), one gets

$$S_t^{ij} = -V S_Z^{ij} + 2G\dot{e}^{ij} - A\dot{W}S^{ij}\,, \tag{3.52}$$

where $\dot{W} = S^{ij}\dot{e}^{ij}$ and

$$\begin{aligned}\dot{e}^{ij} &= \begin{bmatrix} \frac{1}{3\rho}(\rho_t + V\rho_z) & \frac{U_Z}{2} & 0 \\ \frac{U_Z}{2} & V_Z + \frac{1}{3\rho}(\rho_t + V\rho_z) & 0 \\ 0 & 0 & \frac{1}{3\rho}(\rho_t + V\rho_z) \end{bmatrix} \\ &= \begin{bmatrix} -\frac{1}{3}V_Z & \frac{U_Z}{2} & 0 \\ \frac{U_Z}{2} & \frac{2}{3}V_Z & 0 \\ 0 & 0 & -\frac{1}{3}V_Z \end{bmatrix}\end{aligned}$$

As before, Eqs. (3.48), (3.47) and (3.44) can be used with the above strain rate deviator, $\dot{e}^{ij}$, to get $\tilde{S}^{ij}$. Again for programming convenience, we have

$$\begin{aligned}(\dot{e}^{ij}\dot{e}^{ij}) &= \frac{1}{9}V_Z^2 + \frac{1}{4}U_Z^2 + \frac{1}{4}U_Z^2 + \frac{4}{9}V_Z^2 + \frac{1}{9}V_Z^2 \\ &= \frac{2}{3}V_Z^2 + \frac{1}{2}U_Z^2\,, \end{aligned} \tag{3.53}$$

$$(\dot{e}^{ij}S^{ij}) = -\frac{1}{3}V_Z S^{ij} + \left(\frac{1}{2}U_Z + \frac{1}{2}U_Z\right) S^{RZ} + \frac{2}{3}V_Z S^{ZZ} - \frac{1}{3}V_Z S^{\theta\theta}$$
$$= U_Z S^{RZ} + V_Z S^{ZZ} . \tag{3.54}$$

3.3.5 Particle Transport and Remapping

3.3.5.1 Average Velocity for the Particle

After Lagrangian phase calculation, we have $\tilde{U}, \tilde{V}, \tilde{\varepsilon}$ and $\tilde{S}^{ij}$ at each cell. Consistent with these, temporary values of momentum and internal energy are assigned to cell (i) as below.

$$\text{Momentum} = M_i \tilde{U}_i , \tag{3.55a}$$

and

$$\text{internal energy} = M_i \tilde{\varepsilon}_i , \tag{3.55b}$$

where M_i = mass of cell (i).
Then, particle are transported with average velocities, that is, in the radial direction,

$$\bar{U} = \frac{1}{2}\left(U + \tilde{U}\right) . \tag{3.56}$$

Each particle carries with it a fraction of dynamic and state quantities as defined by Eq. (3.55). After the particle transport, total changes in momentum, energy, and stresses are calculated for each cell. This particle transport and subsequent remapping to the original Eulerian mesh are explained in detail in conjunction with the truncation error analysis.

3.3.5.2 Particle Treatment for Void or Multi-material Cell

For each cell, we compute the location, type, and mass for each particle using the following steps:

Step 1. If a cell requires particle treatment or if the cell just spalled, then we move to the next step.

Step 2. If the particle have been previously defined, then the particle information is obtained from the particle flag words. If no particle exists in the cell, then particles will be created for each material, including void.

Step 3. We then move the particles using the interpolated velocities.

Step 4. If a mixed cell has a vacuum interface, then particles are used for the material region and no particle is assigned to the vacuum region.

Step 5. If a void is connected to a vacuum region, then the void region is treated as a vacuum region.

Step 6. After the particles have been moved, we save the information of the new location and the mass for each particle.

Step 7. We then are ready for the next-cycle calculation.

The PIC method is well known for its expensive CPU cost and its horrendous storage requirement. We vectorized the whole code so that it would run four times faster compared to a scalar one. We saved a large amount of storage by using three-packed words that will be described below. Since we were using the CRAY-YMP machine, which contains 64 bits for each word, we divided one word into many groups so that each group possessed its own information. The structures of the material flag word, MFLAG(ij), the particle flag word, PFLAG(ij), and the particle information word, PINDX(NP) are shown in Fig. 3.4. The notations and definitions used in Fig 3.4 are:

- IJ: index of cell at i (R-direction) and j (Z-direction).
- NP: total number of particle in cell IJ.
- INDUV: index of the velocity array, e.g. U(INDUV).
- ISM: flag to indicate spallation and melting for each material (up to maximum of 4).
- IAPL: for applied pressure boundary, IAPL = 1.
- IFULL: 0 for empty cell, 1 for partially filled cell, 2 for full cell.
- MF4-MF1: material number index.
- MF0: the total number of materials in the cell.

Bit Structure for the Packed Array:

MFLAG (I J)

Bits	3	18	4	8	1	2	5	5	5	5	5	3
Name		INDUV		ISM	IAPL	IFULL		MF4	MF3	MF2	MF1	MF0

MFLAG (I J)

Bits	23	21	20
Name	INDP		INDMD

PINDX (NP)

Bits	5	19	19	21
Name	IBT	IBX	IBY	IBM

Figure 3.4. Structure for the packed array MFLAG, PFLAG, and PINDX.

- INDP: pointer for the particle array.
- INDMD: index for the total (summation of all particles) mass, volume, pressure, and internal energy for the first material in cell IJ.
- IBT: material type for the particle (e.g. 1-31).
- IBX: normalized particle location (R-direction) in the cell (positive integer $< 2^{19}$).
- IBY: normalized particle location (Z-direction) in the cell (positive integer $< 2^{19}$).
- IBM: normalized particle mass fraction for one type material (positive integer $< 2^{19}$).
- IBX and IBY are computed from the following (see Fig. 3.5):

$K = 2^{19}$
DO 10 IX = 1, 8
$K = K - 2^{16}$
INXT $= 28 + 8.0 \times$ random (between 0.0 and 1.0) (function of IX)
DO 10 JY = 1, 8
IBX (JY) $= K + 2^{13} \times$ random (between 0 and 7) (function of JY and INXT) $+ 2^{13} \times$ random (between 0 and 1.0)
IBY(JY) $= 2^{16} \times$ (JY $- 1) + 2^{16} \times$ random (between 0 and 1.0) (function of JY)
10 continue

The normalized particle mass fraction IBM is defined by

IBM = (particle mass $\times 2^{21}$)/(Total mass of the same material in the cell)

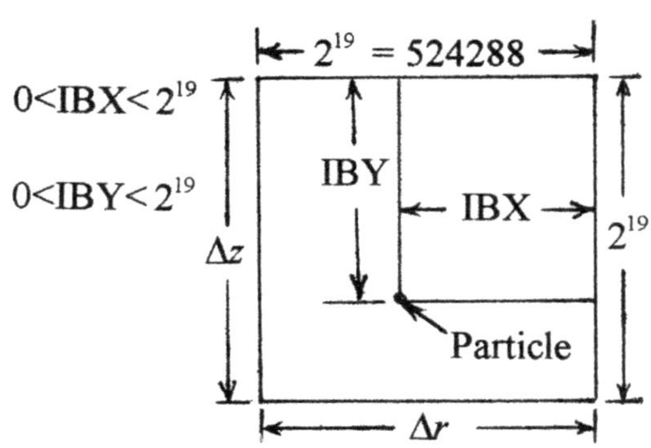

Figure 3.5. The reference point A of any cell (IJ) is defined at the right-top corner with IBX, IBY indicating the position of the particle (Note: IBX and IBY are always positive).

The locations of the particle, i.e. IBX and IBY are illustrated in Fig. 3.5 with the computations:

$$R \text{ coordinate of particle (1)} = R \text{ coordinate of point } A - \text{IBX}/2^{19}.$$
$$Z \text{ coordinate of particle (1)} = Z \text{ coordinate of point } A - \text{IBY}/2^{19}.$$

Mass of particle (1) = mass of material 1/64.

Velocity of particle (1) = interpolated from two velocities, i.e. $U_{i+\frac{1}{2}}$ and $U_{i-\frac{1}{2}}$.

Then, particle (1) is moved to the new position. The logic for creating new particle is as follows:

Step 1. The current calculation is at cell (i).
Step 2. If cell (i) is a mixed cell, then cell (i) should have particles defined from previous cycle calculation.
Step 3. If the velocity $U_{i+\frac{1}{2}}$ is going to the right direction, and if the material at cell ($i+1$) is different from cell (i), then, particles will be created for cell (i).
Step 4. If cell (i) is a pure material cell, i.e. only one material presented, but cell (i) has to spall, then, we will create one type (say blue) of particle for the material and another particle type (say red) for the void space which is defined by the local pressure (i.e. $P = 0.0$). Assign 16 particles for the void volume using random number generator to obtain the locations, i.e. the coordinates (R, Z) of each particle.

3.3.6 Computation for Spall

The computation on or around a spalled cell can vary widely depending on the accuracy desired on spall. Mathematically, the material failure creates a discontinuity within the computational domain. Therefore, for a rigorous treatment, a proper interface condition (possibly jump condition) has to be established across a spalled cell. Basically, spall or fracture has Lagrangian characteristics and to resolve it in an Eulerian mesh system will require a special spall logic. For the present computation, a rather simple scheme is suitable for the particle method.

3.3.6.1 When Spalled

A void region is created and located at the cell center with a minimum number of 16 particles. Then, for subsequent time steps, this is convected

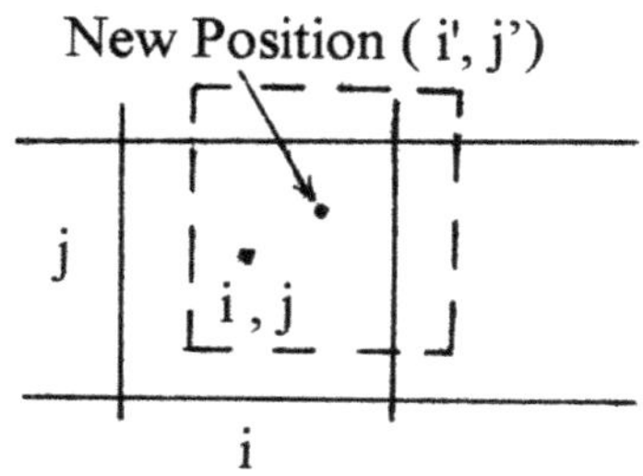

Figure 3.6. Old and new centers of a spalled cell.

with $\bar{U}$ and $\bar{V}$. $P$and S^{ij} are set to be zero at this point. This is essentially the same as introducing a rectangular void at the cell center. The void region has a volume based on the equation of state initially and may grow later.

3.3.6.2 Recombination

If the center of a spalled cell (i, j), as shown in Fig. 3.6 is moved to a new location (i', j'), the specific volume τ for (i', j') is computed by averaging value in the dotted cell. When $\tau \leq \tau_{test}$, the dotted cell is assumed to recombine.

3.3.6.3 Propagation of Fracture Surface

The particle method in its present form will not follow a fracture surface exactly. Rather, the density of the fractured cell will be lowered to behave essentially the same as an empty cell. The propagation of a crack or a material separation will be resolved in this fashion. A possible future option would be to define an interface. Particle can be created in the fractured cell and assigned a different identification number for the fractured void space. (See Fig. 3.7.)

3.3.7 Time Step Control

The time step control in the present particle method is done empirically. The following condition has been applied successfully:

$$N_c \triangleq \frac{(|U| + C)\delta t}{\delta x} < 0.4 \,, \tag{3.57}$$

where $C \triangleq \sqrt{\left(\frac{\partial P}{\partial P}\right)_s}$.

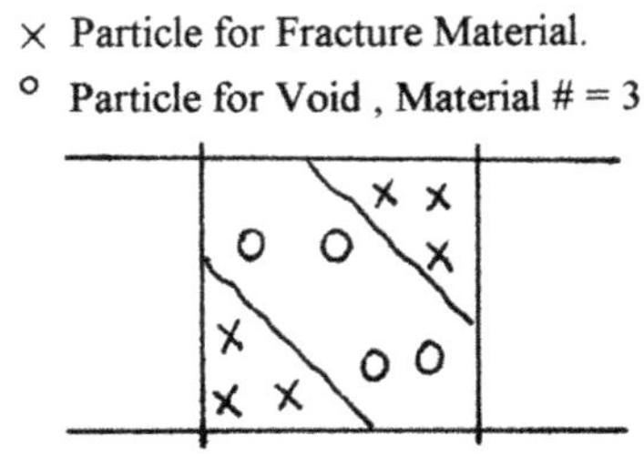

Figure 3.7. Material number 31 is assigned to a spalled or fractured void space.

The above definition of bulk sound speed "C" is rather arbitrary in an elastic material. Physically in an elastic material, wave propagates in a longitudinal or transverse wave form, as shown below. In isotropic elastic media, equations of small motions in the absence of body force are

$$(\lambda + G)\,\frac{\partial e}{\partial x_i} + G\nabla^2 U_i - \rho\frac{\partial^2 U_i}{\partial t^2} = 0\,, \tag{3.58}$$

where e = volume expansion and U_i = displacement in i-direction.

When $e = 0$, the deformation consists of shearing distortion and rotation only, and Eq. (3.58) becomes

$$G\nabla^2 U_i - \rho\frac{\partial^2 U_i}{\partial t^2} = 0\,. \tag{3.59}$$

This is the equation for "wave of distortion" or "transverse waves." Then the wave speed is

$$C_t = \sqrt{\frac{G}{\rho}}\,. \tag{3.60}$$

For the irrotational case, Eq. (3.58) becomes

$$(\lambda + 2G)\Delta^2 U_i - \rho\frac{\partial^2 U_i}{\partial t^2} = 0\,. \tag{3.61}$$

This is the equation for "irrotational waves," "wave of dilatation," or "longitudinal waves." The wave speed is then

$$C_\ell = \sqrt{(\lambda + 2G)/\rho}\,. \tag{3.62a}$$

From thermodynamic relations, one can write the sound speed for an elastic solid as

$$C^2 = \frac{K}{\rho}\left[1 + \frac{\beta^2 KT}{\rho C_v}\right] = \frac{K}{\rho}\,, \tag{3.62b}$$

and

$$C_\ell^2 = C^2 + \frac{4}{3}\frac{G}{\rho}\,, \tag{3.62c}$$

where $\beta =$ thermal expansion coefficient and $K =$ compression modulus.

3.4 Truncation Error Analysis

Here we prove the truncation error of the proposed method and explain details of the particle transport and remapping. Again radial equations are considered, and for convenience, the radial velocity is assumed to be positive. The analysis is done on mass, momentum, energy, and stress equations.

3.4.1 Mass

From Eq. (3.56), a particle moves with

$$\begin{aligned}\bar{U} &= \frac{1}{2}(U^n + \tilde{U}) \\ &= U^n + \frac{\delta t}{2}\left(\frac{\sigma_R^{RR}}{\rho} + \alpha\frac{\Delta\sigma}{R\rho}\right)^{n+1/2} \\ &= \tilde{U} - \frac{\delta t}{2}\left(\frac{\sigma_R^{RR}}{\rho} + \alpha\frac{\Delta\sigma}{R\rho}\right)^{n+1/2}\,,\end{aligned} \tag{3.63}$$

where $\Delta\sigma = \sigma^{RR} - \sigma^{\theta\theta}$.

The maximum distance ℓ that a section of fluid can cross the cell boundary during a time step δt, is shown in Fig. 3.8. Let the $\bar{U}(R_m)$ be the

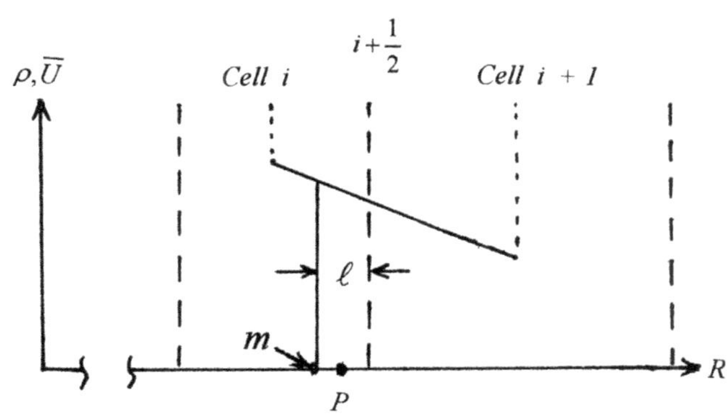

Figure 3.8. A particle originally at position P, can move with a maximum distance ℓ.

average velocity at the location of $R = R_m$, then

$$\bar{U}\,(R_m) = \bar{U}_{i+\frac{1}{2}} - \ell\bar{U}_R = \frac{\bar{U}_i + \bar{U}_{i+1}}{2} + \ell\frac{\bar{U}_i - \bar{U}_{i+1}}{\delta R}\,, \tag{3.64}$$

and

$$\ell = \bar{U}\,(R_m)\,\delta t\,. \tag{3.65}$$

From Eqs. (3.64) and (3.65),

$$\ell = \frac{\bar{U}_{i+\frac{1}{2}}\delta t}{1 + \delta t(\bar{U}_R)_{i+\frac{1}{2}}} = [\bar{U}\delta t(1 - \delta t\bar{U}_R)]_{i+\frac{1}{2}} + O(\Delta^3)\,. \tag{3.66}$$

Then

$$\begin{aligned}
\delta M_{i+\frac{1}{2}} &= \text{mass transported from } (i) \text{ to } (i+1) \text{ during } \delta t \\
&= 2\pi\delta Z \int_{R_{i+1/2}-\ell}^{R_{i+1/2}} \rho(R)RdR \\
&= 2\pi\delta Z \int_{R_{i+1/2}-\ell}^{R_{i+1/2}} [\rho_{i+1/2} - (R_{i+1/2} - R)\rho_R]RdR \\
&= 2\pi\delta Z \left\{ \left[R\rho\ell - (R\rho_R + \rho)\,\frac{\ell^2}{2} \right]_{i+1/2} + \frac{1}{3}\rho_R\ell^3 \right\}.
\end{aligned} \tag{3.67a}$$

Substituting Eqs. (3.63) and (3.65) into Eq. (3.67a), we get

$$\begin{aligned}
\delta M_{i+\frac{1}{2}} &= 2\pi\delta Z \left[R\rho(\bar{U}\delta t - \bar{U}\bar{U}_R\delta t^2) - (R\rho_R + \rho)\frac{\bar{U}^2\delta t^2}{2} \right]_{i+\frac{1}{2}} + O(\Delta^4) \\
&= 2\pi\delta Z\delta t \left\{ R\rho\bar{U} - \delta t \left[R\rho\bar{U}\bar{U}_R + \frac{(R\rho_R + \rho)\bar{U}^2}{2} \right] \right\}_{i+\frac{1}{2}} + O(\Delta^4) \\
&= 2\pi\delta Z\delta t \left\{ R\rho\bar{U} - \frac{1}{2}\delta t \left[\left(R\sigma_R^{RR} + \Delta\sigma \right) + 2R\rho\bar{U}\bar{U}_R \right.\right. \\
&\quad \left.\left. + \,(R\rho_R + \rho)\,\bar{U}^2 \right] \right\}_{i+\frac{1}{2}} + O(\Delta^4)\,.
\end{aligned} \tag{3.67b}$$

The total mass change in cell (i), ΔM_i, is

$$\begin{aligned}
\Delta M_i &= \delta M_{i-\frac{1}{2}} \text{ (in flow)} - \delta M_{i+\frac{1}{2}} \text{ (out flow)} \\
&= -\delta R\frac{\partial}{\partial R}(\delta M) + O(\delta M) \times O(\Delta^3) \\
&= -2\pi(\delta Z)(\delta R)(\delta t)\frac{\partial}{\partial R}\,[\text{from RHS of Eq. (3.67b)}] \\
&\quad + O(\delta M) \times O(\Delta^3)\,.
\end{aligned} \tag{3.67c}$$

Now the density change in cell (i) can be calculated as below.

$$\begin{aligned}\Delta\rho_i &= \rho_i^{n+1} - \rho_i^n \\ &= \frac{\Delta M_i}{2\pi R\delta R\delta Z} = -\frac{\delta t}{R}\left\{R\rho\left[\tilde{U} - \frac{\delta t}{2}\left(\frac{\sigma_R}{\rho} + \frac{\Delta\sigma}{R\rho}\right)\right]\right. \\ &\quad \left. - \,\delta t\left[R\rho U U_R + \frac{(R\rho_R + \rho)U^2}{2}\right]\right\}_R + O(\Delta^3)\,. \end{aligned} \tag{3.68}$$

Now we will prove that Eq. (3.68) satisfies Eq. (3.19) within $O(\Delta^2)$.

From the Taylor series expansion,

$$\rho_t = \frac{\rho^{n+1} - \rho^n}{\delta t} - \frac{\delta t}{2}\rho_{tt} + O(\Delta^2)\,. \tag{3.69}$$

Substituting Eq. (3.68) and ρ_{tt} from Eq. (3.19) into Eq. (3.69) gives

$$\begin{aligned}\rho_t &= -\frac{1}{R}\left\{(R\rho U)_R + \delta t\left[\frac{(R\sigma_R^{RR} + \Delta\sigma)}{2} - R\rho U U_R - \frac{(R\rho_R + \rho)\,U^2}{2}\right]_R\right\} \\ &\quad - \frac{\delta t}{2}\left[-(\rho U)_R - \frac{\rho U}{R}\right]_t + O(\Delta^2) \\ &= -\frac{1}{R}\left\{(R\rho U)_R + \delta t\left[\frac{(R\sigma_R^{RR} + \Delta\sigma)}{2} - R\rho U U_R - \frac{(R\rho_R + \rho)\,U^2}{2}\right]_R\right\} \\ &\quad - \frac{\delta t}{2}\left\{[-(\rho U)_t]_R - \left(\frac{\rho U}{R}\right)_t\right\} + O(\Delta^2) \\ &= (\rho U)_R - \frac{\rho U}{R} + O(\Delta^2)\,. \end{aligned} \tag{3.70}$$

Therefore, Eq. (3.68) satisfies the conservation of mass equation to second-order accuracy.

3.4.2 Radial Momentum

New velocity is calculated based on momentum change during particle transport. The momentum transported from cell (i) to cell $(i+1)$ can be calculated as below.

$$\begin{aligned}\delta\big(M\tilde{U}\big)_{i+\frac{1}{2}} &= 2\pi\delta Z\int_{R_{i+1/2}-\ell}^{R_{i+1/2}} \rho\tilde{U}R dR \\ &= 2\pi\delta Z\int \rho\left[\tilde{U}_{i+1/2} - (R_{i+1/2} - R)\,\tilde{U}_R\right] R dR \\ &= \delta M_{i+\frac{1}{2}}\left(\tilde{U} - \frac{\delta t}{2}\tilde{U}\tilde{U}_R\right)_{i+\frac{1}{2}} + O(\Delta^4)\,. \end{aligned} \tag{3.71}$$

Then the total momentum change in cell (i) is

$$\Delta(M\tilde{U})_i = \delta(M\tilde{U})_{i-\frac{1}{2}} - \delta(M\tilde{U})_{i+\frac{1}{2}} \\ = -\delta R \frac{\partial}{\partial R}[\delta(M\tilde{U})] + O(\Delta^5). \tag{3.72}$$

New velocity U^{n+1} can be calculated as below

$$U^{n+1} = \frac{M\tilde{U} + \Delta(M\tilde{U})}{M + \Delta M} = \frac{\tilde{U} + \Delta(M\tilde{U})/M}{1 + \Delta M/M}. \tag{3.73}$$

Expanding Eq. (3.71) we get

$$\delta(M\tilde{U})_{i+\frac{1}{2}} = 2\pi\delta Z\delta t \Big\{ R\rho\tilde{U}^2 - \frac{\delta t}{2}\left[RU\sigma_R^{RR} + U\Delta\sigma + 3R\rho U^2 U_R \right. \\ \left. + (R\rho_R + \rho)U^3\right]\Big\}_{i+1/2} + O(\Delta^4). \tag{3.73a}$$

Now from Eq. (3.72)

$$\delta(M\tilde{U})_{i+\frac{1}{2}} = 2\pi\delta Z\delta t\delta R \Big\{ (R\rho\tilde{U}^2)_R - \delta t\frac{\partial}{\partial R}\text{ [second term in the RHS} \\ \text{of Eq. (3.73a)]}\Big\} + O(\Delta^5). \tag{3.73b}$$

Then

$$\frac{\Delta(M\tilde{U})}{M} = \frac{\Delta(M\tilde{U})}{(2\pi\delta R\delta Z)\rho} = \underbrace{-\frac{\delta t}{\rho R}(\rho\tilde{U}^2)_R}_{\alpha}\underbrace{-\frac{\delta t^2}{\rho R}\Big\{\frac{\partial}{\partial R}}_{\alpha'}\text{[second term} \\ \text{in the RHS of Eq. (3.73a)]}\Big\} + O(\Delta^5). \tag{3.73c}$$

From Eq. (3.67c),

$$\frac{\Delta M}{M} = \underbrace{-\frac{\delta t}{R\rho}\left(R\rho\tilde{U}^2\right)_R}_{\beta} \\ \underbrace{+\frac{\delta t^2}{R\rho}\left\{\frac{1}{2}(R\sigma_R)_R + \frac{1}{2}(\Delta\sigma)_R + (R\rho UU_R)_R + \frac{1}{2}\left[(R\rho_R + \rho)U^2\right]_R\right\}}_{\beta'} \\ + O(\Delta^5). \tag{3.73d}$$

From Eq. (3.73),

$$\begin{aligned} U^{n+1} &= \frac{\tilde{U} + \Delta(M\tilde{U})/M}{1 + \Delta M/M} \\ &= \frac{\tilde{U} + \alpha + \alpha'}{1 + \beta + \beta'} \\ &= \tilde{U} + (\alpha - \beta\tilde{U}) + (\alpha' - \alpha\beta + \beta^2\tilde{U} - \beta'\tilde{U}) + O(\Delta^3)\,, \end{aligned} \tag{3.73e}$$

$$\begin{aligned} (\alpha - \beta\tilde{U}) &= -\frac{\delta t}{R\rho}(R\rho\tilde{U}^2)_R + \frac{\delta t}{R\rho}(R\rho\tilde{U}^2)_R\tilde{U} \\ &= \delta t\tilde{U}\tilde{U}_R - \delta t^2\left[\left(U\frac{\sigma_R}{\rho}\right)_R + \left(U\frac{\Delta\sigma}{R\rho}\right)_R\right] + O(\Delta^3)\,, \end{aligned} \tag{3.73f}$$

and

$$\begin{aligned} &(\alpha' - \alpha\beta + \beta^2\tilde{U} - \beta'\tilde{U})/\delta t^2 \\ &\quad = \frac{1}{2R\rho}\left[R\sigma_R U_R + U_R\Delta\sigma + R\rho\left(U^2U_R\right)_R\right]. \end{aligned} \tag{3.73g}$$

Now rewriting Eq. (3.73),

$$\begin{aligned} U^{n+1} &= \tilde{U} + [\text{Eq. (3.73f)}] + [\delta t^2 \text{ Eq. (3.73f)}] \\ &= U^n + \delta t\left(\frac{\sigma_R^{RR}}{\rho} + \frac{\Delta\sigma}{R\rho}\right)^{n+1/2} - \delta t U U_R \\ &\quad - \delta t^2\left[\left(\frac{U\sigma_R^{RR}}{\rho}\right)_R + \left(\frac{U\Delta\sigma}{R\rho}\right)_R\right] \\ &\quad + \frac{\delta t^2}{2R\rho}\left[R\sigma_R U_R + U_R\Delta\sigma + R\rho(U^2U_R)_R\right] + O(\Delta^3)\,. \end{aligned} \tag{3.74}$$

Again from the Taylor series expansion,

$$U_t = \frac{U^{n+1} - U^n}{\delta t} - \frac{\delta t}{2}U_{tt} + O(\Delta^2)\,. \tag{3.75}$$

Inserting Eq. (3.74) and U_{tt} from differentiating Eqs. (3.20a) and (3.20b) into Eq. (3.75) and after some algebra, we can get

$$U_t = -UU_R + \left(\frac{\sigma_R^{RR}}{\rho} + \frac{\Delta\sigma}{R\rho}\right) + O(\Delta^2)\,. \tag{3.76}$$

Therefore, the present particle transport and remapping scheme results in a second-order accurate radial momentum equation.

3.4.3 Axial Momentum

Now we will consider the change in axial momentum during a time advancement in radial direction. For a hydrodynamic problem, $\tilde{V} = V^n$, the remapping of the axial momentum is straightforward. For an elastic-plastic flow, deviatoric stress component contributes to Lagrangian phase axial velocity. The truncation error analysis is similar to the radial momentum equation case.

Rewriting Eq. (3.28b),

$$\tilde{V} = V^n + \delta t \frac{dV}{dt} = V^n + \delta t \left(\frac{S_R^{RZ}}{\rho} + \frac{S^{RZ}}{R\rho} \right)^{n+1/2} . \tag{3.77}$$

The axial momentum transported from cell (i) to cell $(i+1)$ during δt is

$$\delta(M\tilde{V})_{i+\frac{1}{2}} = \delta M_{i+\frac{1}{2}} \left(\tilde{V} - \frac{\delta t}{2} \tilde{U} \tilde{V}_R \right)_{i+\frac{1}{2}} + O(\Delta^4) . \tag{3.78}$$

The total axial momentum change in cell (i) is

$$\begin{aligned} \Delta(M\tilde{V})_i &= \delta(M\tilde{V})_{i-\frac{1}{2}} - \delta(M\tilde{V})_{i+\frac{1}{2}} \\ &= -\,\delta R \frac{\partial}{\partial R} \left[\delta \left(M\tilde{V} \right) \right] + O(\Delta^5) , \end{aligned} \tag{3.79}$$

and the new axial velocity is then calculated by

$$V^{n+1} = \frac{M\tilde{V} + \Delta(M\tilde{V})}{M + \Delta M} = \frac{\tilde{V} + \Delta(M\tilde{V})/M}{1 + \Delta M/M} . \tag{3.80}$$

Some algebraic detail:

$$\begin{aligned} \delta(M\tilde{V})_{i+\frac{1}{2}} = (2\pi\delta Z)\delta t \Big\{ R\rho\tilde{U}\tilde{V} - \delta t \Big[\frac{1}{2} RV\sigma_R^{RR} + \frac{1}{2} V\Delta\sigma + R\rho UVU_R \\ + \frac{1}{2}(R\rho_R + \rho)U^2V + \frac{1}{2} R\rho U^2 V_R \Big] \Big\}_{i+\frac{1}{2}} + O(\Delta^4) . \end{aligned} \tag{3.80a}$$

From Eq. (3.79),

$$\Delta(M\tilde{V}) = -2\pi\delta Z(\delta R\delta t)\left\{(R\rho\tilde{U}\tilde{V})_R - \delta t\frac{\partial}{\partial R}\text{ [from the RHS of Eq. (3.80a)]}\right\} + O(\Delta^5)\,. \tag{3.80b}$$

Then

$$\begin{aligned}\frac{\Delta(M\tilde{V})}{M} &= \frac{\Delta(M\tilde{V})}{(2\pi R\delta R\delta Z)\rho} \\ &= \underbrace{-\frac{\delta t}{R\rho}(R\rho\tilde{U}\tilde{V})_R}_{A} + \underbrace{\frac{\delta t^2}{R\rho}\left\{\frac{\partial}{\partial R}\right.}_{A'}\text{ [from the RHS of Eq. (3.80a)]} \\ &\quad + O(\Delta^3)\,.\end{aligned} \tag{3.80c}$$

From Eq. (3.80),

$$\begin{aligned}V^{n+1} &= \frac{\tilde{V} + A + A'}{1 + \beta + \beta'} \\ &= \tilde{V} + \big(A - \beta\tilde{V}\big) + \big(A' - A\beta + \beta^2\tilde{V} - \beta'\tilde{V}\big) + O(\Delta^3)\,,\end{aligned} \tag{3.80d}$$

$$\begin{aligned}\big(A - \beta\tilde{V}\big) &= -\frac{\delta t}{R\rho}\big(R\rho\tilde{U}\tilde{V}\big)_R + \frac{\delta t^2}{R\rho}\big(R\rho\tilde{U}\big)_R\tilde{V} \\ &= -\delta t\left\{UV_R + \delta t\left[U\left(\frac{S_R}{\rho} + \frac{S}{R\rho}\right)_R + V_R\left(\frac{\sigma_R}{\rho} + \frac{\Delta\sigma}{R\rho}\right)\right]\right\} \\ &\quad + O(\Delta^2)\,.\end{aligned} \tag{3.80e}$$

$$\begin{aligned}&\big(A' - A\beta + \beta^2\tilde{V} - \beta'\tilde{V}\big)/\delta t \\ &\quad = \frac{1}{2R\rho}\left[(R\sigma_R V_R) + (\Delta\sigma V_R) + R\rho\left(U^2 V_R\right)_R\right].\end{aligned} \tag{3.80f}$$

Rewriting Eq. (3.80),

$$\begin{aligned}V^{n+1} &= \tilde{V} + \text{RHS of Eq. (3.80e)} + \delta t^2[\text{RHS of Eq. (3.80f)}] \\ &= V^n + \delta t\left(\frac{S_R^{RZ}}{\rho} + \frac{S^{RZ}}{R\rho}\right)^{n+1/2} - \delta t U V_R \\ &\quad - \delta t^2\left[U\left(\frac{S_R}{\rho} + \frac{S}{R\rho}\right)_R + V_R\left(\frac{\sigma_R}{\rho} + \frac{\Delta\sigma}{R\rho}\right)\right] \\ &\quad + \frac{\delta t^2}{2R\rho}\left[R\sigma_R V_R + \Delta\sigma(V_R) + R\rho\big(U^2 V_R\big)_R\right] + O(\Delta^3)\,.\end{aligned} \tag{3.81}$$

From the Taylor series expansion,

$$V_t = \frac{V^{n+1} - V^n}{\delta t} - \frac{\delta t}{2} V_{tt} + O(\Delta^2)\,. \tag{3.82}$$

Inserting Eq. (3.81) and V_{tt} from differentiating Eqs. (3.20a) and (3.20b) into Eq. (3.82) and after some simplification, we get

$$V_t = -UV_R + \frac{1}{\rho}\left(S_R^{RZ} + \frac{S^{RZ}}{R}\right) + O(\Delta^2)\,. \tag{3.83}$$

Therefore, we have a second-order accurate axial momentum equation.

3.4.4 Internal Energy

Following a procedure similar to that in Section 3.4.3, we get

$$\varepsilon^{n+1} = \tilde{\varepsilon} - \delta t \tilde{U} \tilde{\varepsilon}_R + \frac{\delta t^2}{2R\rho}\left[R\sigma_R^{RR}\varepsilon_R + \varepsilon_R \Delta\sigma + R\rho\left(U^2\varepsilon_R\right)_R\right] + O(\Delta^3)\,. \tag{3.84}$$

The truncation error of this depends on the $\tilde{\varepsilon}$ expression. From the Taylor series, we have

$$\varepsilon_t = \frac{\varepsilon^{n+1} - \varepsilon^n}{\delta t} - \frac{\delta t}{2}\varepsilon_{tt} + O(\Delta^2)\,, \tag{3.85}$$

and inserting $\tilde{\varepsilon}$ from Eq. (3.28c) into Eq. (3.84), we get

$$\begin{aligned}
\frac{\varepsilon^{n+1} - \varepsilon^n}{\delta t} = & \left(\frac{\sigma^{RR}U_R}{\rho} + \frac{\sigma^{\theta\theta}U}{R_\rho} + \frac{S^{RZ}V_R}{\rho}\right)^n + \frac{\delta t}{2}\left[\left(\frac{\sigma U_R}{\rho}\right)_t + U\left(\frac{\sigma U_R}{\rho}\right)_R\right. \\
& \left. + \left(\frac{\sigma^{\theta\theta}U}{R\rho}\right)_t + U\left(\frac{\sigma^{\theta\theta}U}{R\rho}\right)_R + \left(\frac{SV_R}{\rho}\right)_t + \left(\frac{SV_R}{\rho}\right)_R\right] - U\varepsilon_R \\
& - \delta t\left[\left(\frac{\sigma_R}{\rho} + \frac{\Delta\sigma}{\rho}\right)\varepsilon_R + U\left(\frac{\sigma U_R}{\rho} + \frac{\sigma^{\theta\theta}U}{R\rho} + \frac{SV_R}{\rho}\right)_R\right] \\
& + \frac{\delta t}{2}\left[\left(\frac{\sigma_R\varepsilon_R}{\rho}\right) + \left(\frac{\varepsilon_R\Delta\sigma}{R\rho}\right) + \left(U^2\varepsilon_R\right)_R\right] + O(\Delta^2)\,.
\end{aligned} \tag{3.86}$$

From Eq. (3.21),

$$\begin{aligned}
\varepsilon_{tt} &= -\left(U\varepsilon_R\right)_t + \left[\frac{1}{\rho}\left(\sigma^{RR}U_R + \frac{\sigma^{\theta\theta}U}{R} + SV_R\right)\right]_t \\
&= -\varepsilon_R U_t - U\varepsilon_{tR} + \left[\frac{1}{\rho}\left(\sigma^{RR}U_R + \frac{\sigma^{\theta\theta}U}{R} + SV_R\right)\right]_t\,.
\end{aligned} \tag{3.87}$$

Inserting Eqs. (3.86) and (3.87) into Eq. (3.85),

$$\begin{aligned}\varepsilon_t &= -U\varepsilon_R + \frac{1}{\rho}\left(\sigma^{RR}U_R + \frac{\sigma^{\theta\theta}U}{R} + SV_R\right)\\ &= -U\varepsilon_R + \frac{1}{\rho}\left(\sigma^{RR}U_R + \sigma^{\theta\theta}\frac{U}{R} + \sigma^{RZ}V_R\right) + O(\Delta^2)\,. \qquad (3.88)\end{aligned}$$

Therefore, we have a second-order scheme for ε.

Note that if the following expression is used instead of Eq. (3.28c), we get a first-order accurate energy equation:

$$\tilde{\varepsilon} = \varepsilon^n + \frac{\delta t}{\rho^n}\left[\left(\sigma^{RR}\right)^{n+1/2}\tilde{U}_R + \left(\sigma^{RZ}\right)^{n+1/2}\tilde{V}_R + \left(\sigma^{\theta\theta}\right)^{n+1/2}\frac{\tilde{U}}{R}\right]. \quad (3.88a)$$

Since we split governing equations in radial and axial directions in transforming the Eulerian to the Lagrangian coordinate system, we have

Jacobian = 1

$$\frac{\partial}{\partial t} = U\frac{\partial}{\partial R} + \frac{\partial}{\partial t}\,.$$

Therefore, Eq. (3.28c) in the Lagrangian phase calculation is a straightforward consequence.

3.4.5 Deviatoric Stresses

In the present scheme, stresses are assigned to each individual material. Therefore, the difference between volume-weighting and mass-weighting scheme in transporting stress components is contributed by the local density gradient. Both schemes result in second-order accurate results. However, for programming, the mass-weighting scheme is more convenient, and the following truncation error analysis is based on that. Similarly to Eq. (3.84), we get for S^{ij} transport

$$(S^{ij})^{n+1} = \tilde{S}^{ij} - \delta t\tilde{U}(\tilde{S})^{ij}_R + \frac{\delta t^2}{2\rho}\left[S_R\sigma_R + \frac{S_R\Delta\sigma}{R} + \left(\rho U^2 SR\right)_R\right] + O(\Delta^3)\,. \quad (3.89)$$

Inserting this expression and $\left(S^{ij}\right)_t$ from Eqs. (3.23a–d) or (3.24a–e) into the following,

$$(S^{ij})_t = \left[(S^{ij})^{n+1} - (S^{ij})^n\right]/\delta t - \frac{\delta t^2}{2}(S^{ij})_{tt} + O(\Delta^2)\,, \quad (3.90)$$

we can obtain the truncation error. We can generalize $\tilde{S}^{ij}$ by Eq. (3.44), that is,

$$\tilde{S} = S^n + \delta t\big(2G\dot{e} - A\dot{W}S\big)^{n+1/2} . \tag{3.91}$$

Then

$$\begin{aligned}\tilde{U}\tilde{S}_R &= \left[U + \delta t\left(\frac{\sigma_R}{\rho} + \frac{\Delta\sigma}{R\rho}\right)\right]\left[S_R + \delta t\left(2G\dot{e} - A\dot{W}S\right)_R\right] \\ &= US_R + \delta t\left[S_R\left(\frac{\sigma_R}{\rho} + \frac{\Delta\sigma}{R\rho}\right) + U\left(2G\dot{e} - A\dot{W}S\right)_R\right] + O(\Delta^2) .\end{aligned} \tag{3.92}$$

Rewriting Eq. (3.43a),

$$S_t = -US_R + 2G\dot{e} - A\dot{W}S .$$

Then

$$S_{tt} = -(US_R)_t + (2G\dot{e} - A\dot{W}S)_t . \tag{3.93}$$

Now from Eq. (3.90), we get

$$S_t^{ij} = -\big(US_R^{ij}\big) + (2G\dot{e}^{ij} - A\dot{W}S^{ij}) + O(\Delta^2) . \tag{3.94}$$

Therefore, mass-weighted transport and remapping of S^{ij} result in second-order accuracy.

3.5 Equivalent Plastic Strain

Total strain rate is

$$\dot{e}^{ij} = (\dot{e}')^{ij} + (\dot{e}'')^{ij} = (\dot{e}')^{ij} + \dot{\alpha}^{ij} , \tag{3.95}$$

where $(\dot{e}')^{ij}$ is the elastic and $(\dot{e}'')^{ij}$ the plastic parts. Since Hook's law is used for elastic phase

$$(\dot{S})^{ij} = 2G(\dot{e}')^{ij} , \tag{3.96}$$

and for plastic part

$$(\dot{\alpha})^{ij} = \frac{\dot{W}}{\frac{2}{3}(Y^0)^2} S^{ij} . \tag{3.97}$$

Combining these will result in a Prandtl-Reuss equation for plastic flow as shown in Eq. (3.7), that is

$$\begin{aligned}(\dot{S})^{ij} &= 2G(\dot{e})^{ij} - 2G(\dot{\alpha})^{ij} \\ &= 2G(\dot{e})^{ij} - \frac{G\dot{W}}{\frac{1}{3}(Y^0)^2}S^{ij}\,. \end{aligned} \tag{3.97a}$$

Then α_p is defined by

$$\alpha_p \triangleq \left(\frac{2}{3}\alpha^{ij}\alpha^{ij}\right)^{1/2} . \tag{3.98}$$

In the present method, the equivalent plastic strain, α_p, is calculated as below:

$$\alpha_p = \int \left(\frac{2}{3}\dot{\alpha}^{ij}\dot{\alpha}^{ij}\right)^{1/2} dt\,, \tag{3.99}$$

where

$$\begin{aligned}\dot{\alpha}^{ij}\dot{\alpha}^{ij} &= \left(\frac{\dot{W}}{\frac{2}{3}(Y^0)^2}\right)^2 S^{ij}S^{ij} \\ &= \left(\frac{\dot{W}}{\frac{2}{3}(Y^0)^2}\right)^2 \left[\left(S^{RR}\right)^2 + \left(S^{ZZ}\right)^2 + \left(S^{\theta\theta}\right)^2 + 2\left(S^{RZ}\right)^2\right] \\ &= \left(\frac{\dot{W}}{\frac{2}{3}(Y^0)^2}\right)^2 \phi\,. \end{aligned} \tag{3.100}$$

It is noted that for a perfect plastic model

$$\phi = \frac{2}{3}(Y^0)^2\,, \tag{3.100a}$$

and then

$$\dot{\alpha}^{ij}\dot{\alpha}^{ij} = (\dot{W})^2\,, \tag{3.100b}$$

and

$$\alpha_p = \sqrt{\frac{2}{3}}\int \dot{W}dt\,. \tag{3.100c}$$

Numerically, Eq. (3.100) is evaluated as

$$\begin{aligned}(\alpha_p)^n &= (\alpha_p)^{n-1} + (\Delta\alpha_p)(\Delta t) \\ &= (\alpha_p)^{n-1} + \sqrt{\frac{2}{3}}\frac{\dot{W}}{\frac{2}{3}(Y^0)^2}\phi^{1/2}(\Delta t)\,, \end{aligned} \qquad (3.101)$$

where

$$\begin{aligned}\dot{W} &= (\dot{e})^{ij}S^{ij} \\ &= U_R S^{RR} + (U_Z + V_R)\,S^{RZ} + V_Z S^{ZZ} + \alpha\frac{U}{R}S^{\theta\theta}\,. \end{aligned} \qquad (3.101a)$$

3.6 FCT Applied to Second-Order PIC

3.6.1 Introduction

The FCT scheme, originally proposed by Boris and Book [3.10], consists of two stages, transport and flux-corrected anti-diffusion. By allowing flux correction, the FTC method generates no local extreme and maintains the positivity of mass and energy density. Thus, steep gradients and shocks can be handled well.

The two stages in FCT algorithm resemble the two computational phases of the PIC scheme, so a slightly modified FCT algorithm can easily be incorporated into the PIC method. Since the positivity of mass is automatically satisfied by PIC, the effect of anti-diffusion is utilized here to handle local extreme. For a semi-continuous version of PIC (see Clark [3.9]), the positivity of mass is maintained by a modified mass transport scheme, which is explained next.

3.6.2 Modified Mass Transport

Around a very steep gradient or local extreme, the mass transport scheme described in Section 3.4.1 may result in negative mass in certain cells. This situation is illustrated in Fig. 3.9.

According to Section 3.4.1, the mass represented by the shaded area will be transported from cell (i) to cell $(i+1)$. When the density gradient is steep locally, this transported mass can be larger than the total mass in cell (i). To prevent this situation, the mass transport scheme is modified as below. Let us consider a linearly interpolated density profile as given in Eq. (3.103) and shown in Fig. 3.10.

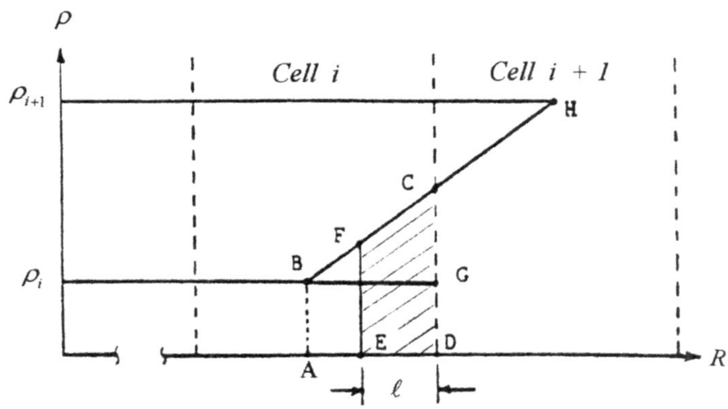

Figure 3.9. A steep gradient of density may result in negative mass during transport.

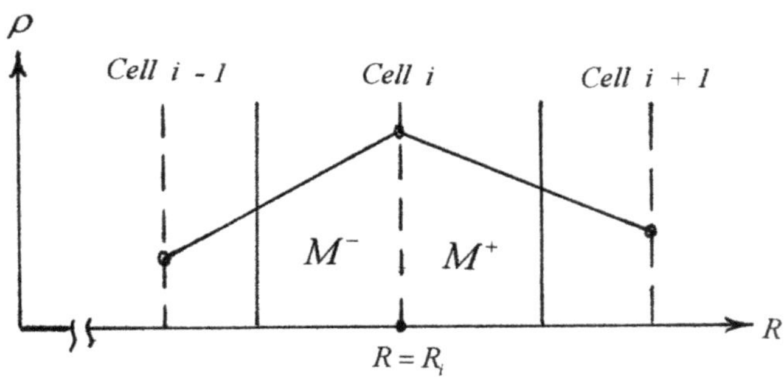

Figure 3.10. Modified mass in cell i based on linear density profile.

Average mass in cell (i):

$$\begin{aligned} M_{avg} &= 2\pi\delta Z \int_{R_i-\Delta R/2}^{R_i+\Delta R/2} \rho_i R' dR' \\ &= (\pi\delta Z\rho_i)\left[\left(R_i + \frac{\Delta R}{2}\right)^2 - \left(R_i - \frac{\Delta R}{2}\right)^2\right] \\ &= 2\pi\delta Z\rho_i(R_i\Delta R)\,. \end{aligned} \tag{3.102}$$

Mass in cell (i) based on linear density profile is:

$$M_{lin} = M^- + M^+\,. \tag{3.103}$$

Let us denote, for convenience, that

$$\begin{aligned}
\rho^- &= \rho_{i-1}, \quad \rho^+ = \rho_{i+1}, \quad R^+ = R_{i+1} \text{ etc}\,. \\
R &= R_i, \quad \rho = \rho_i, \quad \lambda = \frac{\Delta R}{2}\,; \\
M^- &= (2\pi\delta Z)\int_{R-\delta}^{R} \{\rho^- + [R' - (R-\lambda)]\,\rho_R^-\}R'dR' \\
&= (2\pi\delta Z)\left\{[\rho^- + (\lambda - R)\,(\rho-\rho^-)/\lambda]\left(R\lambda - \frac{\lambda^2}{2}\right)\right. \\
&\quad \left. + (\rho-\rho^-)\left(R^2 - R\lambda + \frac{\lambda^2}{3}\right)\right\} \\
&= (2\pi\delta Z)\left[\frac{1}{2}R\rho^-\lambda + \frac{1}{2}R\rho\lambda - \frac{1}{2}\rho\lambda^2 + \frac{\lambda^2}{3}\,(\rho-\rho^-)\right], \qquad (3.103\text{a}) \\
M^+ &= (2\pi\delta Z)\int_{R}^{R+\delta} \{\rho + (R'-R)\rho_R^+\}R'dR' \\
&= (2\pi\delta Z)\left\{[\rho - R\,(\rho^+ - \rho)\,/\lambda]\left(R\lambda + \frac{\lambda^2}{2}\right)\right. \\
&\quad \left. + (\rho^+ - \rho)\left(R^2 + R\lambda + \frac{\lambda^2}{3}\right)\right\} \\
&= (2\pi\delta Z)\left[\rho\left(R\lambda + \frac{\lambda^2}{2}\right) + \frac{1}{2}R\lambda\,(\rho^+-\rho) + \frac{\lambda^2}{3}\,(\rho^+-\rho)\right], \qquad (3.103\text{b})
\end{aligned}$$

and

$$\begin{aligned}
M_{lin} &= M^- + M^+ \\
&= 2\pi\delta Z\left[R\frac{\Delta R}{4}\,(\rho^- + 2\rho + \rho^+) + \frac{(\Delta R)^2}{12}\,(\rho^+ - \rho^-)\right]. \qquad (3.104)
\end{aligned}$$

Linear mass transport:

Mass transported from cell (i) to cell $(i+1)$ is, from Eq. (3.67a),

$$\delta M^+ = 2\pi\delta Z\ell\left\{R\rho - \frac{\ell}{2}\left[\rho + \left(R - \frac{2}{3}\ell\right)\rho_R\right]\right\}_{i+\frac{1}{2}}, \qquad (3.105)$$

where $\rho_R = \frac{\rho^+ - \rho}{(\frac{\Delta R}{2})}$.

Similarly, the mass from cell (i) to cell $(i-1)$ is

$$\delta M^- = 2\pi\delta Z\ell\left\{R\rho + \frac{\ell}{2}\left[\rho + \left(R - \frac{2}{3}\ell\right)\rho_R\right]\right\}_{i-\frac{1}{2}}, \tag{3.106}$$

where $\rho_R = \frac{\rho-\rho^-}{\left(\frac{\Delta R}{2}\right)}$.

Modified mass transport:

To avoid negative mass in a cell, Eqs. (3.105) or (3.106) are weighted as below

$$\left(\delta M_{i+\frac{1}{2}}\right)_{\text{mod}} = \frac{(\delta M^+)\,M_{avg}}{M_{lin}}, \tag{3.107}$$

$$\frac{M_{avg}}{M_{lin}} = \frac{\rho R}{\left[\frac{R}{4}\left(\rho^- + 2\rho + \rho^+\right) + \frac{\Delta R}{12}\left(\rho^+ - \rho^-\right)\right]}, \tag{3.107a}$$

$$\left(\delta M_{i+\frac{1}{2}}\right)_{\text{mod}} = \frac{(2\pi\rho R\Delta R\delta Z)\ell\left\{R_{i+\frac{1}{2}}\rho^+ - \ell\left[\frac{\rho^+}{2} + \left(R_{i+\frac{1}{2}} + \frac{2\ell}{3}\right)\frac{(\rho^+-\rho)}{\Delta R}\right]\right\}}{(\Delta R)^2\left[\frac{R}{4\Delta R}(\rho^+ + 2\rho + \rho^-) + \frac{1}{12}(\rho^+ - \rho^-)\right]}, \tag{3.108}$$

and

$$\left(\delta M_{i-\frac{1}{2}}\right)_{\text{mod}} = \frac{(2\pi\rho R\Delta R\delta Z)\ell\left\{R_{i-\frac{1}{2}}\rho^- - \ell\left[\frac{\rho^-}{2} + \left(R_{i+\frac{1}{2}} + \frac{2\ell}{3}\right)\frac{(\rho-\rho^-)}{\Delta R}\right]\right\}}{(\Delta R)^2\left[\frac{R}{4\Delta R}(\rho^+ + 2\rho + \rho^-) + \frac{1}{12}(\rho^+ - \rho^-)\right]}. \tag{3.109}$$

For the center cell as shown in Fig. 3.11,

$$M_{avg} = \frac{1}{4}\pi\delta Z\rho_i(\Delta R)^2, \tag{3.110}$$

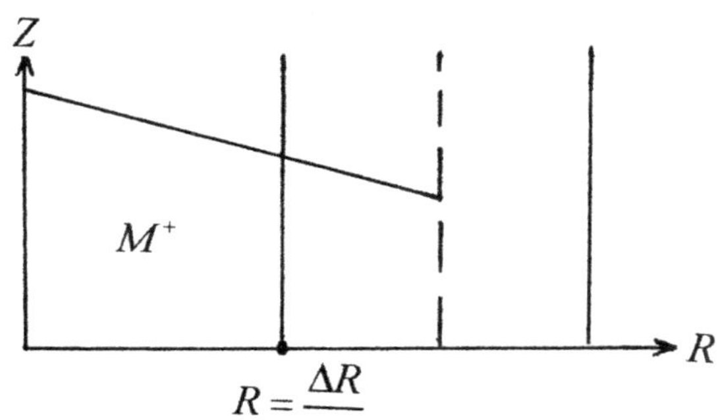

Figure 3.11. Special treatment for the cell along z-axis.

$$M_{lin} = 2\pi\delta Z \int_0^{\Delta R/2} (\rho_0 + \rho_R R')R'dR\,,$$

$$= 2\pi\delta Z(\Delta R)^2 \left[\frac{1}{8}\left(2\rho - \rho^+\right) + \frac{1}{6}\left(\rho^+ - \rho\right)\right], \qquad (3.111)$$

$$\delta M^+ = 2\pi\delta Z\ell\left\{\rho^+ R^+ - \frac{\ell}{2}\left[\rho^+ + \left(R^+ - \frac{2\ell}{3}\right)\rho_R\right]\right\}$$

$$= 2\pi\delta Z\ell\left[\frac{\rho^+(\Delta R - \ell)}{2} - \ell\left(\rho^+ - \rho\right)\left(1 - \frac{4}{3}\frac{\ell}{\Delta R}\right)\right], \qquad (3.112)$$

and

$$(\delta M_{\frac{1}{2}}) = \frac{\frac{\pi\delta Z\ell\rho_i(\Delta R)^2}{4}\left[\frac{\rho^+(\Delta R - \ell)}{2} - \ell\left(1 - \frac{4}{3}\frac{\ell}{\Delta R}\right)\left(\rho^+ - \rho_R\right)\right]}{(\Delta R)^2\left[\frac{1}{8}(2\rho - \rho^+) + \frac{1}{6\Delta R}(\rho^+ - \rho)\right]}. \qquad (3.113)$$

3.6.3 The Modified FCT Analysis

To examine the mass transport scheme applied to the present second-order PIC method, the following simple equation is considered when advancing the quantity f in time.

$$\frac{\partial f}{\partial t} = -\frac{\partial}{\partial x}(fU) - \frac{\partial g}{\partial x}. \qquad (3.114)$$

Assume a positive velocity for convenience, the quantity transported into cell (i) from cell $(i-1)$, $\delta f_{i-1/2}$, and out of cell (i) to cell $(i+1)$, $\delta f_{i+1/2}$, is shown by crosshatched areas in Fig. 3.12.

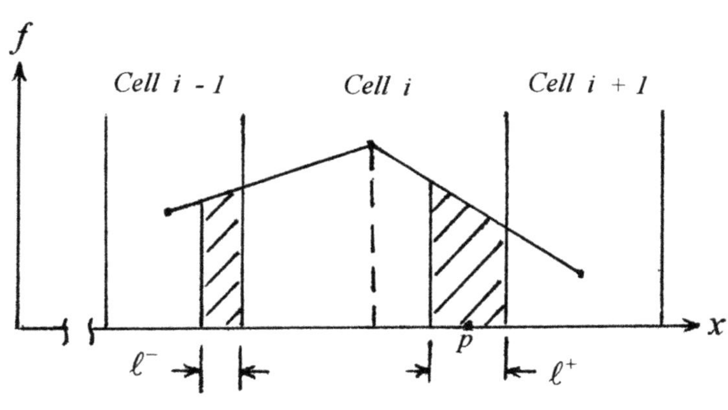

Figure 3.12. Quantity transported from cells $i-1$ to i, and from cells i to $i+1$.

From Eq. (3.66)

$$\ell^{\pm} = \bar{U}_{i\pm\frac{1}{2}}\delta t \Big/ \left[1 + \delta t \left(\bar{U}_x\right)_{i\pm\frac{1}{2}}\right], \tag{3.114a}$$

then

$$\begin{aligned}\delta f_{i+\frac{1}{2}} &= f(p)\ell^{+} \\ &= \frac{1}{2}\left[\left(f_i + f_{i+\frac{1}{2}}\right) + \left(f_i + f_{i-\frac{1}{2}}\right)\frac{\ell^{+}}{\Delta x}\right]\ell^{+}, \end{aligned} \tag{3.114b}$$

and

$$\begin{aligned}\Delta f_i &= \text{total change in cell } (i) \\ &= \delta f_{i-\frac{1}{2}} - \delta f_{i+\frac{1}{2}}. \end{aligned} \tag{3.114c}$$

if f is replaced by $\bar{f}$ (e.g. $\bar{U}$), the effect of $\frac{\partial g}{\partial x}$ will be reflected to $\bar{f}$. To avoid lengthy algebra, a case is considered where $f = \rho$ and $g = 0$ (conservation of mass equation).

$$\left(M_{avg}\right)_i = \rho_i \Delta x, \tag{3.114d}$$

and

$$\left(M_{lin}\right)_i = \frac{\Delta x}{2}\left[\rho_i + \frac{1}{4}\left(\rho_{i+1} + 2\rho_i + \rho_{i-1}\right)\right], \tag{3.114e}$$

According to the modified mass transport scheme, the total mass change in i is

$$\left(\delta M_i\right)_{\text{mod}} = \left(\Delta M_i\right)\left(M_{avg}\right)_i / \left(M_{lin}\right)_i. \tag{3.115}$$

For a uniform velocity, we get simpler equations as below.

$$\ell^{\pm} = U\delta t, \tag{3.115a}$$

$$\Delta M_i = \frac{1}{2}\left[\rho_{i-1} - \rho_{i+1} + \left(\rho_{i-1} - 2\rho_i + \rho_{i+1}\right)\frac{U\delta t}{\Delta x}\right]U\delta t, \tag{3.115b}$$

and

$$\begin{aligned}\left(\delta M_i\right)_{\text{mod}} &= \frac{1}{2}\left[\rho_{i-1} - \rho_{i+1} + \frac{U\delta t}{\Delta x}\left(\rho_{i-1} - 2\rho_i + \rho_{i+1}\right)\right]U\delta t \\ &\quad \cdot \frac{\rho_i}{\frac{1}{2}\left[\rho_i + \frac{1}{4}\left(\rho_{i-1} + 2\rho_i + \rho_{i+1}\right)\right]}. \end{aligned} \tag{3.115c}$$

Then we get

$$\begin{aligned}\rho_i^{n+1} &= \rho_i^n + \frac{(\delta M_i)_{\text{mod}}}{\Delta x} \\ &= \rho_i^n + \left[\frac{-\xi}{2}\left(\rho_{i+1}^n - \rho_{i-1}^n\right) + \frac{\xi^2}{2}\left(\rho_{i-1} - 2\rho_i + \rho_{i+1}\right)\right] \\ &\quad \cdot \frac{2\rho_i}{\left[\rho_i + \frac{1}{4}(\rho_{i-1} + 2\rho_i + \rho_{i+1})\right]} \\ &= \rho_i^n - \frac{\xi}{2}\left(\rho_{i+1}^n - \rho_{i-1}^n\right) + \frac{\xi^2}{2}\left(\rho_{i-1} - 2\rho_i + \rho_{i+1}\right) \\ &\quad + (\text{higher order terms})\,, \end{aligned} \tag{3.116}$$

where $\xi = U\frac{\delta t}{\Delta x}$.

In the original FCT [3.10], the third term on the RHS of Eq. (3.116) looks like

$$\left(\frac{1}{8} + \frac{\xi^2}{2}\right)\left(\rho_{i-1} - 2\rho_i + \rho_{i+1}\right).$$

The coefficient $\frac{1}{8}$ provides a strong diffusion even in the absence of convection. In our scheme, the diffusion occurs only for non-zero velocity. This term plus the additional FCT algorithm for local extremums (see Ref. [3.10]) virtually eliminates the need for artificial viscosity.

References

3.1 Wilkins, ML, Calculation of elastic-plastic flow, Methods in Comput. Physics, 3, pp. 211–264 (1964).
3.2 Swegle, JW, TOODY IV — A computer program for two-dimensional wave propagation, Sandia National Laboratory Report SAND-78-0552 (September 1978).
3.3 Kolsky, HG, A method for the numerical solution of transient hydrodynamic shock problems in two space dimension, Los Alamos Scientific Laboratory Report LA-1867 (March 1955).
3.4 Mader, CL, Numerical modeling of detonations, University of California Press, pp. 333–352 (1979).
3.5 Harlow, FH, The particle-in-cell computing method for fluid dynamics, Method in Comput. Physics, 3 (1964).
3.6 Johnson, WE, Modifications of OIL-type computer programs, Ballistic Research Laboratory Contract Report ARBRL-CR-0476 (January 1982).

3.7 Hageman, LJ, Wilkins, RT, Sedgwick, RT, and Wadell, JL, HELP, A multi-material Eulerian program for compressible fluid and elastic-plastic in two-dimension and time, System, Science, and Software report SSS-75-2654 (July 1975).
3.8 Thompson, SL, CSQII — An Eulerian finite difference program for two-dimensional material response — Part I, material section, Sandia National Laboratory Report SAND-77-1339 (May 1977).
3.9 Clark, RA, Second-order particle-in-cell computing method, Los Alamos Scientific Laboratory Report LA-UR-79-1947 (June 1979).
3.10 Boris, JP, and Book, DL, Flux-corrected transport I, SHASTA, A fluid transport algorithm that works, J. Comput. Physics, 11, pp. 38–69 (1973).

CHAPTER 4

ARTIFICIAL VISCOSITY AND SHOCK CALCULATIONS

Notations

a	some constant
A	area (cm^2)
b	some constant
C_S	sound speed (cm/μs)
C_T	isothermal sound speed (cm/μs)
E	total energy ($\text{Mbar} - \text{cm}^3/\text{cm}^3$)
h_0, h_1	some constants
j	Jacobian of the transformation between (R, Z) and $(k, l)(\text{cm}^2)$
k	Lagrangian coordinate
l	Lagrangian coordinate
M	mass per unit length or a unit of solid angle
P	pressure (Mbar)
Q	artificial viscosity (Mbar)
q	artificial viscosity (Mbar)
R	Eulerian radius (cm)
t	time (μs)
U	particle velocity in the R direction (cm/μs)
u	particle velocity in the R direction (cm/μs)
$\vec{u}$	velocity vector (U, V) (cm/μs)
V	particle velocity in the Z direction (cm/μs)
v	particle velocity in the Z direction (cm/μs)
W	specific volume (cm^2/g)
Z	coordinate (cm)

Greek Letters

γ	ratio of the specific heat, i.e. $\gamma = C_P/C_V$ (no unit)
ε	specific internal energy ($\text{Mbar} - \text{cm}^3$)
η	Lagrangian coordinate, and at $t = 0$, $\eta = Z(\xi, \eta, 0)$
ξ	Lagrangian coordinate, and at $t = 0$, $\xi = R(\xi, \eta, 0)$

ρ density (g/cm^3)
τ specific volume (cm^3/g)

Subscripts

0 initial value
η derivative with respect to η coordinate
ξ derivative with respect to ξ coordinate

Superscripts

n time at n time-step, i.e. $t^n = t_0 + n \cdot \Delta t$
* at the interface (i.e. grid boundary)

4.1 Introduction

Results are shown for the use of several different shock treatments in one- and two-dimensional Lagrangian code calculations of strong shock problems with known solutions. The shock treatments are (a) Richtmyer-von Neumann artificial viscosity; (b) fixed length artificial viscosity; (c) artificial energy diffusivity combined with artificial viscosity; and (d) modified Godunov.

In 1950, von Neumann and Richtmyer [4.1] proposed the use of the artificial viscosity q for calculating shock wave propagation in one-dimensional inviscid flow. It is well known that using the artificial viscosity introduces errors for shock wave propagation through material interfaces or non-uniform meshes [4.2]. At material interfaces, impedence matching reduces the errors. Other errors arise at shock start up and reflection boundaries. The q method was originally derived for steady shock propagation in plane geometry so in taking the method over to curvilinear systems there is a question of whether to use grad U or div U (where U is the particle velocity). In non-tensor codes, grad U should be used.

For shock propagation in variable zoning, Noh [4.3] proposed the use of the fixed length q. This method results in spreading a shock over a fixed physical length rather than a fixed number of zones. A linear term similar to Landshoff's [4.4] is added in this q. This technique computes the thermodynamic properties (e.g. density or pressure) very accurately for a particular direction of shock propagation but very badly for the other direction.

Using an artificial energy diffusivity with the artificial viscosity improves solutions at reflective or non-flow moving boundaries. The other methods tend to give too high internal energy and too low density at boundaries.

The modified Godunov scheme, computes interface velocity and pressure through a Riemann solver. In uniform zoning and a single material, the formulation reduces to that of the regular Godunov scheme. In a variable mesh or multi-material problem, the method has second order features.

4.2 Basic Governing Equations

The mass, momentum, and energy equations for one-dimensional flow in planes, cylinders and spheres are [4.5]:

$$\tau = R^{\alpha-1}\frac{\partial R}{\partial M}\,, \tag{4.1}$$

$$\frac{\partial U}{\partial t} = -R^{\alpha-1}\frac{\partial P}{\partial M}\,, \tag{4.2}$$

and

$$\frac{dE}{dt} = -\frac{\partial PUR^{\alpha-1}}{\partial M}\,. \tag{4.3}$$

In Eqs. (4.1)–(4.3), τ is the specific volume, R the Eulerian radius, M the mass per unit length or unit of solid angle, U the particle velocity, t the time, α (= 1 for plane, = 2 for cylinder, = 3 for sphere), P the pressure, and E the total energy, Also,

$$E = \varepsilon + \frac{1}{2}U^2\,, \tag{4.4}$$

$$\frac{dR}{dt} = U\,, \tag{4.5}$$

and

$$dM = \rho R^{\alpha-1}dR\,, \tag{4.6}$$

where ε is the internal energy, ρ the density $\left(=\frac{1}{\tau}\right)$, and dM the element of mass per unit solid angle (for cylinder or sphere) or of surface (for plane). The two-dimensional Lagrangian calculations, are done in cylindrical coordinates. Define an area Jacobian j as:

$$j = R_\xi Z_\eta - R_\eta Z_\xi\,, \tag{4.7}$$

where
R: Eulerian coordinate in radial direction, $R = R(\xi, \eta, t)$,
Z: Eulerian coordinate in axial direction, $Z = Z(\xi, \eta, t)$,
ξ : Lagrangian coordinate, and at $t = 0$, $\xi = R(\xi, \eta, 0)$,
η : Lagrangian coordinate, and at $t = 0$, $\eta = Z(\xi, \eta, 0)$,

and

$$R_\xi = \frac{\partial R}{\partial \xi}\,, \text{ etc.}$$

The mass, momentum and energy equations in a two-dimensional Lagrangian form can be written as:

$$\rho j = M\,, \tag{4.8}$$

$$\frac{dU}{dt} = -\tau\frac{\partial P}{\partial R}\,, \tag{4.9}$$

$$\frac{dV}{dt} = -\tau\frac{\partial P}{\partial Z}\,, \tag{4.10}$$

and

$$\frac{d\varepsilon}{dt} = -P\frac{\partial \tau}{\partial t}\,. \tag{4.11}$$

In Eqs. (4.9) and (4.10), U and V are the material velocities in the R and Z directions.

4.3 Artificial Viscosity in One-Dimensional Code

In general, the artificial viscosity, q, can be described as combinations of terms linear and quadratic in ΔU. The quadratic term concentrates the artificial viscosity near the shock front while the effect of the linear term is more diffusive. A purely linear form of q will result in a large overshoot in energy behind a shock followed by rapid damping. A purely quadratic form will result in a smaller overshoot followed by undamped oscillations. Some of the q's used for one-dimensional problems are:

(1) Richtmyer-von Neumann [4.1]

$$q = -a^2\rho\Delta U|\Delta U|\,, \quad \text{when } \frac{\partial U}{\partial x} < 0\,, \tag{4.12}$$

where a is some constant,

(2) Rosenbluth [4.6]

$$q = a^2\rho(\Delta U)^2\,, \quad \text{when } \frac{\partial U}{\partial x} < 0\,, \tag{4.13}$$

(3) Landshoff [4.4]

$$q = a^2\rho(\Delta U)^2 - 0.5C_s\Delta U\left[b + (1-b)C_s\frac{\Delta t}{\Delta x}\right], \quad \text{when } \frac{\partial U}{\partial x} < 0\,, \tag{4.14}$$

where $0 < b < 1$ and C_s is the sound speed,

(4) PIC [4.7]

$$q = -a\rho\Delta U\left|\bar{U}\right|, \quad \text{when } \frac{\partial U}{\partial x} < 0\,, \tag{4.15}$$

(5) Schulz [4.8]

$$q = -b\rho\Delta U\left|0.5(\Delta U_{i+1} - \Delta U_{i-1}\right|, \quad \text{when } \frac{\partial U}{\partial x} < 0\,, \tag{4.16}$$

where i is the grid index,

(6) Kuropatenko [4.9]

$$q = 0.5(\gamma+1)\rho(\Delta U)^2 - \rho C_s\Delta U\,, \quad \text{when } \frac{\partial U}{\partial x} < 0\,, \tag{4.17}$$

where $\gamma = \frac{C_p}{C_v}$, C_p = specific heat at constant pressure, and C_v = specific heat at constant volume,

(7) AFWL-PUFF [4.10]

$$q = a^2\rho(\Delta U)^2 - b\rho C_T\Delta U\,, \quad \text{when } \frac{\partial U}{\partial x} < 0\,, \tag{4.18}$$

where C_T is the isothermal sound speed.

(8) QLQ [4.11]

$$q = a^2\rho(\Delta U)^2 - \frac{50.0b^2\rho(\Delta U)^2}{1 + 50.0\left|\Delta U\right|}, \quad \text{when } \frac{\partial U}{\partial x} < 0\,, \tag{4.19}$$

(9) White [4.12]

$$q = a^2\rho(\Delta U)^2\left(\left|\frac{\Delta P}{\rho C_s\Delta U}\right|\right)^{0.5} - b\rho C_s\Delta U\left(\left|\frac{\Delta P}{\rho C_s\Delta U}\right|\right)^{0.25}, \quad \text{when } \frac{\partial U}{\partial x} < 0\,, \tag{4.20}$$

(10) Winkler [4.13]

$$\vec{q} = a^2\rho(\nabla\cdot\vec{U})\left(\nabla U - \frac{\nabla\cdot\vec{U}}{3}\right), \quad \text{when } \nabla\cdot\vec{U} < 0\,, \tag{4.21}$$

and

(11) Noh Fixed Length [4.3]

$$q = a^2 \rho (\Delta x_{\max})^2 \left(\frac{\Delta U}{\Delta x}\right)^2 - b\rho C_s \Delta x_{\max} \frac{\Delta U}{\Delta x}\,, \quad \text{when } \frac{\partial U}{\partial x} < 0\,. \tag{4.22}$$

4.4 Artificial Viscosity in Two-Dimensional Code

In the two-dimensional calculations q is computed only when the cell is compressing. Eq. (4.12) in the two-dimensional calculations is

$$q = 1.4 \rho A \left(\frac{\tau^n - \tau^{n-1}}{\tau^n \Delta t}\right)^2 , \tag{4.23}$$

where A is the area.

Another popular method is extending Eq. (4.12) into a 2D Lagrangian code. If one chooses $a^2 = 2$, then, Eq. (4.12) becomes

$$q = -2\rho \Delta u \left|\Delta u\right| , \tag{4.24}$$

with

$$q = 0 \quad \text{when } \Delta u > 0\,,$$

and

$$q \neq 0 \quad \text{when } \Delta u < 0\,.$$

Referring to Fig. 4.1 that is shown here for convenience, we have

$$\begin{aligned} \Delta \vec{u} &= \Delta(ui + vj) \\ &= (u_{k,\ell} - u_{k,\ell-1})i + (v_{k,\ell} - v_{k,\ell-1})j\,. \end{aligned} \tag{4.25}$$

Therefore

$$|\Delta \vec{u}| = \left[(u_{k,\ell} - u_{k,\ell-1})^2 + (v_{k,\ell} - v_{k,\ell-1})^2\right]^{1/2} , \tag{4.25a}$$

and

$$(\Delta \vec{u})^2 = \left[(u_{k,\ell} - u_{k,\ell-1})^2 + (v_{k,\ell} - v_{k,\ell-1})^2\right] . \tag{4.26}$$

Since ${}_1Q_{k,\ell}$ is along the side between points KML and KL, we need $\Delta v < 0$, or $v_{k,\ell} - v_{k-1,\ell} < 0$. If $\Delta v < 0$, then

$${}_1Q_{k,\ell} = 2\rho_{k,\ell}\left[(u_{k,\ell} - u_{k-1,\ell})^2 + (v_{k,\ell} - v_{k-1,\ell})^2\right] . \tag{4.27}$$

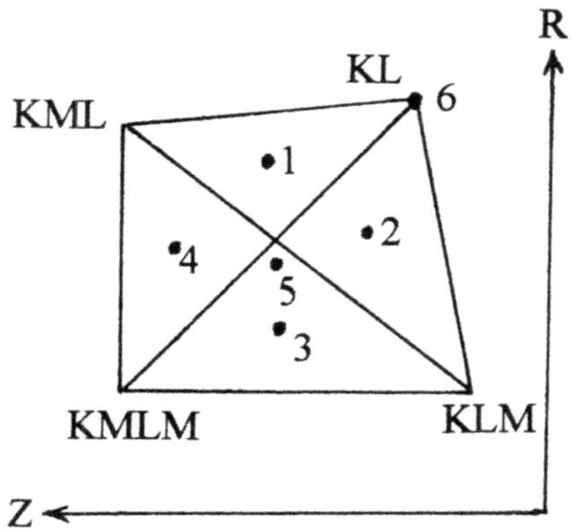

Figure 4.1. M, j, ρ, ε, and S^{ij} are defined at cell center point 5, while U, V, R, Z are defined at vertex point 6, pressure and artificial viscosity are at point 1, 2, 3, and 4.

If $u_{k,\ell} - u_{k,\ell-1} < 0$, then

$$_2Q_{k,\ell} = 2\rho_{k,\ell}\left[(u_{k,\ell} - u_{k,\ell-1})^2 + (v_{k,\ell} - v_{k,\ell-1})^2\right]. \tag{4.28}$$

If $v_{k,\ell-1} - v_{k-1,\ell-1} < 0$, then

$$_3Q_{k,\ell} = 2\rho_{k,\ell}\left[(u_{k,\ell-1} - u_{k-1,\ell-1})^2 + (v_{k,\ell-1} - v_{k-1,\ell-1})^2\right]. \tag{4.29}$$

If $u_{k-1,\ell} - u_{k-1,\ell-1} < 0$, then

$$_4Q_{k,\ell} = 2\rho_{k,\ell}\left[(u_{k-1,\ell} - u_{k-1,\ell-1})^2 + (v_{k-1,\ell} - v_{k-1,\ell-1})^2\right]. \tag{4.30}$$

4.5 With and Without Artificial Viscosity Methods

For methods (A), (B), and (C) with artificial viscosity, the pressure P in Eqs. (4.2), (4.3), (4.9), (4.10) and (4.11) is replaced by $P + q$. (Note that in tensor hydrodynamics, Schulz [4.8] and Winkler [4.13], the equations change in curvilinear coordinates.)

Because it was noticed that the standard artificial viscosity methods produced poor answer when following a shock through a variable mesh, Gee, Kramer, and Noh [4.3] suggested the use of the fixed length q, method

(B), which spreads a shock over a fixed length rather than a fixed number of zones. The coefficients a and b in Eq. (4.22) are empirically related to the zone ratio Z defined as

$$Z = \frac{\Delta X^0_{i-1/2}}{\Delta X^0_{i+1/2}}, \tag{4.31}$$

where i is the grid number defined at cell edge and Z is defined at $t = 0$.

The Richtmyer-von Neumann q was originally derived from considerations of a plane shock running in a mesh where both sides of a cell could respond to the shock. As this condition is not satisfied at a programmed boundary, the entropy production in boundary cells is too large, see Landshoff [4.4]. To correct this, Noh [4.14] suggests the use of an artificial heat diffusivity, method (C). An artificial heat flux H is added to the right hand sides of the energy equations, i.e. Eqs. (4.3) and (4.11). For the two-dimensional calculations,

$$H = C_H \left(\frac{\partial^2 \varepsilon}{\partial R^2} + \frac{\partial^2 \varepsilon}{\partial Z^2} \right). \tag{4.32}$$

The C_H should be at least dependent on coefficient a, e.g. see Eq. (4.12), and zone ratio Z (in case of non-uniform zoning).

For one-dimensional problems, Noh [4.14] suggests that

$$H = h_0^2 \rho \Delta U \Delta \varepsilon + h_1 \rho C_s \Delta \varepsilon, \quad \text{when } \frac{\partial U}{\partial x} < 0, \tag{4.33}$$

and h_0 and h_1 are constants.

In the q-free modified Godunov method (D) discussed here, the one-dimensional Lagrangian equation is solved

$$\frac{\rho d(\vec{W})}{dt} + \frac{\partial [\vec{F}(\vec{W})]}{\partial x} = \vec{S}, \tag{4.34}$$

where $\vec{W}$ represents specific volume τ, velocity U, and the total energy E in the cell $\left(i - \frac{1}{2}\right)$. The flux term $\vec{F}$ represents $-UP$ and UP at the cell boundary (i). $\vec{S}$ is a vector of possible source terms. With initial conditions $\vec{W}^n_{i-\frac{1}{2}}$ at cell center a Riemann problem is solved to get the flux at the cell boundary $\vec{F}_i^{n+1/2}(\vec{W})$. Then

$$\vec{W}_{i-\frac{1}{2}}^{n+1} = \vec{W}_{i-\frac{1}{2}}^{n} - \left(\vec{F}_i^{n+1/2} - \vec{F}_{i-1}^{n+1/2}\right)\frac{\delta t}{\Delta x}, \tag{4.35}$$

where $\delta t = t^{n+1} - t^n$ and $\Delta x = x_i - x_{i-1}$. At $t = t^{n+1/2}$, in compression, the pressure P^* and the velocity U^* at the interface are related by two Hugoniot relations to the adjacent cells:

$$\begin{aligned} P_i^* = {} & P_{i\pm\frac{1}{2}}^n + 0.25\rho_{i\pm\frac{1}{2}}^n(\gamma+1)\left(U_i^* - \bar{U}_{i\pm\frac{1}{2}}^n\right)^2 \pm \left[\left(U_i^* - \bar{U}_{i\pm\frac{1}{2}}^n\right)^2 \gamma P_{i\pm\frac{1}{2}}^n \rho_{i\pm\frac{1}{2}}^n \right. \\ & \left. + \left(\rho_{i\pm\frac{1}{2}}^n\right)^2 (\gamma+1)^2 \left(U_i^* - \bar{U}_{i\pm\frac{1}{2}}^n\right)^4 /16\right]^{1/2} \end{aligned} \tag{4.36}$$

The two unknowns, P_i^* and U_i^*, in these two equations can be solved in any desired manner. We do a three-step iteration. The $\bar{U}^n$ defined at the cell center, is a cell average velocity which we take as a free parameter to partition total energy between the kinetic energy and internal energy. The P_i^* and U_i^* depend on the density, the local velocity and γ. For a non-ideal fluid, γ must be the effective γ.

For non-uniform meshes, a zone ratio Z is defined depending on the direction of wave motion in the mesh.

$$Z_{i-\frac{1}{2}} = \begin{Bmatrix} (\rho\Delta X)_{i-1/2}/(\rho\Delta X)_{i-3/2} & grad(P) < 0 \\ (\rho\Delta X)_{i-1/2}/(\rho\Delta X)_{i+1/2} & grad(P) > 0 \end{Bmatrix} \tag{4.37}$$

By solving the momentum equation for $\delta U = U^{n+1} - U^n$, one gets

$$\delta\bar{U}_{i-\frac{1}{2}} = -\frac{\delta t\left(P_{i-1}^* - P_i^*\right)}{(\rho\Delta X)_{i-\frac{1}{2}}} Z_{i-\frac{1}{2}}^{\alpha}, \tag{4.38}$$

and

$$\varepsilon_{i-\frac{1}{2}}^{n+1} = \varepsilon_{i-\frac{1}{2}}^{n} - div(P^*U^*)/M_{i-\frac{1}{2}} - \left[\frac{(\bar{U}^{n+1} + \bar{U}^n)\delta\bar{U}Z^{\beta-\alpha}}{2}\right]_{i-\frac{1}{2}} \tag{4.39}$$

For the results shown in this chapter, we use $\alpha = 0$ and $\beta = 1/2$.

4.6 Sample Problem Calculations

First, we show comparisons of this q-free Godunov method with the Richtmyer–von Neumann scheme. The calculation is for a plane piston moving from right to left with a velocity of $U = -1.0\,\text{cm}/\mu\text{s}$ against an initially cold ideal gas of density, $\rho_0 = 100\,\text{g/cm}^3$. The left boundary is rigid.

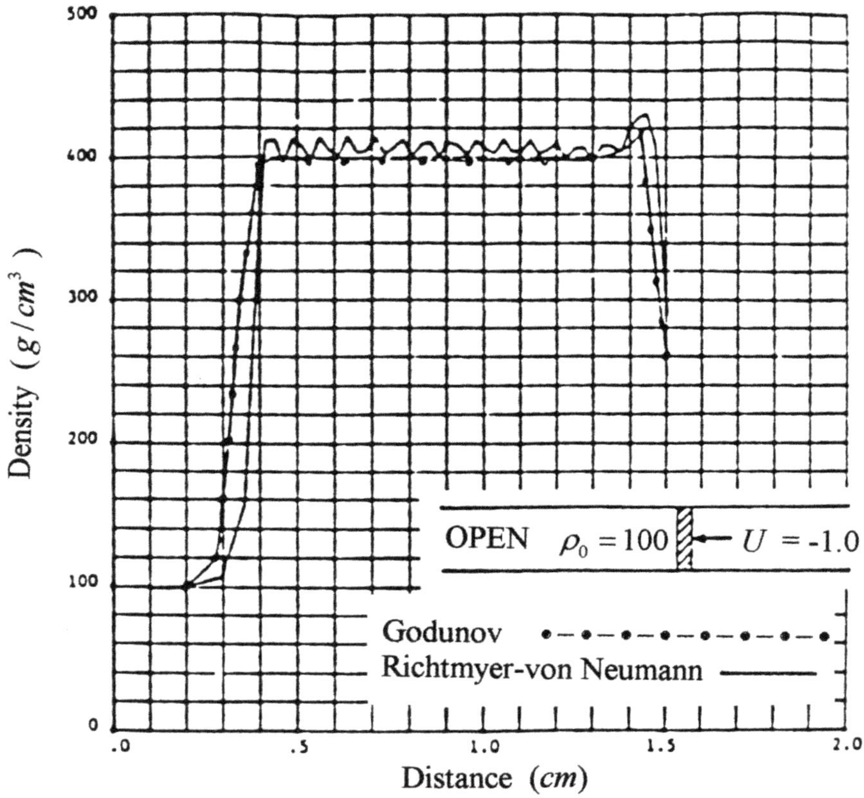

Figure 4.1-a. Density profile for the open-end cylinder calculated with uniform mesh at time $t = 0.035\,\mu$s.

Figure 4.1-a shows the piston calculation before shock reflection. The initial density of $100\,\text{g/cm}^3$ has quadrupled behind the shock. The Richtmyer–von Neumann method with $a = 2$ and no linear term has oscillations behind the shock which are much reduced in the Godunov scheme.

Figure 4.1-b shows the density profile after shock reflection. The figure again shows that the modified Godunov scheme calculates a smoother density, closer to the analytical solution of $1000\,\text{g/cm}^3$. The Richtmyer–von Neumann front is a little behind. In general, the various methods calculate pressure well. Therefore, the errors in density are reflected in the calculated energy.

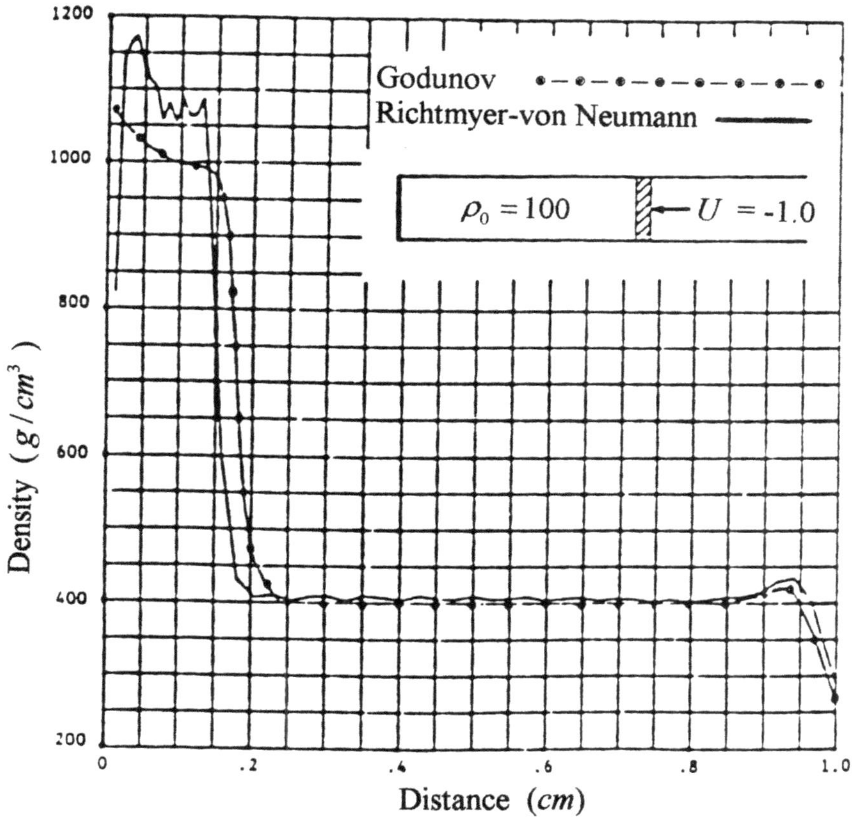

Figure 4.1-b. Density profile after the first shock reflected from the closed end at time $t = 0.004\,\mu$s.

Figures 4.2-a and 4.2-b show the piston problem in a non-uniform mesh known as Noh's problem [4.14]. The initial zoning was coarse at the boundaries and, then, decreasing uniformly ($Z = 1.15$) to the center. In Fig. 4.2-a, the shock has just passed through the minimum zone area. Neither methods (A) or (D) does well (the density behind the shock should be $400\,\mathrm{g/cm^3}$), although the modified Godunov scheme does a little better. The shock velocity calculated with the modified Godunov scheme agrees with the analytic solution, but the shock position calculated with the Richtmyer–von Neumann scheme is a little behind the analytical position again.

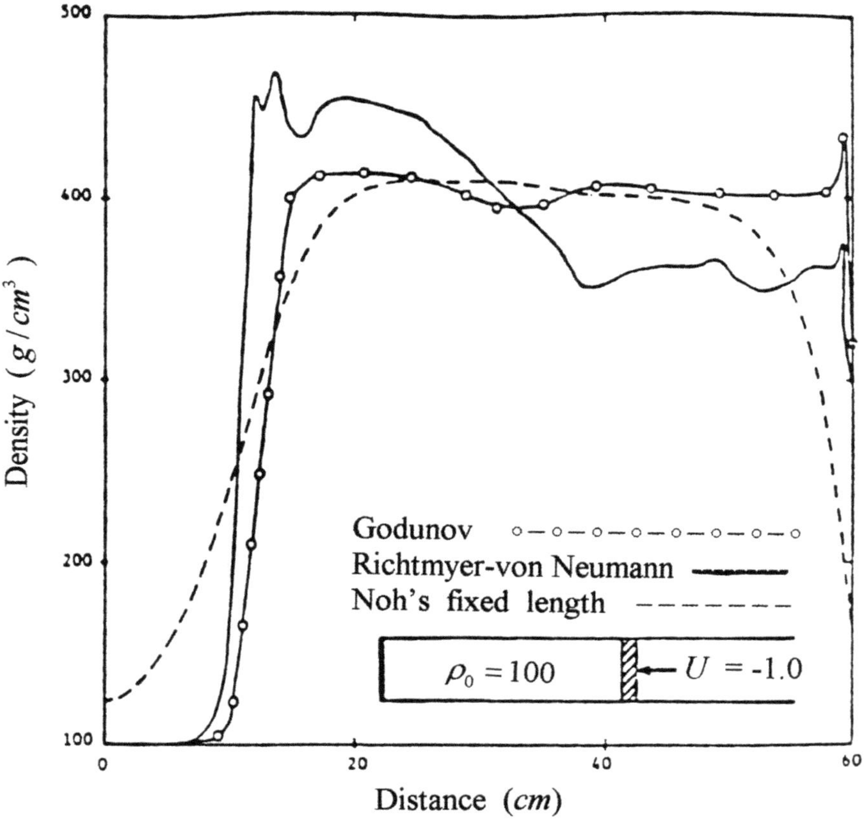

Figure 4.2-a. Density profile with variable mesh before reflection.

The calculation using Noh's fixed length q agrees very well with the analytic solution at this time. Figure 4.2-b shows the density profile after shock reflection. The fixed length q method has become a diffusion solution for U and has lost all relation to the analytic solution $\frac{\rho}{\rho_0} = 10$.

The next problem shown is a spherical convergent shock reflecting from the origin for ideal gas with $\gamma = \frac{5}{3}$; initial conditions are, $\rho_0 = 1$, $E_0 = 0$, $P_0 = 0$, and $U_0 = 0$. Boundary conditions are in the form of $R_B(t)$ and $\dot{R}_B(t)$ which are given in Table 4.1. Shock collapse occurs at $t = 440\,\mu$s, while the analytic solutions are obtained by the similarity method.

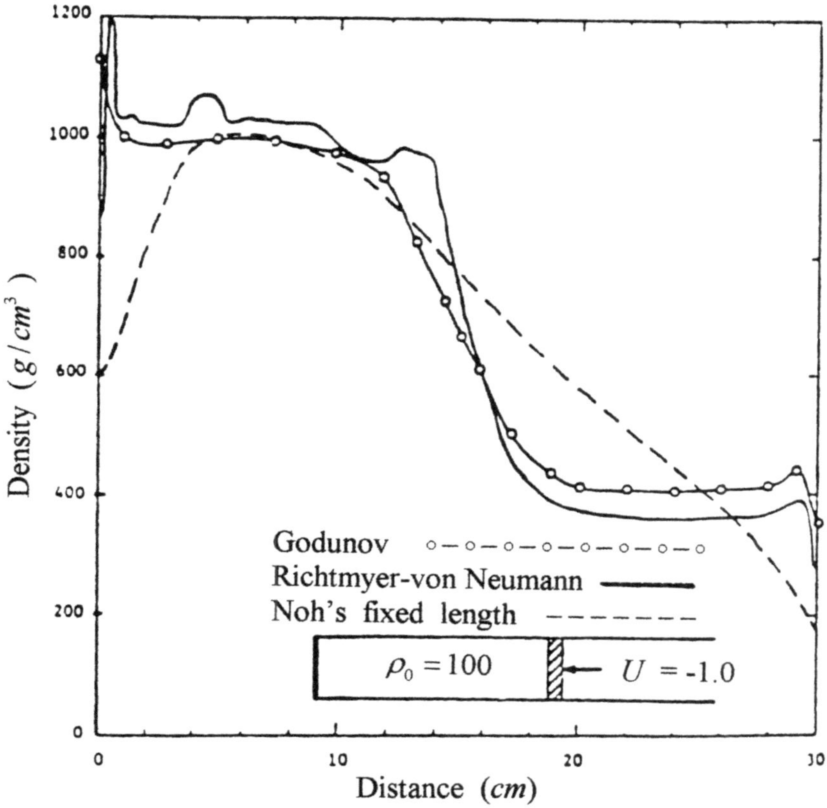

Figure 4.2-b. Density profile with variable mesh after reflection.

Figure 4.3 shows the pressure profiles at time $t = 420\,\mu s$ as calculated by the two-dimensional code before shock collapse showing that the calculation with heat conduction is much better than the one without heat conduction. However, the code calculations give higher pressure in the shocked region as compared with the similarity solution.

Figures 4.4 and 4.5 show the pressure profiles at time $t = 460\,\mu s$ after the shock has reflected from the center in the two-dimensional Lagrangian calculations. Without artificial heat diffusivity, the maximum pressure may be off by more than 133 percent; with the artificial heat diffusivity, it is in error by 20 percent. In addition, the shock front position is much closer to the similarity solution.

Table 4.1. Similarity boundary conditions.

Time (μsec)	Radius (cm)	Velocity (cm/μsec)
0.	6.0000	−.00705
20.	5.8585	−.00708
40.	5.7167	−.00711
60.	5.5743	−.00714
80.	5.4308	−.00717
100.	5.2869	−.00721
120.	5.1423	−.00724
140.	4.9972	−.00727
160.	4.8515	−.00730
180.	4.7055	−.00732
200.	4.5588	−.00734
220.	4.4121	−.00735
240.	4.2649	−.00736
260.	4.1177	−.00736
280.	3.9707	−.00735
300.	3.8738	−.00733
320.	3.6775	−.00730
340.	3.5319	−.00726
360.	3.3874	−.00720
380.	3.2438	−.00712
400.	3.1024	−.00702
420.	2.9633	−.00689
440.	2.8273	−.00672
460.	2.6951	−.00650
480.	2.5677	−.00622
500.	2.4471	−.00578
520.	2.3329	−.00544
540.	2.2278	−.00492
560.	2.1344	−.00430
580.	2.0560	−.00350

4.7 Conclusions and Discussions

In solving problems of shock generation and propagation by numerical integration, the most popular method has been to add artificial viscosity to smear the shock front (or any discontinuity). For 50 years, many researchers have tried to invent new forms of q with magical properties for their problems. Others have tried to solve shock problems without an entropy generation mechanism. Tests of the myriad forms of q with claimed magical

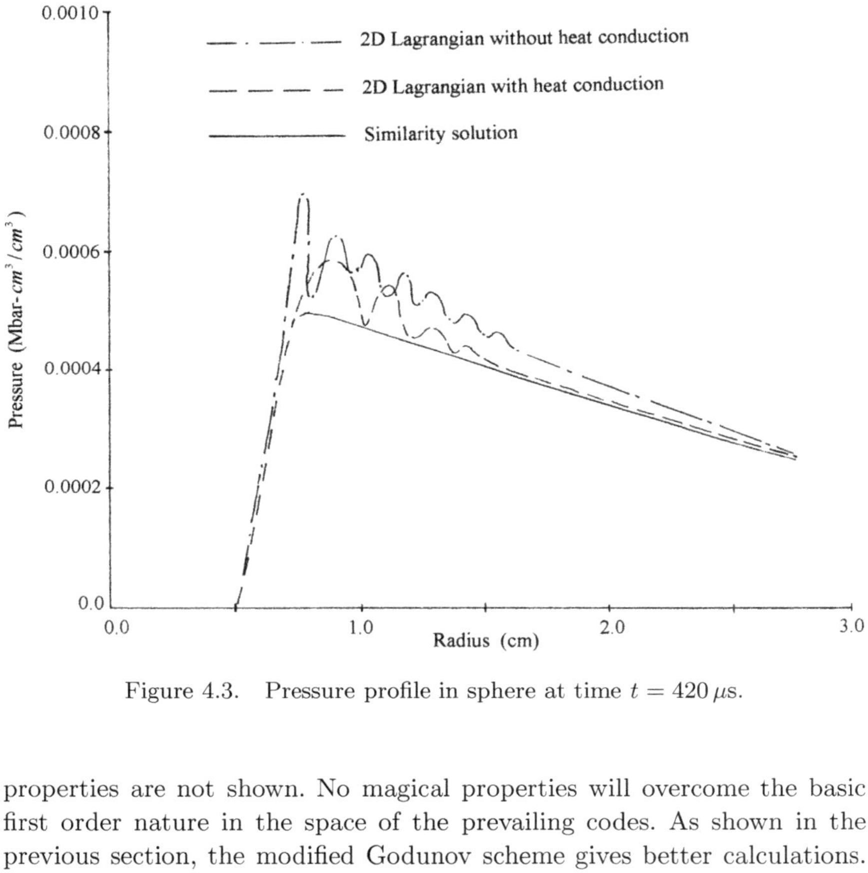

Figure 4.3. Pressure profile in sphere at time $t = 420\,\mu$s.

properties are not shown. No magical properties will overcome the basic first order nature in the space of the prevailing codes. As shown in the previous section, the modified Godunov scheme gives better calculations. For non-uniformly zoned problems, a clever choice of coefficients a and b of Eq. (4.22) may produce good results for the first shock passage but not for multiple shocks or reflected shocks. However, since artificial viscosity is still very popular, we recommend the artificial heat flux mechanism to couple with the q term. The appropriate relation between q and the flux can be taken from statistical mechanics.

Recently developed methods such as adaptive grid, moving finite element, Glimm's Riemann method, piecewise-parabolic method, and local mesh refinement are all of q-free type methods. Most of these methods show good results for one-dimensional shock problems and some of them may have a practical application for two-dimensional geometry especially when solving problems with multiple materials. We recommend pursuing techniques such as the fully second-order Godunov scheme.

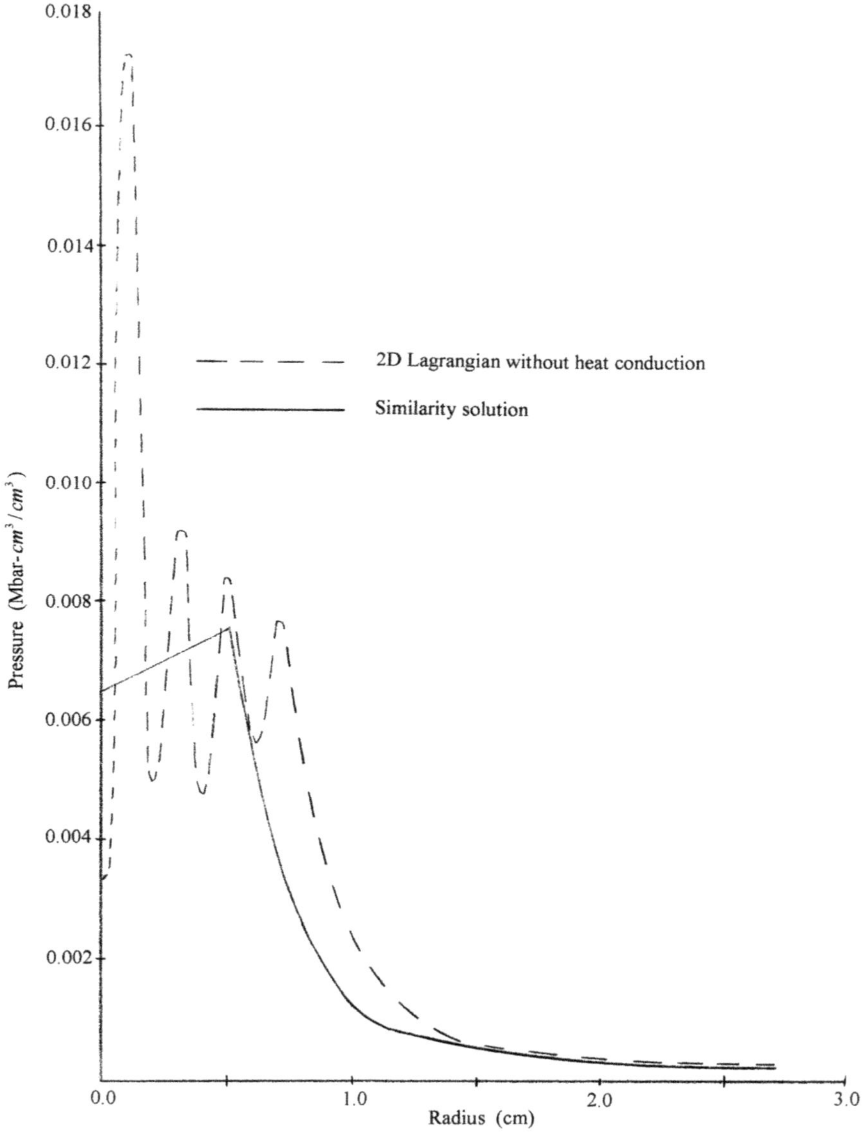

Figure 4.4. Pressure profile at time $t = 460\,\mu s$ for 2D Lagrangian code without heat conduction.

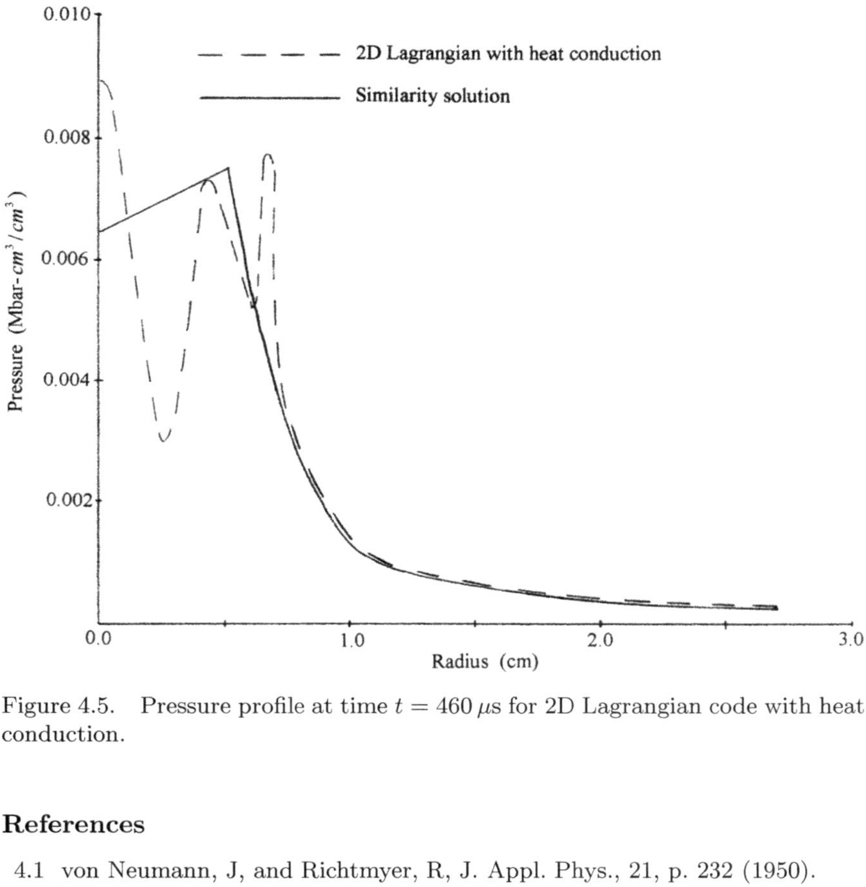

Figure 4.5. Pressure profile at time $t = 460\,\mu s$ for 2D Lagrangian code with heat conduction.

References

4.1 von Neumann, J, and Richtmyer, R, J. Appl. Phys., 21, p. 232 (1950).

4.2 Cameron, I, An analysis of the errors caused by using artificial viscosity terms to represent steady-state shock waves, J. Comp. Phys. 1, pp. 1-20 (1966).

4.3 Gee, M, Kramer, G, and Noh, W, Fixed and non-fixed length Q studies using a one-dimensional Lagrangian code on the variable zoned one-dimensional hydro shock test problems, Lawrence Livermore Laboratory report, UCID-18515 (1979).

4.4 Landshoff, R, Numerical method for tracing fluid flow in the presence of shocks, Los Alamos National Laboratory report, LA-1930 (1955).

4.5 Mader, C, Numerical modeling of explosives and propellants, CRC Press, second edition (1998).

4.6 Rosenbluth, M, unpublished work, Los Alamos National Laboratory (1950).

4.7 Harlow, F, and Evans, M, The particle-in-cell method for hydrodynamic calculations, Los Alamos National Laboratory report, LA-2139 (1957).

4.8 Schulz, W, Two-dimensional Lagrangian hydrodynamic difference equations, Lawrence Radiation Laboratory report, UCRL-667 (1963).
4.9 Kuropatenko, V, in Difference methods for solutions of problems of mathematical physics, I, (N.N. Janeko, Ed.), p116, Amer. Math. Soc., Providence, R.I. (1967).
4.10 Hicks, D, and Pelzl, R, Comparison between a von Neumann/Richtmyer hydrocode (AFWL's PUFF) and Lax-Wendroff hydrocode, Air Force Weapon Laboratory report, AFWL-TR-68-112 (1968).
4.11 Meyer, K, Artificial viscosities considered for use in the CIRCE Lagrangian hydrocode, Los Alamos National Laboratory report, LA-4383-MS (1970).
4.12 White, J, A new form of artificial viscosity postscript, J. Comp. Phy., 12, pp. 553-556 (1973).
4.13 Winkler, K, A method for computing selfgravitating gas flows with radiation, MPI/PAE-Astro 163 (1978).
4.14 Noh, W, private communication.

CHAPTER 5

NUMERICAL SIMULATIONS OF OBLIQUE SHOCK-WAVE REFLECTION

Notations

a^2	cell area used in von Neumann artificial viscosity (cm^2)
B_1, B_2	calculating domain dimension (cm)
c	constant coefficient in von Neumann artificial viscosity (no unit)
c_1, c_2	linear and nonlinear constant coefficients in artificial viscosity (no unit)
c_s	sound speed (cm/μs)
e	specific internal energy ($\text{Mbar} - \text{cm}^3$)
f_p^z, f_z^p	corner force ($\text{g} \cdot \text{cm}/(\mu\text{s})^2$)
H_1, H_2	calculating domain dimension (cm)
m_p^z, m_z^p	corner mass (g)
M	mass (g)
M_s	mach number of shock wave (no unit)
NX, NY	grid number (no unit)
P	pressure (Mbar)
q	scalar artificial viscosity (Mbar)
$\overline{\overline{q}}$	tensor form of artificial viscosity (Mbar)
S	edge area vector (cm^2)
t	time (μs)
T_0	initial temperature (°K)
u, v	velocities in the x and y directions respectively (cm/μs)
$\vec{v}$	velocity vector (cm/μs)
V_z^p	subzonal volume (cm^3)

Greek Letters

γ	ratio of the specific heat; $\gamma = C_p/C_v$ (no unit)
ζ	step function
θ	wedge angle (degree)

ρ density (g/cm^3)
ψ artificial viscosity limiter (no unit)

Superscripts and Subscripts

i the edge of the zone in logic domain in Lagrangian method
p main grid point
z main zone

5.1 Introduction

When a planar shock wave encounters a wedge, a complex shock-wave diffraction phenomenon will occur. This phenomenon has long been noticed and studied by many researchers. As early as 1878, Mach [5.1] had observed shock regular reflection and single Mach reflection. But only in this century has this phenomenon been studied extensively. In the 1940's, von Neumann [5.2, 5.3] used two and three shock theories to predict these two reflections quite well. Smith [5.4], White [5.5], and Colella and Henderson [5.6] have found more basic physics in the shock wave reflection. Today, it is widely known that, depending on the Mach number of incident shock wave, wedge angle, and gas property, there exist three types of reflection: regular reflection (RR), von Neumann reflection (vNR), and Mach reflection (MR). MR can be further divided into simple-Mach reflection (SMR), transitional-Mach reflection (TMR), and double-Mach reflection (DMR). Glass and many other researchers [5.7, 5.8] have done much work in determining the transition conditions among these reflections.

Since the complete set of Euler equations in these situations cannot be solved analytically, experimental work has been an effective way to find new information and to benchmark different kinds of numerical approximations. Much experimental work on shock wave reflection has been conducted. In our current study area, a series of experiments done by a group of people under leadership of Glass [5.7, 5.8] is very useful. Their extensive experimental work supplies reliable shock behavior that could be used to check the new theory and the performance of numerical methods.

In the past few decades, a number of shock-capturing numerical techniques have been developed to solve Euler equations for both the Eulerian and Lagrangian codes. In view of treating discontinuity, one can use the artificial viscosity method, Godunov's method or linear hybridization method. In the case of mesh layout, one may choose simple uniform rectangular mesh in Eulerian code or unstructural, adaptive mesh refinement for both Eulerian and Lagrangian codes. In this investigation, artificial viscosity is used to smooth out the discontinuities in both the Eulerian and Lagrangian codes. It is well known that the accuracy of numerical simulation of shock waves is heavily dependent on the method of treating the discontinuity. Details about the form of artificial viscosity used and its effects on the results will be discussed. Case (1) and (3) in Ref. [5.9] were chosen to demonstrate the capability of these Eulerian and Lagrangian codes and to evaluate our results because there are other numerical work available for these two cases.

In case (1), planar shock with Mach number 2.05 propagates in the initial still air with $P = 250\,\text{torr}$ and $T_0 = 298.4°\text{K}$ and then encounters a wedge with $\theta = 63.4°$. This results in regular reflection. Its configuration consists of an incident and a reflected shock wave. These two shock waves intersect at the reflection point, which is located on the reflecting surface. In case (3), parameters are $M_s = 2.03$, $P = 250\,\text{torr}$, $T_0 = 299.2°\text{K}$ and $\theta = 27.0°$, and simple-Mach reflection will occur. Besides the incident and the reflected shock waves, a Mach stem also presents and produces a triple point in the shock wave configuration. In present numerical simulation, the same field parameters are used except that the ideal gas equation of state is assumed here. For air, this assumption is valid under room temperature and moderate Mach number. In the calculations, some necessary boundary conditions were obtained theoretically.

Sections 5.2 and 5.3 will introduce the Eulerian and the Lagrangian method, respectively; then Section 5.4 will discuss the calculating procedure and its results; and in Section 5.5, some conclusions will be summarized.

5.2 Eulerian Method

The conservation equations of mass, momentum, and internal energy for this problem are solved in a planar coordinate system with a staggered mesh. The pressure P, specific internal energy E, and density ρ are defined at the cell center along with the velocity vector $\vec{v}$ (where $\vec{v} = u\mathbf{i} + v\mathbf{j}$ with u, v the velocity components in x, y directions respectively), and coordinates x and y are defined at the vertices. Explicit finite differences are used for solving the dependent variables.

An operator splitting scheme is used for the present Eulerian code, in which the whole calculation is divided into two phases. The first phase solves the Lagrangian part in both the x and y directions simultaneously.

The momentum and energy equations solved in the Lagrangian phase are

$$\frac{\partial v}{\partial t} = -\frac{1}{\rho}\nabla(P+q)\,, \tag{5.1}$$

and

$$\frac{\partial e}{\partial t} = \frac{(P+q)}{\rho}\nabla\cdot\vec{v}\,. \tag{5.2}$$

The artificial viscosity used in Eqs. (5.1) and (5.2) is the von Neumann-Richtmyer [5.10] quadratic q which is

$$q = -ca^2\rho\Delta v|\Delta v|\,, \tag{5.3}$$

where $c = 2.0$ and a^2 = area of the cell.

At the end of phase one, we have the temporary values of velocity $\tilde{u}$ and $\tilde{v}$, and energy $\tilde{e}$, momentum $M\tilde{u}$ and $M\tilde{v}$ (where M is the mass in the cell), and internal energy $M\tilde{e}$. In phase two or the remap phase, the average velocities $\bar{u}$ and $\bar{v}$ are calculated by

$$\bar{u} = \frac{1}{2}(u+\tilde{u})\,, \tag{5.4a}$$

and

$$\bar{v} = \frac{1}{2}(v+\tilde{v})\,, \tag{5.4b}$$

where u and v are the velocities at the beginning of each time step. These average velocities are used for computing the mass, the momentum, and the energy flux across the cell boundaries. The governing equations are split into x and y directions. Therefore, in the x direction, we solve

$$\frac{\partial \rho}{\partial t} = -\bar{u}\frac{\partial \rho}{\partial x}\,, \tag{5.5}$$

$$\frac{\partial u}{\partial t} = -\bar{u}\frac{\partial u}{\partial x}\,, \tag{5.6a}$$

$$\frac{\partial v}{\partial t} = -\bar{u}\frac{\partial v}{\partial x}\,, \tag{5.6b}$$

and

$$\frac{\partial e}{\partial t} = -\bar{u}\frac{\partial e}{\partial x}\,, \tag{5.7}$$

and, in the y direction, we have

$$\frac{\partial \rho}{\partial t} = -\bar{v}\frac{\partial \rho}{\partial y}, \tag{5.8}$$

$$\frac{\partial u}{\partial t} = -\bar{v}\frac{\partial u}{\partial y}, \tag{5.9a}$$

$$\frac{\partial v}{\partial t} = -\bar{v}\frac{\partial v}{\partial y}, \tag{5.9b}$$

and

$$\frac{\partial e}{\partial t} = -\bar{v}\frac{\partial e}{\partial y}. \tag{5.10}$$

Then calculation is performed in each separated direction. The order of this calculation is alternated for each time advance, i.e. $x-y-y-x$, to maintain the accuracy of the one-dimensional procedure. For flux calculations, the monotonic advection methods of van Leer [5.11] are used to avoid the non-physical oscillations and negative densities.

5.3 Lagrangian Method

The governing equations in the Lagrangian description are the same as those used in the first phase of the Eulerian method described in the previous section except that a tensor form of artificial viscosity, $\bar{\bar{q}}$, is used. The momentum and the energy equations are

$$\frac{\partial \vec{v}}{\partial t} = -\frac{1}{\rho}\nabla P - \frac{1}{\rho}\nabla \cdot \bar{\bar{q}}, \tag{5.11}$$

and

$$\frac{\partial e}{\partial t} = -\frac{P}{\rho}\nabla \cdot \vec{v} + \frac{1}{\rho}\bar{\bar{q}} : \nabla \vec{v}. \tag{5.12}$$

Unlike the Eulerian method where mesh and zone are fixed in space, the zone represents fixed mass and the mesh moves with time in the Lagrangian method. The disadvantage of the Lagrangian method is that the mesh may tangle or collapse and may cause calculation to terminate. Mesh tangling is physically correct in some cases such as turbulence where the Lagrangian code is not effective or suitable. But sometimes, it is fictitious such as hour-glass phenomena and spurious vorticity. So a good code should have the ability to filter all spurious distortions. As in the Eulerian code, special attention must be paid in introducing artificial viscosity. The desired form should be able to capture the shock front and should not effect the calculations of the other part of the flow too much. Preventing the spurious mesh distortion and selecting the artificial viscosity are two important factors in simulation. The 2D Lagrangian code used here is based on Caramana's work [5.12, 5.13, 5.14].

As in the Eulerian code, staggered grid formulation is used with position and velocity defined at the grid point and the density, internal energy, and pressure defined in the zone center. Finite volume method is used, and the momentum equation is discretized in the control volume formed by the median mesh around point p, and the energy equation is discretized in each main zone. Consequently, the momentum and the energy equations can be expressed in the following semi-discretized form,

$$M_p \frac{d\vec{v}_p}{dt} = \sum_z \vec{f}_z^p \,, \tag{5.13}$$

and

$$M_z \frac{de_z}{dt} = -\sum_p \vec{f}_p^z \cdot \vec{v}_p \,, \tag{5.14}$$

where p denotes a main grid point and z denotes zone. In this code, every main zone is divided into four subzones by median mesh. In every subzone, corner mass (m_z^p or m_p^z) and corner force (f_z^p or f_p^z) are defined. Here $[\ ]_z^p$ or $[\ ]_p^z$ means the quantity in the subzone associated with point "p" and "z". The nodal and zonal masses are calculated as $M_p = \sum_z m_z^p$, and $M_z = \sum_p m_p^z$. The corner mass is treated as a Lagrangian variable. Thus there exists subzonal pressure that arises from the subzonal corner density as computed from $\rho_z^p = m_z^p / V_z^p(t)$. The subzonal pressure may be different from the mean zone pressure and the difference is defined as perturbed corner pressure. This perturbed corner pressure stabilizes the Lagrangian grid and eliminates spurious distortions. Every corner force consists of the forces that arise from mean zone pressure, perturbed pressure and artificial viscosity tensor. The force discretizations are constructed by the support-operator-type of method, and the total momentum and energy are conserved. More detailed descriptions of the method are given in [5.12, 5.13, 5.14].

A tensor form of artificial viscosity is used, where the artificial viscosity force is calculated at the edge of each subzone and exerts in $\Delta\vec{v}_i$ direction.

$$\overline{\overline{q_{i,z}}} = q_{sc,i}(1 - \psi_i)\zeta[-\Delta\vec{v}_i \cdot \vec{S}_{i,z}]\Delta\hat{v}_i \Delta\hat{v}_i \,, \tag{5.15}$$

$$q_{sc,i} = c_2 \frac{\gamma+1}{4} \rho (\Delta \vec{v}_i)^2 + \rho |\Delta \vec{v}_i| \sqrt{c_2^2 \left(\frac{\gamma+1}{4}\right)^2 (\Delta \vec{v}_i)^2 + c_1^2 c_s^2}\,, \tag{5.16}$$

and

$$\vec{f}_{m,i} = \vec{q}_{i,z} \cdot \vec{S}_{i,z} = q_{sc,i}(1-\psi_i)(\Delta \vec{v}_i \cdot \vec{S}_{i,z})\zeta[-\Delta \vec{v}_i \cdot \vec{S}_{i,z}]\Delta \vec{v}_i\,, \tag{5.17}$$

where $\Delta \hat{v}_i$ is the unit vector of the velocity change along the edge i of a subzone; ζ is a step function that turns the viscosity off for expansion; $\vec{S}_{i,z}$ is the edge area vector; c_1 and c_2 are constant coefficients of linear and nonlinear parts of artificial viscosity; ρ and c_s are density and sound speed at the shock front; and ψ is a limiter to distinguish the uniform compression. The form of ψ is similar to Christiansen's limiter [5.15] in one-dimension,

$$\psi_i = \max\{0.0, \min[0.5(r_{l,i} + r_{r,i}), 2r_{l,i}, 2r_{r,i}, 1.0\}\,, \tag{5.18}$$

$$r_{l,i} = \frac{\dfrac{\Delta \vec{v}_{i+1} \cdot \Delta \vec{v}_i}{\Delta \vec{x}_{i+1} \cdot \Delta \vec{x}_i}}{\dfrac{|\Delta \vec{v}_i|^2}{|\Delta \vec{x}_i|^2}}\,, \tag{5.19}$$

and

$$r_{r,i} = \frac{\dfrac{\Delta \vec{v}_{i-1} \cdot \Delta \vec{v}_i}{\Delta \vec{x}_{i-1} \cdot \Delta \vec{x}_i}}{\dfrac{|\Delta \vec{v}_i|^2}{|\Delta \vec{x}_i|^2}}\,, \tag{5.20}$$

where $r_{l,i}$ and $r_{r,i}$ are two indicators that distinguish the uniform contraction flow field, the subscripts "$i-1$" and "$i+1$" denote two neighboring edges of the edge i in the logic domain. This limiter will turn the artificial viscosity off in the case of uniform compression.

5.4 Numerical Calculations and Results

In the Eulerian calculation, the grid and mesh are fixed in space; therefore, one can use a smaller domain between the inlet region and the wedge section for computations. Figure 5.1 is a schematic diagram of the problem setup and zoning for both regular and Mach reflections. The shocked air enters at the left boundary, and its states are calculated theoretically. The top and bottom boundaries are treated as impermeable whereas the wedge surface is treated as a fixed boundary. Orthogonal mesh is used and the number of

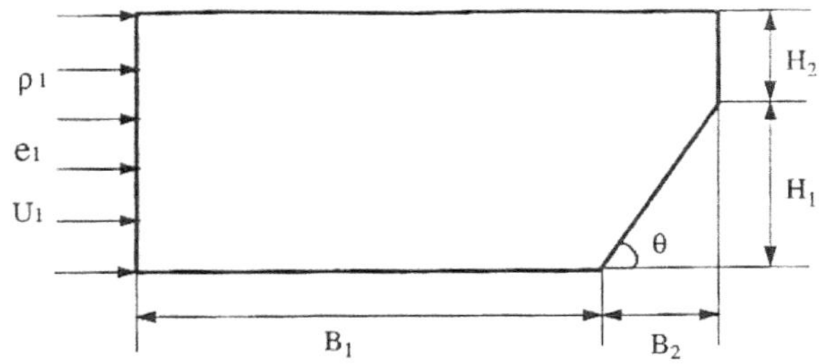

Figure 5.1. Schematic diagram of problem setup in Eulerian method.

Table 5.1. Dimensions of calculating domain and properties of shocked air.

	B_1 (cm)	B_2 (cm)	H_1 (cm)	H_2 (cm)	NX	NY	ρ_1 (Kg/m^3)	e_1 (10^5J)	v_1 (m/s)
Regular Reflection	4	8	16.0	0	480	640	1.060	3.685	450.76
Mach Reflection	8	16	8.15	3.85	960	480	1.049	3.650	444.22

this mesh is $NX \times NY$. The dimensions of the calculating domain, mesh number, and the properties of the shocked air are given in Table 5.1.

In the Lagrangian code, the overall dimension used is the same as that in the Eulerian code except longer B_1 is needed because the left boundary is now moving toward the wedge and B_1 must be long enough to make sure that the reflected shocked does not interfere with the inlet boundary. In the calculation, $B_1 = 14\,\text{cm}$ is used in regular reflection and $B_1 = 30\,\text{cm}$ in Mach reflection. The left boundary in the Lagrangian method works like a piston, and only the velocity boundary condition is needed. The other boundary conditions are the same as those in the Eulerian code.

One advantage of Lagrangian code is its flexibility in defining the mesh. Lagrangian meshes are not necessarily orthogonal, and non-uniform meshes may be structured for a specific problem. For instance, in this problem three mesh layouts could be used as shown in Fig. 5.2. Every layout has its own advantage. Generally speaking, skew mesh is liable to spurious motions. In practice, we should choose the best one depending on a specific situation. When the Mach number is not too high (that means that the computed domain is relatively short), and the wedge angle is large, mesh layout (a) is the best due to its simplicity. When the Mach number is large, the required

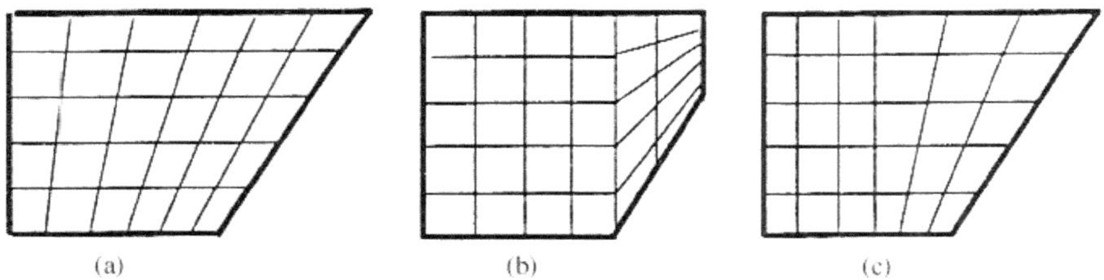

Figure 5.2. Different mesh layouts in Lagrangian method.

straight segment, B_1, is longer. Mesh layout (b) or (c) may be the best since a large aspect ratio in straight segment can be used without causing spurious motions. This allows us to obtain good results with a relatively small number of grids. In our calculation, layout (a) is used for regular reflection and the mesh number is 80×150; and layout (c) is used for simple-Mach reflection and the mesh number is 200×80.

In the following, experimental results and two different numerical simulations are presented for the regular and the simple-Mach reflection, respectively. For each case, experimental isopycnics, computed isopycnics, wall-density distribution, and the isobars will be discussed. All the densities and pressures in the figures are normalized by initial values. Since the solution is self-similar with time, the time scale is not important and will not be mentioned.

5.4.1 Regular Reflection

In this case, $M_s = 2.05$, $\theta = 63.4°$, and the regular reflection occurs. Figure 5.3-a shows isopycnics of experimental data, and computed results by Eulerian code. The Lagrangian calculation is shown in Fig. 5.3-b. Good agreement of overall shock structure between experiment and calculation is obtained. From Fig. 5.3-a (top), the reflected shock makes an angle of 16° from the wedge. The calculated angles are 16.2° and 16° by the Eulerian and Lagrangian codes, respectively. If an overlay of the three isopycnics (two calculations and one experimental data) were made by matching at the reflecting point, the computed reflected shock location by the Lagrangian code would coincide with the experimental data, whereas that of the Eulerian code would fall outside the other two.

Figures 5.3-a (top-left) and 5.3-b also show the excellent agreement of computed isopycnics with experimental data. The wall-density distribution of experimental data and the Eulerian code calculation is shown in

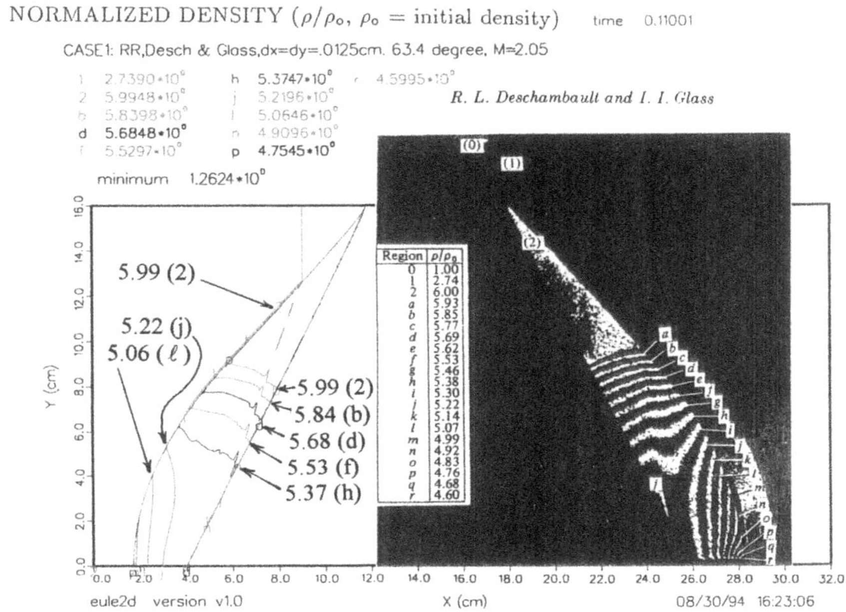

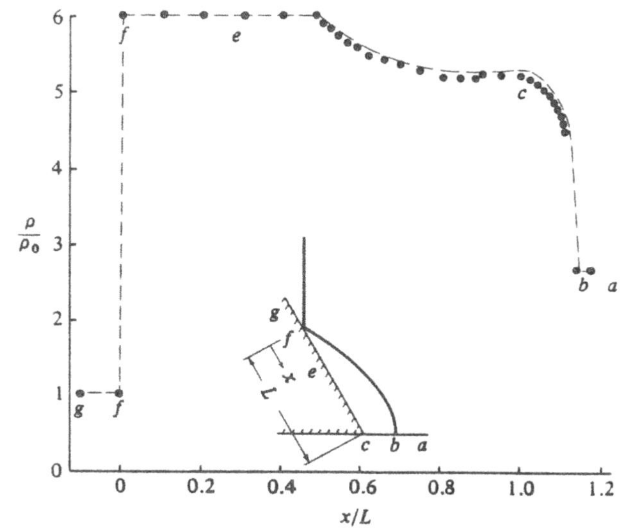

Figure 5.3-a. Isopycnics for regular reflection with Mach number 2.05. The Eulerian code calculation (top-left), experimental results (top-right), and the wall-density distribution of experimental data, '•', and code calculations (– – ––).

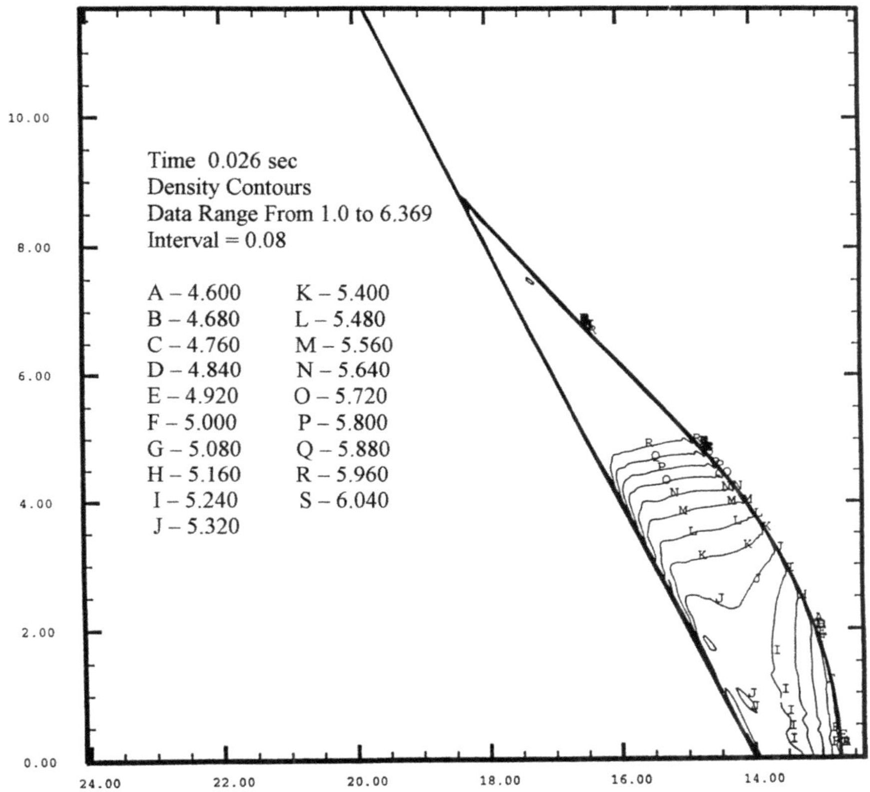

Figure 5.3-b. Computed isopycnics of regular reflection by Lagrangian code.

Fig. 5.3-a (bottom). Eulerian code gives isopycnics location with an error of about one fringe respect to the experimental data, whereas Lagrangian code gives about half a fringe in the whole reflected shock region except those near the wedge surface. The Lagrangian method also gives smoother isopycnics than those of the Eulerian method.

From the above comparisons, the Lagrangian method is seen to be more accurate in calculating isopycnics even though coarse meshes are used. There are two main reasons for this. First, Lagrangian code is generally more accurate than the Eulerian code because of its lack of an advective term. The discretization of advective term could bring additional numerical errors into the Eulerian code. Second, the Lagrangian code has a better way of handling the artificial viscosity. In the Eulerian code, the effect of the artificial viscosity is added to the cell center; therefore, the artificial viscosity is a scalar quantity. In the Lagrangian code, a tensor form of artificial viscosity is used, and the artificial viscosity force is added in the normal direction of the shock front. The limiter also turns unnecessary artificial viscosity off in the uniform compression region. All of these efforts eliminate spurious diffusion of artificial viscosity and make for better results. In Fig. 5.3-b, one may observe that the isopycnics curve sharply along the wedge surface. This error is known as the "wall heating" and it is inherent in the artificial viscosity method. The overheating and the small oscillation shown in Fig. 5.3-a (top-left) are probably produced from the combination of the "wall heating" phenomenon and the treatment of artificial viscosity. Wall heating affects only the specific energy and density, but not the pressure, which can be seen from the computed isobars shown in Fig. 5.4.

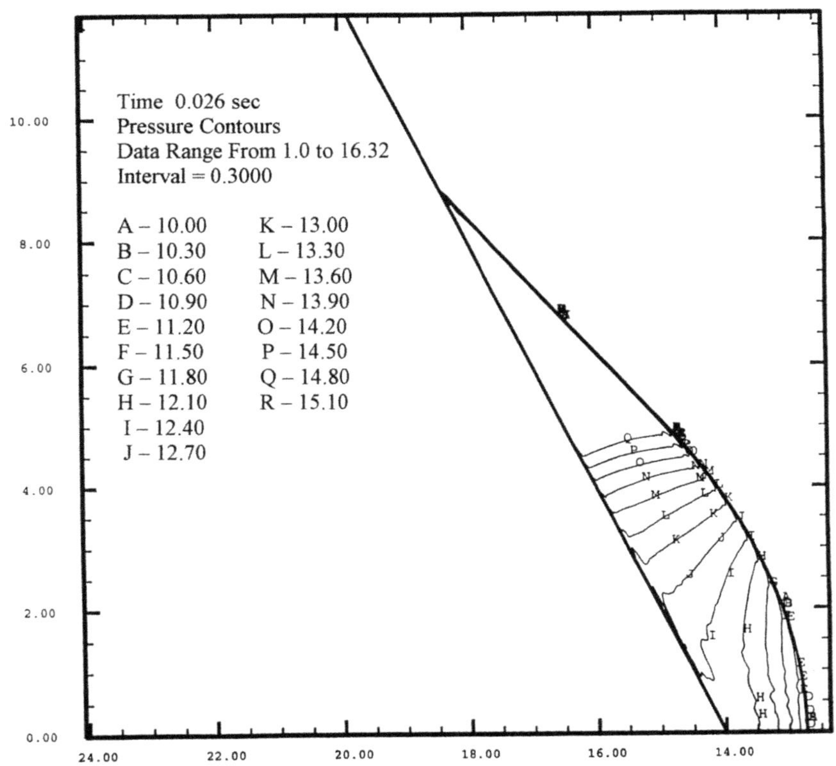

Figure 5.4. Computed isobars of regular reflection by Lagrangian code.

In 1975, Schneyer [5.16] used an Eulerian code and was able to predict the overall shock reflection structure. However, the calculated isopycnics were really poor. In 1985, Glaz, *et al.* [5.17] used Eulerian code with a second-order Godunov scheme and obtained better results with some oscillations near the reflected shock front. In their methods, a procedure to eliminate the "starting error" is added, and they argue that this procedure is very important. This special treatment is a little unnatural and is not required in both of our present codes. In all of these calculations, the reflected shock locations always fall outside the experimental data.

Experimental and numerical wall-density distributions along the wedge are shown in Fig. 5.5. Eulerian code predicts wall density excellently. The Lagrangian code gives a little lower density along the wedge surface. In fact, this discrepancy is expected because of the "wall heating" phenomenon. The good results in the Eulerian method are largely attributed to the fine mesh used. In the Lagrangian method, the mesh is not fine enough to have high resolution, and its aspect ratio is high along the wedge surface. All these deteriorate the calculation. An adaptive mesh structure may improve the results greatly.

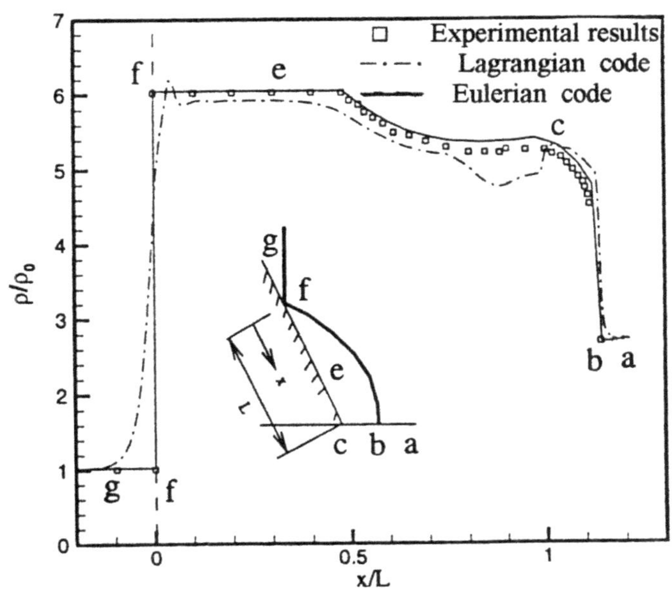

Figure 5.5. Comparison of experimental and computed wall-density distributions.

5.4.2 Mach Reflection

In this case, $M_s = 2.03$, $\theta = 27.0°$, and a single-Mach reflection occurs. The four discontinuities (incident shock wave, reflected shock wave, Mach stem, and slip line) meet at the triple point, which moves along a straight line originating at the toe of the wedge and forming an angle χ with the reflecting surface. Figure 5.6 shows the isopycnics from the experimental data, the Eulerian, and the Lagrangian calculations, respectively. Both codes predict the overall shock reflection structure well. From Fig. 5.6 (top), the trajectory angle χ is equal to 9° which is the same as the computed value from both the Eulerian and Lagrangian codes. If one overlays those three figures by matching them at the triple point, then, the computed shock location computed by the Lagrangian code would coincide with the experimental location, and the shock location by the Eulerian code (see Fig. 5.6 center plot) would fall outside of the other two.

As for the isopycnics, the Eulerian code does predict the value in the reasonable range, but it has large oscillations. The error is caused by the same problem as in the regular reflection. Figure 5.6 (bottom) shows the isopycnics computed by the Lagrangian code. The results are really good. The isopycnics are smooth and are in good agreement with experimental data with an error of about half a fringe.

The little spurious expansion curve "A" along the top-horizontal line in Fig. 5.6 (bottom) is the interface of the fine mesh and the coarse mesh. Since mesh layout like that in Fig. 5.2-c is used in this case, there exists a great variation in grid spacing. When shock wave propagates across a grid whose initial spacing is variable, errors in solution will occur. This error is inherent in the artificial viscosity method. Schneyer [5.16] observed much larger spurious expansion of this kind in his studies and had interpreted them mistakenly as the result of the improper initial velocity profile.

In the Lagrangian code, the meshes may move as time changes, but the mass in one cell remain constant. The mesh plot itself can also show some characteristics of the flow field. Figure 5.7 is the final mesh plot. It shows clearly the structure of the whole shock reflection. The top boundary of the darkened region near the reflection surface starting from the triple point is the real slip line. Slip line exhibits discontinuities in density and shear velocity. Lagrangian code would break down if too much distortion or shear motion appeared. Due to its robustness, the present code can run to the end successfully without interruption. However, pressure is continuous at the slip line as shown in Fig. 5.8. Also, it can be seen that, unlike the

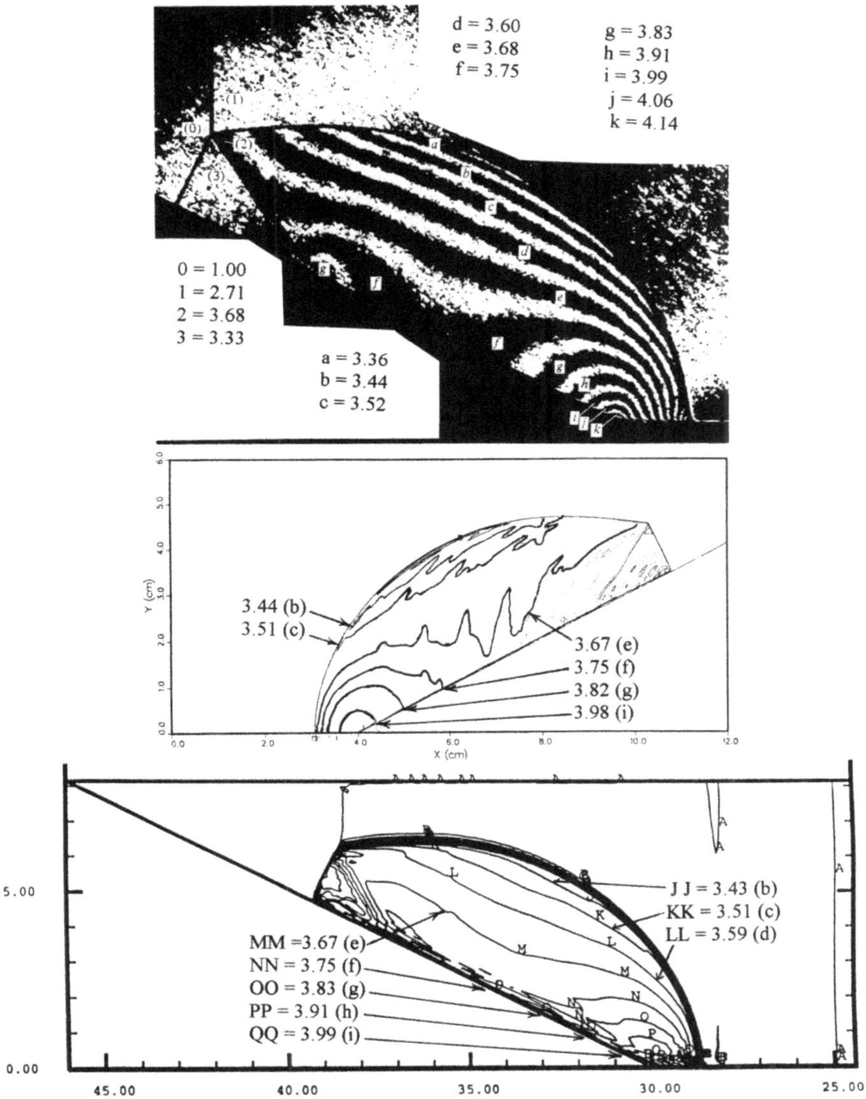

Figure 5.6. Isopycnics of Mach reflection: (top) experimental results, (center) Eulerian results, (bottom) Lagrangian results.

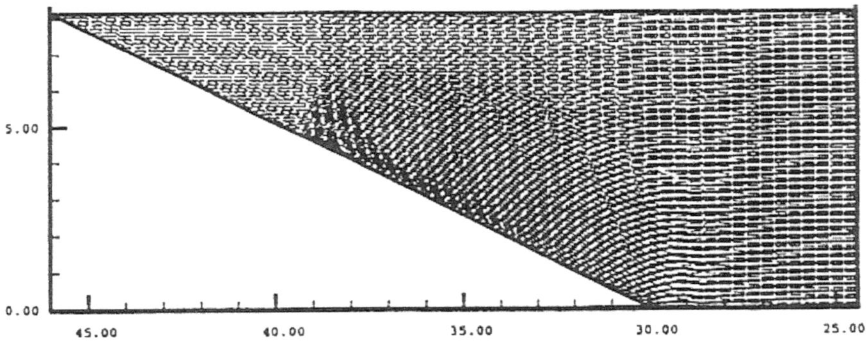

Figure 5.7. Plot of final mesh.

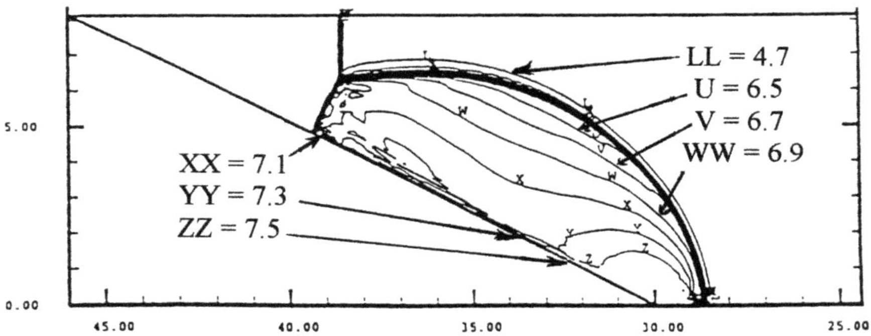

Figure 5.8. Computed isobars of Mach reflection by Lagrangian code.

isopycnics, the isobars are smooth near the wall since "wall-heating" errors only affect internal energy and density, but not the pressure.

5.5 Conclusions

Both the Eulerian and Lagrangian codes are used to simulate the regular and the simple-Mach reflections. The Eulerian code uses the operator splitting scheme and the normal von Neumann artificial viscosity. In the Lagrangian code, Lagrangian subzonal masses and pressure are used to eliminate the spurious grid distortion; the edge-centered artificial viscosity is used to smear the discontinuity. Both codes capture the shock reflections very well and give quantitatively accurate results including the sensitive isopycnics.

Lagrangian method gives the best numerical simulations of the regular and the simple-Mach reflections up to date. Its computed overall shock structure completely coincides with its experimental counterpart. And the computed isopycnics are smooth and are in good agreement with experimental data with an error of about half a fringe. This result is due to the robustness of the code, the better treatment of discontinuity, and the careful problem setup.

An Eulerian method is also used. Although its results are not as accurate as the Lagrangian method, it is used for handling shock problems with large distortion.

It is shown that the accuracy of the simulated results is largely due to the numerical method itself such as the discretization method, the treatment of discontinuity, and the robustness of the code. The perfect gas equation of state assumption in these two cases is not the main reason for the discrepany between the simulated results and the experimental data.

In our future work, we intended to modify the code to reduce the effect due to the "wall heating" phenomenon and try to investigate other Mach reflection cases.

References

5.1 Mach, E, Uber der Vevlauf des Funkenwellen in der Ebene und im Raume, Sitzungsbr. Akad. Wiss. Wien, 78, pp. 819–838 (1878).
5.2 von Neumann, J, Oblique reflection of shocks, Explos. Res. Rep. 12, Navy Department, Bureau of Ordinance, Washington, DC, USA (1943a).
5.3 von Neumann, J, Refraction, intersection and reflection of shock waves, NAVORD Rept. 203-45, Navy Department, Bureau of Ordinance, Washington, DC, USA (1943b).
5.4 Smith, LG, Photographic investigation of the reflection of plane shock in air, NDRC Rept. A-350 (1945).
5.5 White, DR, An experimental survey of the Mach reflection of shock waves, Princeton University, Department of Physics. Rept. II-10, Princeton, NJ, USA (1951).
5.6 Colella, P, and Henderson, LF, The von Neumann paradox for the diffraction of weak shock waves, J. Fluid Mech., 213, pp. 71–94 (1990).
5.7 Glass, II, and Sislian, JP, Nonstationary flow and shock waves, Oxford University Press, Walton Street, Oxford (1994).
5.8 Ben-dor, G, Shock wave reflection phenomena, Springer-Verlag, NY (1992).
5.9 Deschambault, RL, and Glass, II, An update on non-stationary oblique shock-wave reflection: Actual isopycnics and numerical experiments, J. Fluid Mech., 131, pp. 27–57 (1983).
5.10 Von Neumann, J, and Richtmyer, RD, A method for the numerical calculation of hydrodynamic shocks, J. Appl. Phys., 21, pp. 232–237 (1950).

5.11 Van Leer, B, Towards the ultimate conservative difference scheme, IV. A new approach to numerical convection, J. Comput. Phys., 23, pp. 276–299 (1977).
5.12 Caramana, EJ, and Shashkov, MJ, Elimination of artificial grid distortion and hourglass-type motion by means of Lagrangian subzonal masses and pressures, Los Alamos National Laboratory report, LA-UR-97-1631 (1997).
5.13 Caramana, EJ, Shashkov, MJ, and Whalen, PP, Formulations of artificial viscosity for multi-dimensional shock wave problems, Los Alamos National Laboratory report, LA-UR-97-2146 (1997).
5.14 Caramana, EJ, Burton, DE, Shashkov, MJ, and Whalen, PP, The construction of compatible, mimetic hydrodynamics algorithm utilizing conservation of total energy, Las Alamos National Laboratory report, LA-UR-97-3097 (1997).
5.15 Christiansen, RB, Godunov method in a staggered mesh — An improved artificial viscosity, Lawrence Livermore National Laboratory report, UCRL-JC-105269 (1991).
5.16 Schneyer, GP, Numerical simulation of regular and Mach reflections, The Physics of Fluids, 18(9), pp. 1119–1124 (1975).
5.17 Glaz, HM, Colella, P, Glass, II, and Deschambault, RL, A numerical study of oblique shock-wave reflections with experimental comparisons, Proc. R. Soc., A398, pp. 117–140 (1985).

CHAPTER 6

COMPARISONS BETWEEN THE CELL-CENTERED AND STAGGERED MESH LAGRANGIAN HYDRODYNAMICS

Notations

a	constant speed
A	potential
c	speed of light (3×10^{10} cm/s)
D	the cell-face velocity divergence
E	total energy per mass (Mbar-cm^3/g)
$\tilde{E}$	temporary value of energy per mass (Mbar-cm^3/g)
f^*	function defined by Eq. (6.3)
F_j	mass flux across interface j (g/cm^2 μs)
j	Jacobian of the transformation between (R, Z) and (k, l) (cm^2)
K	Lagrangian coordinate used in figures and difference equation
k	Lagrangian coordinate
L	Lagrangian coordinate used in figures and difference equation
ℓ	Lagrangian coordinate
m	mass (g)
M	mass constant (g)
$\hat{n}$	normal unit vector
p	pressure (Mbar)
P	pressure (Mbar)
p^*	flux of momentum (Mbar)
Q	artificial viscosity (Mbar)
Q_A	artificial viscosity (Mbar)
Q_B	artificial viscosity (Mbar)
r, R	radial coordinate (cm)
$\vec{r}$	position vector (R, Z) (cm)
s	surface vector = unit normal vector times area (cm^2)
S, S_C, S_L, S_R	slopes as defined by Eqs. (6.39), (6.38), (6.36), and (6.37)

t	time (μs)
u	velocity in r direction (cm/μs)
u^*	flux of volume
$\vec{u}$	velocity vector (U, V) (cm/μs)
$\vec{u}_v$	vertex velocity vector (cm/μs)
$\tilde{u}$	temporary value of u (cm/μs)
$\bar{u}$	average velocity (cm/μs)
v	velocity in z direction (cm/μs)
$\tilde{v}$	temporary value of v (cm/μs)
$\bar{v}$	average velocity (cm/μs)
V	volume (cm^3)
x	rectangular coordinate (cm)
z, Z	axial coordinate (cm)

Greek Letters

α	dimensionless parameter defined by Eq. (6.4)
ε	specific internal energy per mass (Mbar-cm^3/g)
η	defined by Eq. (6.3a)
ρ	density (g/cm^3)
σ	area (cm^2)
ϕ, ϕ_1, ϕ_2	TVD limiters [see Eqs. (6.13), (6.14) and (6.17)]
ψ	potential defined by Eq. (6.19)
ω	vorticity (1/μs)

Subscripts

0	initial value
k	derivative with respect to k coordinate
ℓ	derivative with respect to ℓ coordinate
r	derivative with respect to r coordinate
t	derivative with respect to time
z	derivative with respect to z coordinate

Superscripts

n	time at n time-step, i.e. $t^n = t_0 + n \cdot \Delta t$

6.1 Introduction

The classical approach to numerical Lagrangian hydrodynamics is to divide the problem domain into control volumes (cells) containing a fixed mass of material. The volume of these cells increases or decreases according to the dynamical motion of their vertices that is determined by the fluid velocity at the cell vertex. The vertex velocity is naturally determined by conservation of linear momentum in another set of control volumes centered at the mass control volume vertices. This well-known staggered mesh approach is shown schematically in Fig. 6.1, and has been troubled with two main difficulties since its first use. The first problem is that artificial terms corresponding to a bulk viscosity must be added to the equation in order to obtain reasonably smooth variation in density near shock wave [6.1]. The second difficulty is that spurious fluid motion can develop as a result of a purely non-physical circumstance brought on by the staggered mesh [6.2].

Both of these difficulties have been the subject of numerous prior investigations [6.1, 6.2]. Despite the efforts of many great investigators, satisfactory general solutions to the non-monotonicity and hour-glassing problems have not been found. We are involved in an ongoing effort to develop general methods for dealing with these classical difficulties. An earlier report in this same forum presented our progress on the monotonicity problem [6.3]. This is the main contribution made by the CAVEAT [6.4] code, in which a single control volume approach is used. When mass, momentum, and energy are all conserved on the same control volume, the variables are said to be all cell-centered. This is the central feature of Godunov's method, which has been highly successful for problem in one dimension. The hitch with this is in the generalization to two and three dimensions. The rub lies in the definition of the vertex velocity u_v used for moving the mesh, as shown in Fig. 6.2.

In one dimension, the definition is natural for cell-centered schemes like that of Godunov. However, in higher dimensions, the definition is not

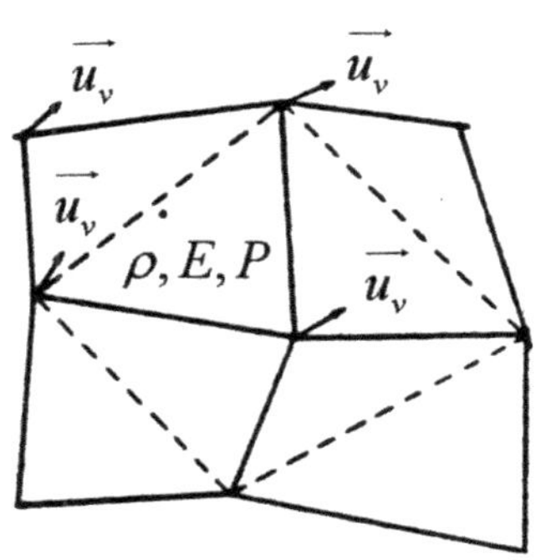

Figure 6.1. Classical staggered mesh. Solid lines define mass control volumes and dashed lines define momentum control volumes.

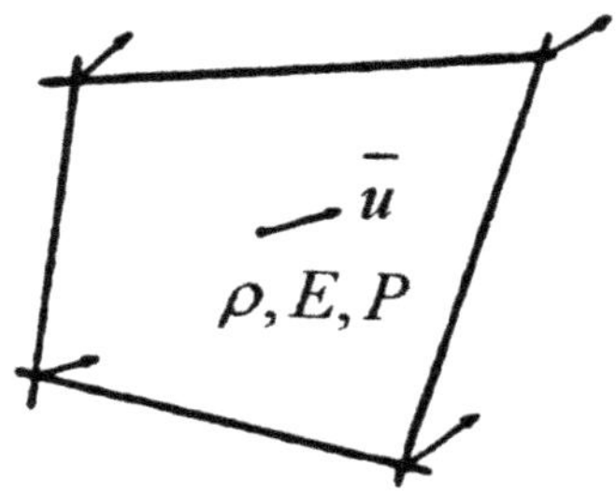

Figure 6.2. Non-staggered mesh with cell-centered velocity.

natural so that some latitude exists for defining the vertex velocity. We find that vertex velocity definition is a crucial matter for cell-centered schemes insofar as determining the robustness of the method.

The subject of this chapter is two-fold. The first is an alternative to the CAVEAT method for handling the monotonicity problem and the new development in the definition of the vertex velocity for cell-centered Lagrangian methods. The second is to investigate the accuracy and stability of both the cell-centered and the staggered mesh methods. After describing these developments, results on some of the well-known test problems are presented.

6.2 A TVD Lagrangian Scheme

Here we describe the general ideas behind our TVD Lagrangian scheme. To accomplish this we first provide a simplified discussion of a single hyperbolic equation. Then, we indicate the manner in which this relates to the more general case of the hydrodynamic equation system.

Consider the wave equation with constant speed a

$$\frac{\partial u}{\partial t} + \frac{\partial au}{\partial x} = 0\,. \tag{6.1}$$

Let $\Delta u = u^{n+1} - u^n$, $\Delta t = t^{n+1} - t^n$ and $\Delta x_j = x_{j+\frac{1}{2}} - x_{j-\frac{1}{2}}$. A linearly stable, forward-time discretization is

$$\frac{\Delta u}{\Delta t} + \frac{1}{\Delta x_j}\left(f^*_{j+\frac{1}{2}} - f^*_{j-\frac{1}{2}}\right) = 0\,, \tag{6.2}$$

where

$$
\begin{aligned}
f^*_{j+\frac{1}{2}} &= \frac{a}{2}(u^n_{j+1} + u^n_j) - \frac{\alpha}{\Delta x_{j+\frac{1}{2}}}(u^n_{j+1} - u^n_j) \\
&= a\left[\frac{1}{2}\left(u^n_{j+1} + u^n_j\right) - \frac{\alpha}{a\Delta x_{j+\frac{1}{2}}}\left(u^n_{j+1} - u^n_j\right)\right].
\end{aligned} \tag{6.3}
$$

Now note that for

$$\alpha = \frac{1}{2}a^2\Delta t \quad \text{then } f^*_{j+\frac{1}{2}} = au^{n+1/2}_{j+\frac{1}{2}},$$

and the method is known as Lax-Wendroff which is second-order, but non-monotone. Notice also that for

$$\alpha = \frac{1}{2}|a|\,\Delta x, \quad \text{then } f^*_{j+\frac{1}{2}} = au^{n+1/2\eta}_{j+\frac{1}{2}}$$

where

$$\eta = \frac{|a|\,\Delta t}{\Delta x}, \tag{6.3a}$$

the method is known as upwind and is monotone, but first order.

A higher-order monotone scheme uses the solution itself to generate a function $\phi(u)$, such that $0 < \phi < 1$, to define the highest order possible, while avoiding development of overshoots and undershoots (new extrema):

$$\alpha = \frac{1}{2}\left[\phi\,|a|\,\Delta x + (1-\phi)a^2\Delta t\right]. \tag{6.4}$$

In one dimension, one can use a precise rule known as a Total-Variation-Diminishing (TVD) test to derive an expression for $\phi(u)$. For systems of equations, the task is extremely difficult and, in multiple dimensions, TVD rules have not been developed. Our approach is to use the basic idea of TVD rules to develop a simple generalization of the foregoing flux calculation. That is, we define fluxes using the conservation equations themselves to compute time-advanced values at the cell faces. The flux time level varies at least $n + \frac{1}{2}$ to $n + \frac{1}{2\eta}$, depending on the local value of the limiter ϕ.

Now consider the system of conservation equations for a compressible, non-viscous fluid in surface integral form

$$\frac{dV}{dt} = \int_s u \cdot \hat{n}\,d\sigma\,, \tag{6.5}$$

$$m\frac{du}{dt} = -\int_s p\hat{n}d\sigma\,, \tag{6.6}$$

and

$$m\frac{dE}{dt} = -\int_s pu\cdot\hat{n}d\sigma\,. \tag{6.7}$$

where V is an arbitrary volume containing a fixed mass m with velocity u and total energy per mass E, and $\hat{n}$ is the normal unit vector. Here $d\sigma$ represents a differential area. A state relation between the pressure p and the other dependent variables closes the system. For each control volume, a discrete form for these is given by

$$\frac{\Delta V}{\Delta t} = \sum u^*\cdot s\,, \tag{6.8}$$

$$m\frac{\Delta u}{\Delta t} = -\sum p^* s\,, \tag{6.9}$$

and

$$m\frac{\Delta E}{\Delta t} = -\sum p^* u^*\cdot s\,, \tag{6.10}$$

where the fluxes are identified by the * and the sum is over the sides of the control volume with outward surface vector s (normal times area). The flux of volume is u^*, the flux of momentum is p^* and the flux energy is the product of the two. Again, the main idea for defining the fluxes is to use the equation system itself for relations that give advanced time quantities. The appropriate equations for defining the fluxes are the momentum and energy equations written

$$\frac{du}{dt} = -\frac{1}{\rho}\nabla p\,, \tag{6.11}$$

and

$$\frac{dp}{dt} = -\rho c^2\nabla\cdot u\,, \tag{6.12}$$

where c is the sound speed. The discrete forms, after identifying the limiter ϕ in Section 6.3 and choosing $\phi_1 = \phi_2 = \phi$, are

$$u^* = u_0^* - \phi_1\left[\frac{\Delta x\nabla p}{(\rho c)_+ + (\rho c)_-}\right]^n - (1-\phi_1)\left(\frac{\Delta t\nabla p}{2\rho}\right)^n\,, \tag{6.13}$$

and

$$p^* = p_0^* - \phi_2 \left[\frac{\Delta x \nabla \cdot u}{(\rho c)_+ + (\rho c)_-} \right]^n - (1 - \phi_2) \left(\frac{\Delta t \rho c^2 \nabla \cdot u}{2} \right)^n , \qquad (6.14)$$

where u_0^* and p_0^* will be given shortly. For the "initial" face values of velocity and pressure, we make use of a result from the method of characteristics. For a uniform mesh, we simply take the leading terms from the corresponding fluxes. These are

$$u_0^* = \frac{(\rho c)_- u_- + (\rho c)_+ u_+}{(\rho c)_- + (\rho c)_+} , \qquad (6.15)$$

and

$$p_0^* = \frac{p_-/(\rho c)_- + p_+/(\rho c)_+}{1/(\rho c)_- + 1/(\rho c)_+} , \qquad (6.16)$$

where the subscript plus and minus refer to the cell-centered values on opposite sides of the face for which the flux is desired.

6.3 The DIVU Limiter

The limiter used in our TVD method is described here. This limiter can be considered to be a generalization of the results due to Davis [6.5] who considered the case of Eulerian coordinates. At the present, we consider the divergence of the velocity to be an essential means by which to measure the smoothness of the solution. The limiter is required at the cell face and is a function of the cell-face velocity divergence D and the cell-centered velocity divergence on either side of the face D_+ and D_-. Therefore, $\phi = \phi'$, if $D \leq 0$; and $\phi = 0$, otherwise, where the trial value of the limiter is

$$\phi' = 1 - \max \left[0, \min \left(\frac{D}{D_-}, \frac{D}{D_+}, \frac{D_-}{D}, \frac{D_+}{D} \right) \right] . \qquad (6.17)$$

Note that if any of the D's change sign, then the limiter is unity, making the diffusion coefficient first-order. Also, if the velocity divergence is positive (indicating expansion), then the limiter is always zero, so that the finite difference is second-order.

6.4 Vertex Velocity Definition

One of the main challenges in cell-centered Lagrangian numerical hydrodynamics is the definition of the vertex velocity. Since the vertex velocity is not computed directly, it must be interpolated from the flow field or otherwise found from the other dependent variables. The only criterion available in formulating the definition of the vertex motion is that the volume change should be preserved. That is, the volume conservation equation should be satisfied by whatever vertex velocity is determined. In general the normal motion at each cell face is sufficient to define the motion everywhere, but our attempts to use this condition alone have failed to produce a scheme that is satisfactory in terms of robustness. Basically what we observe with the standard scheme used in the CAVEAT [6.4] code is the tendency for vertex crossing in problem with significant rotation. In general, we seek a Lagrangian scheme that computes problems with rotation for as long as the mesh permits, without vertex crossing.

One scheme has emerged recently that holds real promise for improving the robustness of our cell-centered methods. In this the velocity at the cell vertex is defined as

$$u_v = \nabla\psi + \nabla \times A\,, \tag{6.18}$$

where the potential ψ and A are defined in the standard fashion by

$$\nabla^2\psi = \nabla \cdot u\,, \tag{6.19}$$

and

$$-\nabla^2 A = \nabla \times u\,, \tag{6.20}$$

in which the vector potential is chosen to be solenoidal. The divergence of the velocity field is known from the Lagrangian equations themselves ($u = u^*$), so the only remaining item needed is the curl of the velocity. In principle the curl is known also, at least at the n time level, but we choose instead to integrate the vorticity equation

$$\frac{d\omega}{dt} + \omega\nabla \cdot u = \nabla p \times \nabla\left(\frac{1}{\rho}\right), \tag{6.21}$$

where ω is the vorticity vector (which in two dimensions has only one non-zero component). The reason for going to all this trouble is purely heuristic, and goes as follows. We know that in general a multi-dimensional flow field can develop rotation locally as part of the dynamics of the problem. We also know that a distorted computational mesh can be responsible for introducing a rotation in the flow that is completely artificial. If we take the position that a practical resolution of the physical rotation is accomplished by "capturing" the rotation in a manner analogous to the way we are "capturing" shocks, then we require a mechanism for smearing out physical rotation in the problem. It turns out to be quite convenient to incorporate the smearing mechanism in the process of solving the vorticity equation itself. A stable, forward-time differencing procedure is given by

$$\frac{\Delta\omega}{\Delta t} + \omega^{n+1}\nabla \cdot u^* = \nabla p^n \times \nabla\left(\frac{1}{\rho^n}\right) + \nabla \cdot \alpha \nabla \omega^n\,, \tag{6.22}$$

where α is already defined. The diffusion term is required because the forward-time difference used in the temporal term introduces a negative diffusion effect that must cancel to ensure stability.

All that is required now is the specification of boundary conditions on the potentials, which is simply given by the normal velocity on the exterior of the domain. Specifically, we use the conditions on the problem boundary

$$\nabla\psi = u^* \cdot \hat{n}\,, \tag{6.23}$$

and

$$A = 0\,. \tag{6.24}$$

Clearly this formulation is not the simplest, because we are now faced with the difficulty of solving two elliptic boundary value problems at each time step. We find, however, that conjugate gradient procedure works sufficiently fast to make the method practical, even when the remainder of the computation is purely explicit.

6.5 Staggered Mesh

The staggered mesh uses the two-dimensional Lagrangian method by Schulz [6.6] as described in Chapter 2. The basic equations in finite difference form are given in Eqs. (2.45)–(2.48). For the present test problems, we do not have any stress term, therefore, the governing equations can be simplified to

Mass

$$\rho R(R_k Z_\ell - R_\ell Z_k) = M\,. \tag{6.25}$$

Momentum

$$u_t = -\frac{1}{\rho j}\left(Z_\ell \frac{\Delta P}{\Delta k} - Z_k \frac{\Delta P}{\Delta \ell}\right) - \frac{1}{M}\frac{\Delta}{\Delta k}\left(RQ_A \frac{R_\ell u_k + Z_\ell v_k}{u_k^2 + v_k^2} u_k\right)$$
$$+ \frac{1}{M}\frac{\Delta}{\Delta \ell}\left(RQ_B \frac{R_k u_\ell + Z_k v_\ell}{u_\ell^2 + v_\ell^2} u_\ell\right), \tag{6.26}$$

$$v_t = \frac{1}{\rho j}\left(R_\ell \frac{\Delta P}{\Delta k} - R_k \frac{\Delta P}{\Delta \ell}\right) + \frac{1}{M}\frac{\Delta}{\Delta k}\left(RQ_A \frac{R_\ell u_k + Z_\ell v_k}{u_k^2 + v_k^2} v_k\right)$$
$$- \frac{1}{M}\frac{\Delta}{\Delta \ell}\left(RQ_B \frac{R_k u_\ell + Z_k v_\ell}{u_\ell^2 + v_\ell^2} v_\ell\right), \tag{6.27}$$

and Energy

$$\rho \varepsilon_t = -P\nabla \cdot \bar{u} - \frac{Q_A}{\rho j}(Z_\ell u_k - R_\ell v_k) + \frac{Q_B}{\rho j}(Z_k u_\ell - R_k v_\ell), \tag{6.28}$$

with all of the variables defined in Chapter 2 or in the nomenclature section.

6.6 The Mesa and Unicorn Method

Mesa and Unicorn are two-dimensional Eulerian codes. In the first phase, they solve the Lagrangian part in both the r and z directions simultaneously. The momentum and energy equations solved in the Lagrangian phase are

$$\frac{\partial u}{\partial t} = -\frac{1}{\rho}\frac{\partial p}{\partial r}, \tag{6.29}$$
$$\frac{\partial v}{\partial t} = -\frac{1}{\rho}\frac{\partial p}{\partial z}, \tag{6.30}$$

and

$$\frac{\partial E}{\partial t} = -\frac{p}{\rho r}\frac{\partial (ru)}{\partial r} - \frac{p}{\rho}\frac{\partial v}{\partial z}. \tag{6.31}$$

At the end of phase one, we have the temporary values of velocity $\tilde{u}$ and $\tilde{v}$, energy $\tilde{E}$, momentum $M\tilde{u}$ and $M\tilde{v}$ (where M is the mass in the cell), and total internal energy in the cell $M\tilde{E}$. In phase two or remap phase, the average velocities $\bar{u}$ and $\bar{v}$ are calculated by

$$\bar{u} = \frac{1}{2}(u + \tilde{u}), \tag{6.32}$$

and

$$\bar{v} = \frac{1}{2}(v + \tilde{v}), \tag{6.33}$$

where u and v are the old velocities. These average velocities are used for computing the momentum and energy flux across the cell boundaries. The governing equations are split in radial and axial directions. Then calculation is performed in each separated direction. The order of this calculation is alternated for each time advancement to maintain the accuracy of one-dimensional procedure. Mesa uses the monotonic advection methods of Van Leer [6.7] to avoid the non-physical oscillations and negative density. Unicorn code uses LeBlanc limiting [6.8] for the remapped phase, that is a second order monotonic interpolation scheme. To describe the method, let us start with the radial mass conservation equation

$$\frac{\partial \rho}{\partial t} = -\frac{1}{r}\frac{\partial (r\rho u)}{\partial r}\,. \tag{6.34}$$

The finite difference of Eq. (6.34) is

$$\rho^{n+1}_{j+\frac{1}{2}} = \rho^{n}_{j+\frac{1}{2}} - \frac{\Delta t^{n+\frac{1}{2}}}{r_{j+\frac{1}{2}}\Delta r_{j+\frac{1}{2}}}\left[(rF)^{n+\frac{1}{2}}_{j+1} - (rF)^{n+\frac{1}{2}}_{j}\right]. \tag{6.35}$$

where n is the discrete time level and $n+1/2$ means $n+\Delta t/2$, j is the zone index in r direction, and F_j the mass flux across interface j. Here $F_j = \rho^*_j \bar{u}_j$. Unicorn method uses slope S_L, S_R, and S_C to obtain the optimum value of ρ^*. The expressions of S_L, S_R, and S_C are

$$S_L = \frac{\rho_{j+\frac{1}{2}} - \rho_{j-\frac{1}{2}}}{0.5\Delta r_{j+\frac{1}{2}}}\,, \tag{6.36}$$

$$S_R = \frac{\rho_{j+\frac{3}{2}} - \rho_{j+\frac{1}{2}}}{0.5\Delta r_{j+\frac{1}{2}}}\,, \tag{6.37}$$

and

$$S_C = \frac{1}{\Delta r_{j+\frac{1}{2}}}\left(\frac{\rho_{j+\frac{1}{2}}\Delta r_{j+\frac{3}{2}} + \rho_{j+\frac{3}{2}}\Delta r_{j+\frac{1}{2}}}{\Delta r_{j+\frac{3}{2}} + \Delta r_{j+\frac{1}{2}}} - \frac{\rho_{j-\frac{1}{2}}\Delta r_{j+\frac{1}{2}} + \rho_{j+\frac{1}{2}}\Delta r_{j-\frac{1}{2}}}{\Delta r_{j+\frac{1}{2}} + \Delta r_{j-\frac{1}{2}}}\right). \tag{6.38}$$

The minimum value of the slope S is

$$S = \min\{|S_L|, |S_C|, |S_R|\}\,. \tag{6.39}$$

If the slope S_L has an opposite sign of the slope S_R then $S = 0$. Finally, we interpolate backward along the slope S by a distance $\frac{1}{2}\Delta t^{n+1/2}\bar{u}_j$ from the zone boundary j. This location gives the desired value of the interpolated mass density ρ^*. Unicorn method is second order accurate, and it greatly reduces the numerical diffusion.

6.7 Sample Results

Here we display the results of three test problems known respectively as Saltzman, blast wave, and strong shock tube, each computed using the classical Schulz hydrodynamic scheme and the modified TVD scheme described here. They all use ideal gas with γ-law as the equation of state. The Saltzman and the blast wave problems are treated as two-dimensional problems while the strong shock tube is calculated with one dimension only. In the two-dimensional problem, the TVD method, using the velocity potential scheme for determining the vertex velocity, far exceeds the Schulz (staggered mesh) method.

The initial skewed mesh of the Saltzman problem is shown in Fig. 6.3 with 100 meshes in the axial direction (z-axis) and 10 meshes in the radial direction (r-axis). The moving piston boundary is on the left-hand side and the right-hand side is a non-flow boundary. The Schulz method shows some bad meshes near the shock front along the z-axis at time $t = 0.7\,\mathrm{s}$ (see Fig. 6.4, lower portion), while the modified TVD method gives a very nice shock front definition (see Fig. 6.4, upper portion). The final configuration

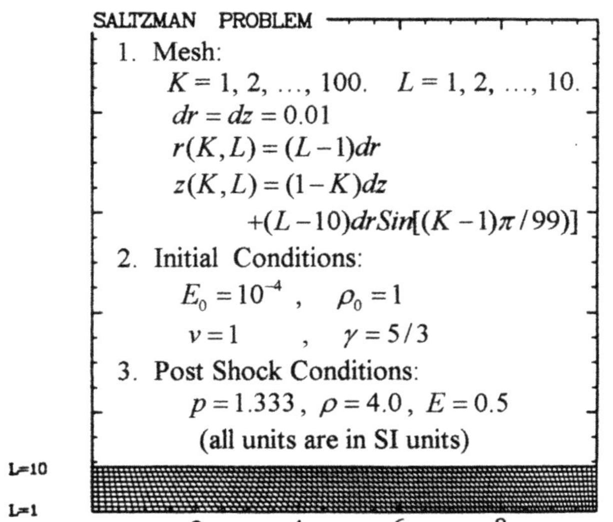

Figure 6.3. The initial skewed mesh of the Saltzman problem with meshes 10×100 (radial by axial direction).

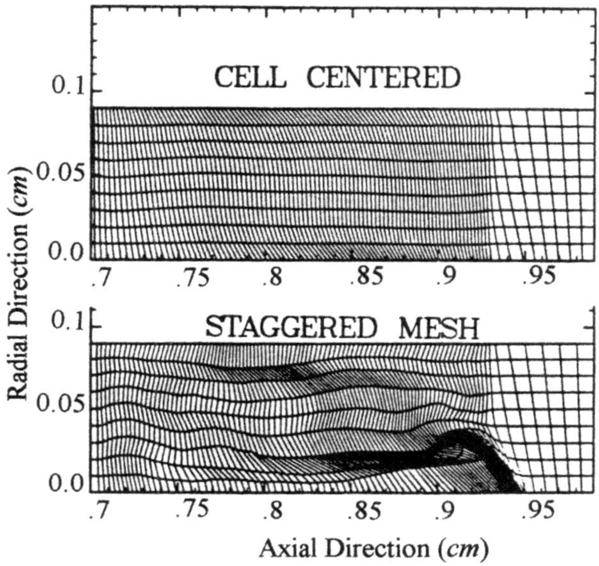

Figure 6.4. Saltzman problem at time 0.7 sec. by cell centered method (upper) and by staggered mesh method (lower).

calculated by the TVD method is shown in Fig. 6.5, after the shock has reflected numerous times between the wall and the piston.

The second problem, called the blast wave, has initial density $\rho_0 = 1\,\mathrm{g/cm^3}$, energy $E_0 = 0$, and pressure $P_0 = 0$. The problem is computed for a region of $0 \le r \le 1.125\,\mathrm{cm}$ and $-1.125\,\mathrm{cm} \le z \le 1.125\,\mathrm{cm}$ with mesh 46×91. The uniform square zone size is $0.025\,\mathrm{cm}$ (i.e. $dr = dz = 0.025\,\mathrm{cm}$). At time $t = 0\,\mathrm{s}$, a high pressure, $p = 3.3053 \times 10^7$ Mbar, is set for four zones near the origin (i.e. $k = 45, 46, 47$, and $\ell = 1, 2$) as shown in Fig. 6.6. Due to the high pressure, a spherical wave starts to propagate outward. At time $t = 1\,\mathrm{s}$, the wave front should reach radius $r = 1\,\mathrm{cm}$. Figure 6.7 contains the meshes and wave fronts as computed by Schulz's hydro (top) and the TVD (bottom) methods. The TVD method not only shows smoother meshes, it also obtains the correct wave front position. The artificial viscosity used in Schulz's hydro has smeared the wave front to a wrong position (approximately at $r = 0.97\,\mathrm{cm}$). The density distributions at time $t = 1.0\,\mathrm{s}$ are shown in Fig. 6.8. Again, the TVD result is much closer to the analytical solution and has no fluctuation.

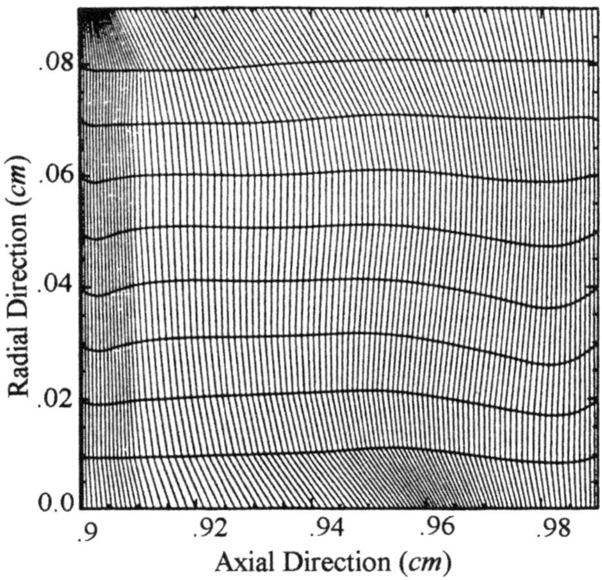

Figure 6.5. The configuration calculated by modified TVD method (cell centered) at time 0.9 sec.

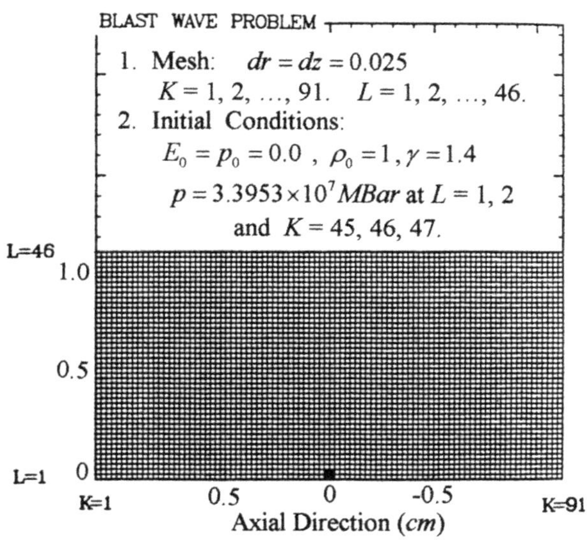

Figure 6.6. The initial meshes of 46 × 91 (radial by axial direction) and the initial conditions of the blast wave problem (all units are in SI units).

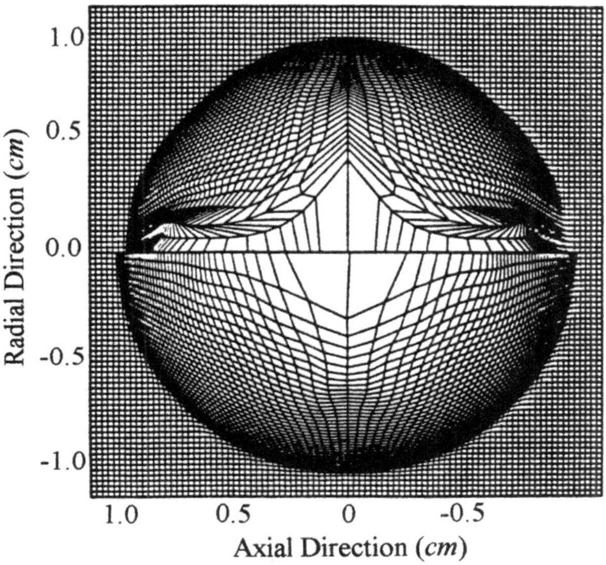

Figure 6.7. The meshes and wave front as computed by Schulz's hydro (upper) and the modified TVD method (lower).

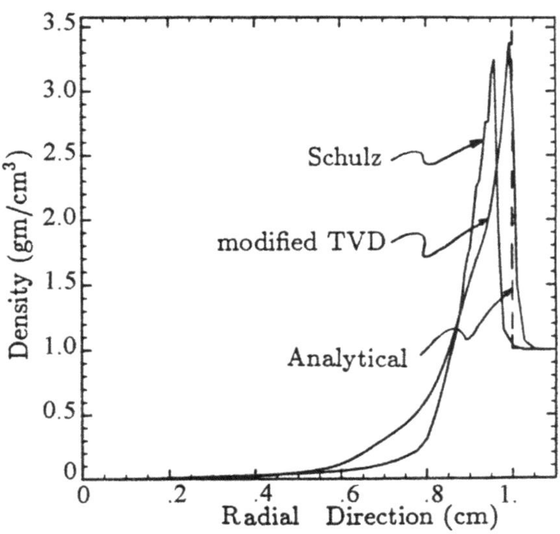

Figure 6.8. The density vs. radial direction at time 1 s. for analytical, Schulz, and modified TVD solutions.

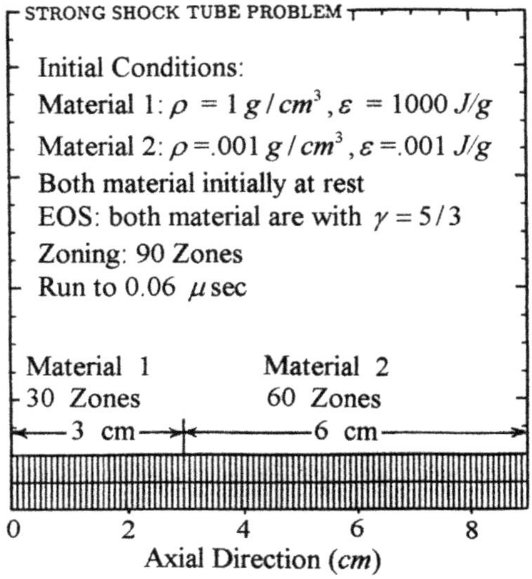

Figure 6.9. The initial uniformed meshes (material 1 has 30 zones and material 2 has 60 zones) and initial conditions for the strong shock tube problem.

The last problem is known as strong shock tube problem that consists of two materials. The initial configuration and conditions are described in Fig. 6.9 with a total of 90 zones. The density distribution obtained by analytical method at time $0.06\,\mu$s, and the initial mesh for mass matching are shown in Fig. 6.11. The density distributions at time $t = 0.06\,\mu$s are shown in Fig. 6.10 for no mass matching and Fig. 6.12 for mass matching (i.e. $\rho\Delta x = 10^{-4}\,\text{g/cm}^2$). For no mass matching case, Schulz's hydro is better than the TVD scheme. In the case of mass matching, there is no difference between them.

The computed results of the density along the axial direction are presented with the analytical solutions. Figure 6.13 shows the density distributions as simulated by Mesa and Unicorn codes. The Unicorn calculations for the shock front and contact discontinuity are better than those of Mesa, but overall they are almost the same.

6.8 Conclusions

We conclude from these studies that the new TVD method sketched here, together with the velocity potential scheme for u_v, may provide a significant improvement over existing staggered mesh methods for multi-dimentional

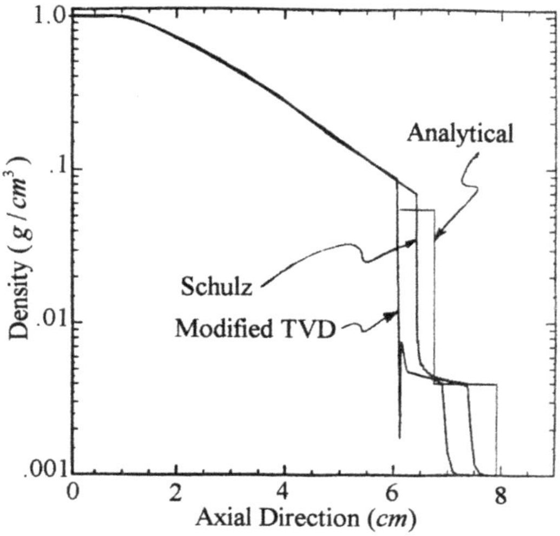

Figure 6.10. The density distributions at time 0.06 μs for analytical, Schulz and modified TVD solutions.

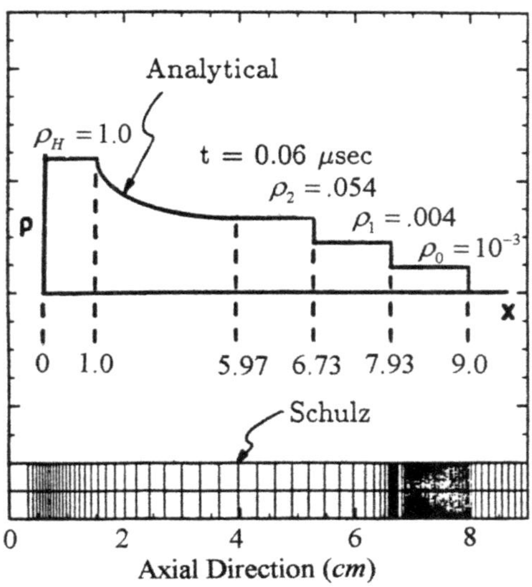

Figure 6.11. The density distribution for analytical solution at time 0.06 μs, (upper portion) and the final mesh for mass matching (lower portion).

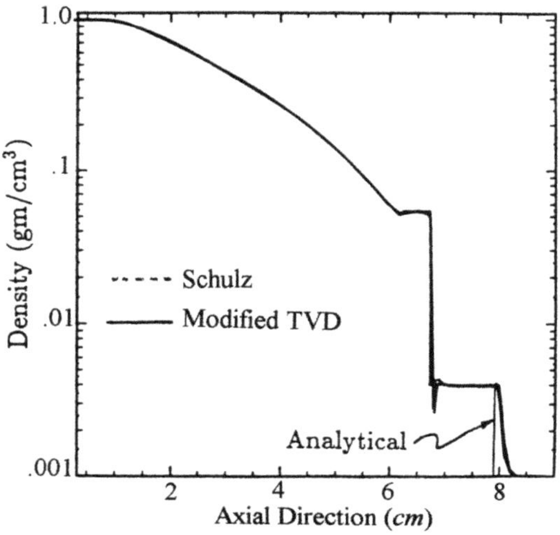

Figure 6.12. The density distribution at time 0.06 μs for analytical, Schulz and modified TVD solutions with mass matching.

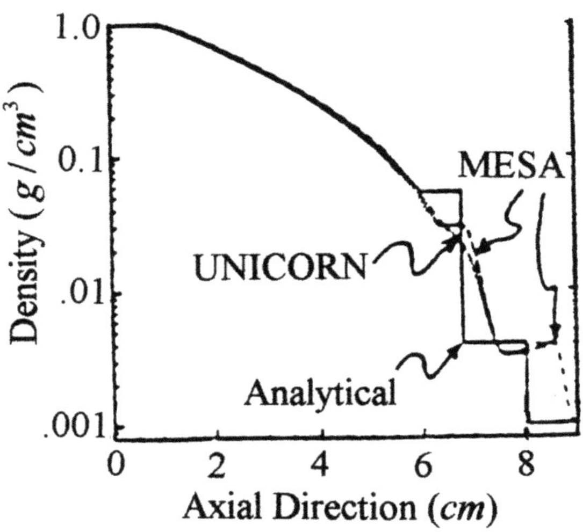

Figure 6.13. The density distributions as calculated by MESA and UNICORN codes.

problems. Additional testing of these schemes in practical problems is clearly warranted.

References

6.1 von Neumann, J, and Richtmyer, R, J. Appl. Phys., 21, pp. 232 (1950).
6.2 Margolin, L, and Pyun, J, Fourth Int. Conf. in Numerical Methods in Laminar and Turbulent Flow, Montreal, Quebec, Canada, July 6–10 (1987).
6.3 Baumgardner, J, *et al.*, Nuclear Explosives Code Developers Conference, San Diego, Ca., November 6–8 (1984).
6.4 Addessio, F, *et al.*, Los Alamos National Laboratory Report LA-10613-MS (1986).
6.5 Davis, S, SIAM J. Sci. Stat. Comput., 8, pp. 1–18 (1987).
6.6 Schulz, W, Two-Dimensional Lagrangian Hydrodynamic Difference Equation. Method in Computational Physics, 3, pp. 1–45 (1964).
6.7 van Leer, B, J. Comput. Phys., 23, pp. 276–299 (1977).
6.8 Clancy, S, Ph.D. Thesis, University Texas, Austin, Texas (1989).

CHAPTER 7

MULTIPHASE FLOW TREATMENT

Notations

C_d	standard sphere drag coefficient
C_p	dynamic pressure coefficient
C^*	defined by Eq. (7.19a)
g	gravitational acceleration
h	specific enthalpy
K	drag function
k	heat conduction coefficient
P	pressure
R	exchange function describing heat transfer between fields
r_p	radius of particle
t	time
T	temperature
$\vec{U}$	velocity vector, i.e. $\vec{U} = (u, v)$
u_m	mean velocity in x direction defined by Eq. (7.9)
v_m	mean velocity in y direction defined by Eq. (7.10)
u	velocity in x direction
v	velocity in y direction
x	coordinate direction
y	coordinate direction

Greek Letters

α	volume fraction
β	iteration step
Γ_c	condensation rate
Γ_e	evaporation rate
ζ	cell index in x direction
η	cell index in y direction
$\lambda_1, \lambda_2, \lambda_3, \lambda_4$	eigenvalues used in Section 7.3
λ_c, λ_e	relaxation parameters used in Eqs. (7.40)–(7.41)
μ	dynamic viscosity

ρ microscopic material density
ρ' macroscopic material density, e.g. $\rho'_i = \alpha_i \rho_i$
ρ_m mean density defined by Eq. (7.11)
ρ^* pseudo density defined by Eq. (7.19-b)
ω relaxation factor

Subscripts

c continuous phase
d discontinuous phase
i phase $i =$ solid or liquid
j phase $j =$ gas or vapor
m mean value
s saturation state

7.1 Introduction

During the collapsing process of the shaped charge liner, particles may be ejected into the air (or into an attached foam) due to the spallation of the liner. Usually, no particles will be ejected into the high explosive side. Therefore, the mixture of the liner and high explosive burnt gas is not considered. In the late time of the liner jet formation, the liner jet may break into separated drops which will mix with the air, e.g. see Section 9.4.8. The broken liner jet particle is called dispersed fluid, and the air is continuous fluid. For a more accurate calculation of this situation, one should use the so called two-phase flow model. However, the two-phase flow model equations are not well defined. Some of these equation systems possess complex characteristics. The full implications of the complex characteristics are still not completely agreed upon, but many investigators agree that the equations are probably ill-posed as an initial-value or Cauchy problem. Certainly they are not completely hyperbolic, which causes numerical difficulties because the von Neumann analysis for the linear system predicts exponential growth. *Ad hoc* numerical techniques using space increments or interfacial drag sufficiently large to stabilize some unstable modes have recently been suggested.

It now appears that the modeling difficulties arise from constitutive models and assumptions concerning the treatment of the pressure, stress, and transient flow forces at the interface between the phases or inconsistent approximations. Stuhmiller [7.1] presented an analysis which tended to support the importance of the interfacial pressure modeling by obtaining the averaged equation of motion for a single accelerating sphere that has real characteristics. Banerjee *et al.* [7.2] admitted two phase pressures into their model by adopting an interphase pressure difference constitutive equation and also obtained all real characteristics. Gidaspow [7.3], on the other hand, showed that a consistent thermodynamic approach would also produce two-phase model equations having all real characteristics.

The objective of this chapter is to present the computational results of five-transient two-phase flow models all solved using the same finite-differencing technique, numerical scheme, and computer code structure [7.4]. The problems analyzed are:

(1) One-dimensional fluidized bed.
(2) One-dimensional vertical batch settling of a two-phase mixture.
(3) Horizontal two-phase jet impingement on a vertical wall.

The one-dimensional fluidized bed problem was chosen because data are available concerning the frequency of voidage oscillations. The batch settling problem demonstrates the propagation of density or kinematic waves and phase separation which has been analyzed by Soo [7.5]. The two-phase jet-impingement problem can be compared with experimental data [7.6]. The five sets of equations under consideration are:

(1) Soo's model [7.7].
(2) Gidaspow's model [7.3].
(3) Rudinger-Chang model [7.8] as modified by Lyczkowski [7.9].
(4) The Hancox *et al.* model [7.10].
(5) The ill-posed model [7.11] which is the same as the TRAC [7.12] code vessel model.

These five two-phase models represent a diversified and representative cross section of active research. The first four models were chosen because they all explicitly attempted to develop well-posed two-phase models to remedy the ill-posed model which is widely used in two-phase analysis [7.11]. Gidaspow's model was developed for fluidized beds. Arastoopour and Gidaspow analyzed four two-phase models for steady-state one-dimensional vertical pneumatic conveying of solids [7.13]. The results of the computations from each of the five models are compared with the available analytical results, with available data and against each other.

7.2 Two-phase Flow Model Equations

This section presents the five different two-phase flow equation sets under consideration. In the interest of compactness, it is best to begin this section with the equation set, which we will refer to as the "basic" or "ill-posed" set [7.11]. In this section, when we refer to an equation set as being ill-posed or well-posed, we refer to whether the characteristics for one-dimensional, incompressible (ρ_i, ρ_j), "potential type" [7.11], and isothermal flow are either complex-valued or real. They are summarized in Section 7.3. When the phases are compressible, the polynomial for the characteristics does not factor in general. In this case, much more effort is needed to determine the characteristics numerically [7.11]. Modifications of this basic set will then be given for each of the subsequent four sets. The governing equations of mass and momentum in a two-dimensional Cartesian geometry are given by:

7.2.1 Basic Equation Set

Continuity

$$\frac{\partial}{\partial t}(\alpha_i \rho_i) + \frac{\partial}{\partial x}(\alpha_i \rho_i u_i) + \frac{\partial}{\partial y}(\alpha_i \rho_i v_i) = (\Gamma_c - \Gamma_e)\,, \tag{7.1}$$

$$\frac{\partial}{\partial t}(\alpha_j \rho_j) + \frac{\partial}{\partial x}(\alpha_j \rho_j u_j) + \frac{\partial}{\partial y}(\alpha_j \rho_j v_j) = (\Gamma_e - \Gamma_c)\,, \tag{7.1a}$$

x-Direction Momentum

$$\frac{\partial}{\partial t}(\alpha_i \rho_i u_i) + \frac{\partial}{\partial x}(\alpha_i \rho_i u_i u_i) + \frac{\partial}{\partial y}(\alpha_i \rho_i v_i u_i) + \alpha_i \frac{\partial P}{\partial x} - K(u_j - u_i) - (\Gamma_c u_j - \Gamma_e u_i) = 0\,, \tag{7.2}$$

$$\frac{\partial}{\partial t}(\alpha_j \rho_j u_j) + \frac{\partial}{\partial x}(\alpha_j \rho_j u_j u_j) + \frac{\partial}{\partial y}(\alpha_j \rho_j v_j u_j) + \alpha_j \frac{\partial P}{\partial x} - K(u_i - u_j) - (\Gamma_e u_i - \Gamma_c u_j) = 0\,, \tag{7.2a}$$

y-Direction Momentum

$$\frac{\partial}{\partial t}(\alpha_i \rho_i v_i) + \frac{\partial}{\partial x}(\alpha_i \rho_i u_i v_i) + \frac{\partial}{\partial y}(\alpha_i \rho_i v_i v_i) + \alpha_i \frac{\partial P}{\partial y} - K(v_j - v_i) - (\Gamma_c v_j - \Gamma_e v_i) + \alpha_i \rho_i g = 0\,, \tag{7.3}$$

$$\frac{\partial}{\partial t}(\alpha_j \rho_j v_j) + \frac{\partial}{\partial x}(\alpha_j \rho_j u_j v_j) + \frac{\partial}{\partial y}(\alpha_j \rho_j v_j v_j)$$
$$+ \alpha_j \frac{\partial P}{\partial y} - K(v_i - v_j) - (\Gamma_e v_i - \Gamma_c v_j) + \alpha_j \rho_j g = 0 \,. \qquad (7.3a)$$

In Eqs. (7.1) through (7.3a) above, i refers to the liquid or solid phase and j refers to the gas or vapor phase. Gravity is assumed to act in the y-direction. The volume fractions are constrained by:

$$\alpha_i + \alpha_j = 1 \,, \qquad (7.4)$$

since there are never more than two co-existent phases. The drag function K and evaporation and condensation rates are flow regime dependent. They are given in Section 7.4.

The energy equations are given by:

$$\frac{\partial}{\partial t}(\alpha_i \rho_i h_i) + \frac{\partial}{\partial x}(\alpha_i \rho_i u_i h_i) + \frac{\partial}{\partial y}(\alpha_i \rho_i v_i h_i)$$
$$= \alpha_i \left(\frac{\partial P}{\partial t} + u_i \frac{\partial P}{\partial x} + v_i \frac{\partial P}{\partial y} \right) + R_i(T_j - T_i) + (\Gamma_c - \Gamma_e) h_j \,, \qquad (7.5)$$

$$\frac{\partial}{\partial t}(\alpha_j \rho_j h_j) + \frac{\partial}{\partial x}(\alpha_j \rho_j u_j h_j) + \frac{\partial}{\partial y}(\alpha_j \rho_j v_j h_j)$$
$$= \alpha_j \left(\frac{\partial P}{\partial t} + u_j \frac{\partial P}{\partial x} + v_j \frac{\partial P}{\partial y} \right) + R_j(T_i - T_j) + (\Gamma_e - \Gamma_c) h_j$$
$$+ \left(K + \frac{\Gamma_e - \Gamma_c}{2} \right) [(u_i - u_j)(v_i - v_j)] \,. \qquad (7.5a)$$

The continuity and energy equations will be taken to be common to all five of the equation sets. Only the momentum equations change from set to set. The internal shear and thermal conductivity have been taken to be zero so that we have "potential type" flow [7.11]. The densities for each phase are given by:

$$\rho_{i,j} = \rho_{i,j}(P, h_{i,j}) \,, \qquad (7.6)$$

or

$$\rho_{i,j} = \text{constant} \,. \qquad (7.6a)$$

depending on the problem. Equations (7.1)–(7.3a) possess a complex conjugate pair of characteristics in one dimension given by Eq. (7.22) and are therefore ill-posed as an initial-value problem. The usual assumption has been made that the phase pressures are equal [7.11]. The momentum equation set, involving pressure times vapor volume fraction gradients and no additional terms, is not under consideration; however, it is discussed by Lyczkowski *et al.* [7.14]. Arastoopour and Gidaspow found this set to be physically unacceptable for one-dimensional steady flows [7.13].

7.2.2 Soo's Momentum Equations

The two-dimensional dispersed Brownian motion flow form of Soo's [7.7] momentum equations is given by:

x-Direction Momentum

R.H.S. of Eq. (7.2)

$$-P\frac{\partial \alpha_i}{\partial x} + \frac{\partial}{\partial x}[\rho_i \alpha_i (u_m - u_i)^2] + \frac{\partial}{\partial y}[\alpha_i \rho_i (u_m - u_i)(v_m - v_i)]\,, \tag{7.7}$$

R.H.S. of Eq. (7.2a)

$$-P\frac{\partial \alpha_j}{\partial x} + \frac{\partial}{\partial x}[\alpha_j \rho_j (u_m - u_j)^2] + \frac{\partial}{\partial y}[\alpha_j \rho_j (u_m - u_j)(v_m - v_j)]\,. \tag{7.7a}$$

y-Direction Momentum

R.H.S. of Eq. (7.3)

$$-P\frac{\partial \alpha_j}{\partial y} + \frac{\partial}{\partial x}[\rho_i \alpha_i (v_m - v_i)(u_m - u_i) + \frac{\partial}{\partial y}[\alpha_i \rho_i (v_m - v_i)^2]\,, \tag{7.8}$$

R.H.S. of Eq. (7.3a)

$$-P\frac{\partial \alpha_j}{\partial y} + \frac{\partial}{\partial x}[\rho_j \alpha_j (v_m - v_j)(u_m - u_j) + \frac{\partial}{\partial y}[\alpha_j \rho_j (v_m - v_j)^2]\,, \tag{7.8a}$$

where

$$\rho_m u_m = \alpha_i \rho_i u_i + \alpha_j \rho_j u_j\,, \tag{7.9}$$

$$\rho_m v_m = \alpha_i \rho_i v_i + \alpha_j \rho_j v_j\,, \tag{7.10}$$

and

$$\rho_m = \alpha_i \rho_i + \alpha_j \rho_j\,. \tag{7.11}$$

When these equations are added together for each direction they become:

Mixture Momentum Equations

$$\frac{\partial}{\partial t}(\rho_m u_m) + \frac{\partial}{\partial x}(\rho_m u_m u_m) + \frac{\partial}{\partial y}(\rho_m v_m u_m) + \frac{\partial P}{\partial x} = 0\,, \tag{7.12}$$

and

$$\frac{\partial}{\partial t}(\rho_m v_m) + \frac{\partial}{\partial x}(\rho_m u_m v_m) + \frac{\partial}{\partial y}(\rho_m v_m v_m) + \frac{\partial P}{\partial y} = -\rho_m \mathrm{g}\,. \tag{7.12a}$$

The mixture momentum equations given by Eqs. (7.12) and (7.12a) are identical in form to the inviscid single-phase momentum equations. This set of equations is hyperbolic [7.7]. The characteristics are given by Eqs. (7.23)–(7.27).

7.2.3 Extended 'Rudinger-Chang' Momentum Equations

The Rudinger-Chang [7.8] momentum equations as modified by Lyczkowski [7.9] are given by:

x-Direction Momentum

$$\text{R.H.S. of Eq. (7.2)} \qquad \begin{cases} -\alpha_j \dfrac{\partial P}{\partial x} & 0 < \alpha_j < 0.5 \\ +\alpha_i \dfrac{\partial P}{\partial x} & 1 > \alpha_j \geq 0.5 \end{cases} \tag{7.13}$$

$$\text{R.H.S. of Eq. (7.2a)} \qquad \begin{cases} +\alpha_j \dfrac{\partial P}{\partial x} & 0 < \alpha_j < 0.5 \\ -\alpha_i \dfrac{\partial P}{\partial x} & 1 > \alpha_j \geq 0.5 \end{cases} \tag{7.13a}$$

y-Direction Momentum

$$\text{R.H.S. of Eq. (7.3)} \qquad \begin{cases} -\alpha_j \dfrac{\partial P}{\partial y} & 0 < \alpha_j < 0.5 \\ +\alpha_i \dfrac{\partial P}{\partial y} & 1 > \alpha_j \geq 0.5 \end{cases} \tag{7.14}$$

$$\text{R.H.S. of Eq. (7.3a)} \qquad \begin{cases} +\alpha_j \dfrac{\partial P}{\partial y} & 0 < \alpha_j < 0.5 \\ -\alpha_i \dfrac{\partial P}{\partial y} & 1 > \alpha_j \geq 0.5 \end{cases} \tag{7.14a}$$

These equations are hyperbolic. The characteristics are given by Eqs. (7.28)–(7.29).

7.2.4 Hancox *et al.* Momentum Equations

These momentum equations were developed for so-called mixed flow by extending analytical relations derived by Stuhmiller [7.1] for a single accelerating sphere [7.10]. The basic assumption is that the average phase pressures are equal but different from the interfacial pressure. These equations are, in two dimensions:

<u>x-Direction</u>

R.H.S. of Eq. (7.2)

$$-C_p\rho_c(u_d-u_c)^2\frac{\partial\alpha_i}{\partial x}+\frac{1}{2}\alpha_d\rho_c\left[\frac{\partial}{\partial t}(u_c-u_d)+u_d\frac{\partial}{\partial x}(u_c-u_d)\right.$$
$$\left.+v_d\frac{\partial}{\partial y}(u_c-u_d)\right], \tag{7.15}$$

R.H.S. of Eq. (7.2a)

$$-C_p\rho_c(u_d-u_c)^2\frac{\partial\alpha_j}{\partial x}+\frac{1}{2}\alpha_d\rho_c\left[\frac{\partial}{\partial t}(u_c-u_d)+u_d\frac{\partial}{\partial x}(u_c-u_d)\right.$$
$$\left.+v_d\frac{\partial}{\partial y}(u_c-u_d)\right]. \tag{7.15a}$$

<u>y-Direction</u>

R.H.S. of Eq. (7.3)

$$-C_p\rho_c(v_d-v_c)^2\frac{\partial\alpha_i}{\partial y}+\frac{1}{2}\alpha_d\rho_c\left[\frac{\partial}{\partial t}(v_c-v_d)+u_d\frac{\partial}{\partial x}(v_c-v_d)\right.$$
$$\left.+v_d\frac{\partial}{\partial y}(v_c-v_d)\right], \tag{7.16}$$

R.H.S. of Eq. (7.3a)

$$-C_p\rho_c(v_d-v_c)^2\frac{\partial\alpha_j}{\partial y}-\frac{1}{2}\alpha_d\rho_c\left[\frac{\partial}{\partial t}(v_c-v_d)+u_d\frac{\partial}{\partial x}(v_c-v_d)\right.$$
$$\left.+v_d\frac{\partial}{\partial y}(v_c-v_d)\right]. \tag{7.16a}$$

In Eqs. (7.15)–(7.16a), the subscripts c and d denote continuous and dispersed phases respectively with the identification

$$\left.\begin{array}{l} c = i \\ d = j \end{array}\right\} \text{ for } 0 < \alpha_j < 0.5\,, \tag{7.17}$$

and

$$\left.\begin{array}{l} c = j \\ d = i \end{array}\right\} \text{ for } 1 > \alpha_j \geq 0.5\,.$$

The model used for C_p is from Stuhmiller [7.1] and is given by

$$C_p = 0.37\, C_d\,, \tag{7.18}$$

where C_d is the single sphere drag function. We used $C_d = 0.42$ (fully turbulent flow) so that the value of $C_p = 0.155$. This set of equations is hyperbolic for incompressible flow as long as [7.10]

$$\frac{4C_p\rho_c(\rho^* + c^*) + \alpha_d^2 C^{*2}}{4\alpha_c\alpha_d\rho_c(\rho_d + C^*))} > 1\,, \tag{7.19}$$

where

$$C^* = \frac{1}{2}\rho_c/\alpha_c\,, \tag{7.19a}$$

and

$$\rho^* = \alpha_c\rho_d + \alpha_d\rho_c\,. \tag{7.19b}$$

The characteristics are given by Eqs. (7.30)–(7.33).

7.2.5 Gidaspow's Momentum Equations

These equations, given by a mixture momentum equation and a relative velocity constitutive equation [7.3], were extended to two dimensions by Shih and Arastoopour [7.14]. They may be cast into the forms given by Eqs. (7.2) and (7.3) by subtracting the relative velocity equation times $\alpha_j\rho_j$ from the mixture momentum for phase i and using the mixture momentum equation for phase j. This manipulation results in the following, including mass transfer:

<u>x-Direction</u>

R.H.S. of Eq. (7.2)

$$-\alpha_j\frac{\partial P}{\partial x} - \alpha_j\rho_j\left[\frac{\partial u_j}{\partial t} + (u_j - u_i)\frac{\partial u_i}{\partial x} + u_i\frac{\partial u_j}{\partial x} + (v_j - v_i)\frac{\partial u_i}{\partial y} + v_i\frac{\partial u_j}{\partial y}\right]$$
$$- u_j(\Gamma_e - \Gamma_c)\,, \tag{7.20}$$

R.H.S. of Eq. (7.2a)

$$-\alpha_i \frac{\partial P}{\partial x} - \alpha_i \rho_i \left[\frac{\partial u_i}{\partial t} + u_i \frac{\partial u_i}{\partial x} + v_i \frac{\partial u_i}{\partial y} \right] - u_i(\Gamma_c - \Gamma_e) - K(u_i - u_j) . \tag{7.20a}$$

y-Direction

R.H.S. of Eq. (7.3)

$$-\alpha_j \frac{\partial P}{\partial y} - \alpha_j \rho_j \left[\frac{\partial v_j}{\partial t} + (u_j - u_i) \frac{\partial v_i}{\partial x} + u_i \frac{\partial v_j}{\partial x} + (v_j - v_i) \frac{\partial v_i}{\partial y} + v_i \frac{\partial v_j}{\partial y} \right] + v_j(\Gamma_e - \Gamma_c) , \tag{7.21}$$

R.H.S. of Eq. (7.3a)

$$-\alpha_i \frac{\partial P}{\partial y} - \alpha_i \rho_i \left[\frac{\partial v_i}{\partial t} + u_i \frac{\partial v_i}{\partial x} + v_i \frac{\partial v_i}{\partial y} \right] - v_i(\Gamma_c - \Gamma_e) - K(v_i - v_j) . \tag{7.21a}$$

These equations are hyperbolic [7.3, 7.15]. The characteristics are given by Eqs. (7.34)–(7.35).

7.3 Characterictics for the Five Two-phase Flow Equation Sets

(1) For the basic equation set [7.19], the eigenvalues are

$$\lambda_{1,2} = -\frac{u_i \rho_i \alpha_j + u_j \rho_j \alpha_i}{\rho_i \alpha_j + \rho_j \alpha_i} \pm I \left[\frac{\rho_j \alpha_j \rho_i \alpha_i (u_i - u_j)^2}{(\rho_i \alpha_j + \rho_j \alpha_i)^2} \right]^{\frac{1}{2}} , \tag{7.22}$$

and

$$\lambda_3^{-1} = \lambda_4^{-1} = 0 ,$$

where

$$I = \sqrt{-1} .$$

(2) For Soo's momentum equations [7.7] with $c = 1$, and $b = 0$, the eigenvalues are

$$\lambda_{1,2}^{-1} = \frac{-b \pm (b^2 - 4ac)^{\frac{1}{2}}}{2a} , \tag{7.23}$$

and

$$\lambda_3^{-1} = \lambda_4^{-1} = 0\,, \tag{7.24}$$

where

$$\begin{aligned} a &= 1 + \frac{\alpha_i \rho_j}{\alpha_j \rho_i}\,, \\ b &= 2u_j \left(\frac{\alpha_i \rho_j}{\alpha_j \rho_i} + \frac{1}{\alpha_j} - \frac{u_i \alpha_j}{u_j \alpha_i} \right), \end{aligned} \tag{7.25}$$

and

$$c = u_j^2 \left(\frac{\alpha_i \rho_j}{\alpha_j \rho_i} + \frac{1 + \alpha_i}{\alpha_j} - \frac{2u_i \alpha_i}{u_j \alpha_j} - \frac{P}{\alpha_j \rho_j u_j^2} \right). \tag{7.26}$$

The characteristics are real since

$$b^2 - 4ac = 4u_j^2 \left[\left(\frac{\alpha_i}{\alpha_j} \right)^2 \left(\frac{u_i}{u_j} - 1 \right)^2 + \frac{P}{\alpha_j \rho_i u_j^2} \left(1 + \frac{\alpha_i \rho_j}{\alpha_j \rho_i} \right) \right] > 0\,. \tag{7.27}$$

(3) For the extended Rudinger and Chang momentum equations [7.9], the eigenvalues are

$$\lambda_1 = -u_j\,, \tag{7.28}$$

and

$$\lambda_2 = \lambda_3 = \lambda_4 = -u_i\,. \tag{7.29}$$

(4) For the Hancox *et al.* momentum equations [7.10], the eigenvalues are

$$\lambda_{1,2} = u_m \pm v_m\,, \tag{7.30}$$

and

$$\lambda_3 = \lambda_4 = 0\,. \tag{7.31}$$

Equation (7.30) holds when u_d approaching u_c and the propagation velocities of interfacial waves are defined as

$$u_m = \{\alpha_c \rho_d u_d + \alpha_d \rho_c u_c + [2u_d - \alpha_d(u_d - u_c)]C^*/2\}/\rho^*\,, \tag{7.32}$$

and

$$v_m^2 = [(\rho^* + C^*)C_p \rho_c + \alpha_d^2 (C^*)^2/4] - \alpha_c \alpha_d \rho_c (\rho_d + C^*)](u_c - u_d)^2/(\rho^*)^2\,, \tag{7.33}$$

where C_p, C^*, and ρ^* are defined by Eqs. (7.18), (7.19a), and (7.19b), respectively.

(5) For the Gidaspow's momentum equations [7.3], the eigenvalues are

$$\lambda_1^{-1} = \lambda_2^{-1} = \lambda_3^{-1} = 0\,, \tag{7.34}$$

and

$$\lambda_4^{-1} = \frac{\alpha_j \rho_i - \alpha_i \rho_j}{\alpha_i \rho_j u_j - \alpha_j \rho_i u_i}\,. \tag{7.35}$$

7.4 The Drag Function and Phase Change Rates

In order to close the set of equations, one needs the additional information, i.e. the three interface jump conditions (interfacial mass, momentum, and heat transfers), and the additional boundary conditions.

The above mass, momentum, and energy equations are solved as an initial and boundary value problem to obtain $\alpha_i, \alpha_j, P, u_i, v_i, u_j, v_j, h_i$, and h_j. The densities, ρ_i and ρ_j, are provided through the use of the equation of state, Eqs. (7.6) or (7.6a).

In general, the drag function K is dependent on the flow regime, local vapor volume fraction, vapor and liquid density, Reynolds' number, and phase velocity. There are two models of K function used in the computer program. One is given by [7.15]

$$K = Cf(\alpha_i)\,,$$

where C is some coefficient, and

$$\mathrm{f}(\alpha_i) = \begin{cases} 1 + 10^{2000(0.01-\alpha_i)}, & \text{when } 0 < \alpha_i \leq 0.01 \\ 1, & \text{when } 0.01 < \alpha_i < 0.99 \\ 1 + 10^{2000(\alpha_i - 0.99)}, & \text{when } 1 > \alpha_i \geq 0.99 \end{cases} \tag{7.36}$$

A more elaborate form of K was used in this study given by [7.4, 7.16]

$$K = 0.375(\rho_i + \rho_j)\left\{ C_d \left|\vec{U_i} - \vec{U_j}\right| + \frac{12[\alpha_i \mu_i/\rho_i + (1-\alpha_i)\mu_j/\rho_j]}{r_p} \right\} A(\alpha_i, N)\,, \tag{7.37}$$

where

$$A = \begin{cases} \alpha_i^{2/3}\left(\dfrac{4\pi N}{3}\right)^{1/3}, & \text{when } \alpha_i \le 1/2 \\ (1-\alpha_i)^{2/3}\left(\dfrac{4\pi N}{3}\right)^{1/3}, & \text{when } \alpha_i > 1/2 \end{cases} \tag{7.38}$$

$$r_p = \begin{cases} \left(\dfrac{3\alpha_i}{4\pi N}\right)^{1/3}, & \text{when } \alpha_i \le 1/2 \\ \left[\dfrac{3(1-\alpha_i)}{4\pi N}\right]^{1/3}, & \text{when } \alpha_i > 1/2 \end{cases} \tag{7.39}$$

The evaporation and condensation rates, Γ_e and Γ_c are determined from [7.4]

$$\begin{aligned} \Gamma_e &= \lambda_e \rho_j (T_j - T_s)/T_s, && \text{for } T_j \ge T_s \\ &= 0, && \text{for } T_j < T_s \end{aligned} \tag{7.40}$$

$$\begin{aligned} \Gamma_c &= \lambda_c \rho_i (T_s - T_i)/T_s, && \text{for } T_i \le T_s \\ &= 0, && \text{for } T_i > T_s \end{aligned} \tag{7.41}$$

where λ_e and λ_c are time relaxation parameters with unit 1/s. For the tested problem described in this chapter, λ_e and λ_c are set equal to 0.1.

For the dispersed flow regime, the interfacial heat transfer coefficients, R_i and R_j are calculated from [7.17]

$$R_i = 8.067\frac{k_i}{r_p} \tag{7.42}$$

for the liquid phase and

$$R_j = \frac{1}{r_p}\left(1 + 0.37 Re^{0.5} Pr_j^{0.33}\right) \tag{7.43}$$

for the vapor phase, where

$$Re = \frac{2\rho_j |\vec{U}_i - \vec{U}_j| r_p}{\mu_j}, \tag{7.44}$$

and

$$Pr = \frac{C_{p_j}\mu_j}{k_j}. \tag{7.45}$$

The correlations R_j and R_i are obtained with r_p defined by Eq. (7.39).

7.5 Numerical Solution Procedure

The finite difference equations of mass, momentum, and energy are solved as an initial value problem in time and a boundary value problem in space. The procedure is iterative and the main steps necessary to determine the flow conditions at time $t + \Delta t$ from those at time t are as follows:

(1) The pressure distribution at time $t + \Delta t$ is guessed.
(2) The momentum equations are solved to compute the first approximation of the two velocity components for each of the two phases.
(3) Partial energy equations are solved for approximating enthalpies of both phases.
(4) Densities and temperatures of both phases are calculated using equation of state and estimated enthalpies and pressures.
(5) Liquid volume fraction α_i is estimated from the continuity equation of liquid. Gas volume fraction α_j is then calculated from the relation $\alpha_j = 1 - \alpha_i$.
(6) It is now checked to see if the gas continuity equation is satisfied. If the equation is not satisfied, then pressure is corrected, the amount of correction being dependent on the mass residual, and steps (2) to (6) are repeated. This iterative procedure is continued until the mass residual is less than the specified value.
(7) Energy equations are solved for enthalpies. Temperatures and densities are then calculated using the equations of state.
(8) A new time step is chosen and steps (1) to (7) are repeated.

The numerical scheme used in the pressure calculation is:

$$P^{\beta+1} = P^{\beta} - \omega_j \frac{D_j^{\beta}}{\left(\dfrac{\partial D_j}{\partial P}\right)^{\beta}} \quad \text{when } \alpha_j \geq \alpha^* , \tag{7.46}$$

and

$$P^{\beta+1} = P^{\beta} - \omega_i \frac{D_i^{\beta}}{\left(\dfrac{\partial D_i}{\partial P}\right)^{\beta}} \quad \text{when } \alpha_j < \alpha^* , \tag{7.47}$$

where α^* is an input constant, e.g. $\alpha^* = 0.001$. In the pressure iteration loop, i.e. Eqs. (7.46) and (7.47), the relaxation factors, ω_i and ω_j, are set to 0.95 since we use Jacobi's method for the relaxation procedure which requires that $0 < \omega \leq 1.0$. D_j^{β} and D_i^{β} are the residue masses from the continuity equations for j and i phases at iteration step β. The definitions of D_j and D_i are given by

$$\begin{aligned}(D_j)^{\beta}_{\zeta,\eta} &= \left(\frac{1}{\delta t}\right)\left[(\rho'_j)^{\beta}_{\zeta,\eta} - (\rho'_j)^{\beta-1}_{\zeta,\eta}\right] \\ &+ \left(\frac{1}{\delta x}\right)\left[(\rho'_j u_j)^{\beta}_{\zeta+\frac{1}{2},\eta} - (\rho'_j u_j)^{\beta}_{\zeta-\frac{1}{2},\eta}\right] \\ &+ \left(\frac{1}{\delta y}\right)\left[(\rho'_j v_j)^{\beta}_{\zeta,\eta+\frac{1}{2}} - (\rho'_j v_j)^{\beta}_{\zeta,\eta-\frac{1}{2}}\right], \end{aligned} \tag{7.48}$$

and

$$\begin{aligned}(D_i)^{\beta}_{\zeta,\eta} &= \left(\frac{1}{\delta t}\right)\left[(\rho'_i)^{\beta}_{\zeta,\eta} - (\rho'_i)^{\beta-1}_{\zeta,\eta}\right] \\ &+ \left(\frac{1}{\delta x}\right)\left[(\rho'_i u_i)^{\beta}_{\zeta+\frac{1}{2},\eta} - (\rho'_i u_i)^{\beta}_{\zeta-\frac{1}{2},\eta}\right] \\ &+ \left(\frac{1}{\delta y}\right)\left[(\rho'_i v_i)^{\beta}_{\zeta,\eta+\frac{1}{2}} - (\rho'_i v_i)^{\beta}_{\zeta,\eta-\frac{1}{2}}\right]. \end{aligned} \tag{7.49}$$

The momentum exchange terms, $K(\vec{U}_i - \vec{U}_j)$ and $K(\vec{U}_j - \vec{U}_i)$ in the liquid and vapor momentum equations, respectively, are neglected in calculating Eqs. (7.46) and (7.47). It is found that these formulations result in greater stability for the pressure calculations of the problems that have been investigated so far. The expressions of these two derivatives with the finite difference grid shown in Fig. 7.1 are

$$\begin{aligned}\frac{\partial D_i}{\partial P} &= \frac{1}{\left(\dfrac{\partial P}{\partial \rho'_i}\right)_{\zeta,\eta}\delta t} + \frac{\delta t\left[(\alpha_i)_{\zeta+\frac{1}{2},\eta} + (\alpha_i)_{\zeta-\frac{1}{2},\eta}\right]}{(\partial x)^2} \\ &+ \frac{\delta t\left[(\alpha_i)_{\zeta,\eta+\frac{1}{2}} + (\alpha_i)_{\zeta,\eta-\frac{1}{2}}\right]}{(\partial y)^2}\end{aligned} \tag{7.50}$$

and

$$\begin{aligned}\frac{\partial D_j}{\partial P} &= \frac{1}{\left(\dfrac{\partial P}{\partial \rho_j}\right)_{\zeta,\eta}\delta t} + \frac{\delta t\left[(\alpha_j)_{\zeta+\frac{1}{2},\eta} + (\alpha_j)_{\zeta-\frac{1}{2},\eta}\right]}{(\partial x)^2} \\ &+ \frac{\delta t\left[(\alpha_j)_{\zeta,\eta-\frac{1}{2}} + (\alpha_j)_{\zeta,\eta+\frac{1}{2}}\right]}{(\partial y)^2}\end{aligned} \tag{7.51}$$

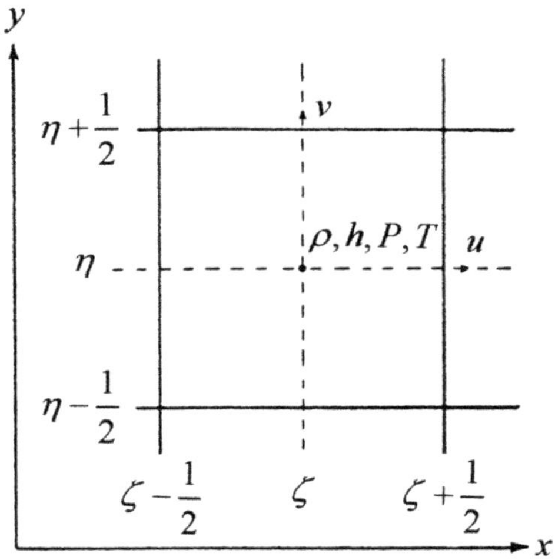

Figure 7.1. Finite difference grid notations and the locations of variables; ρ, h, P, and T are located at the cell center; velocities u, v, and flux terms are at cell boundaries.

The enthalpy equations were divided into two groups. The first group, including the heat transfer due to mass transfer, the interfacial sensible heat transfer, and the energy dissipation due to interfacial friction, is solved inside the pressure iteration loop. The second group, which contains the pressure compression work, shear stress dissipation energy, and the thermal diffusion, is solved outside the pressure iteration loop. A more detailed description of the numerical solution procedure is given in Lee [7.4]. An outline of a more stable, semi-implicit numerical scheme which solves the two-fluid, six-equation model is briefly described in the next section and generalized the above description. The jet impingement problem described in Section 7.7 was computed with both numerical schemes, and found to produce the same results.

7.6 A More Stable Numerical Scheme

This section gives an outline of a stable, semi-implicit numerical scheme which calculates the pressure P and the vapor volume fraction α_j based on the following procedures:

(1) Consider both vapor and liquid phases to be compressible.
(2) Calculate the pressure P and vapor volume fraction α_j inside the iteration loop.

(3) Compute the residue masses $(D_j)^\beta$ and $(D_i)^\beta$ from the continuity equations for vapor and liquid respectively, based on the iteration step β.

(4) Use the relationships $D_j = D_j(P, \alpha_j)$, $D_i = D_i(P, \alpha_i)$ and the momentum equations to obtain

$$\frac{\partial D_j}{\partial P^\beta}, \frac{\partial D_i}{\partial P^\beta}, \frac{\partial D_j}{\partial \alpha_j^\beta}, \text{ and } \frac{\partial D_i}{\partial \alpha_j^\beta}.$$

(5) Solve $\partial P^{\beta+1} = P^{\beta+1} - P^\beta$ and $\partial \alpha_j^{\beta+1} = \alpha_j^{\beta+1} - \alpha_j^\beta$ from the following matrix

$$\begin{bmatrix} \dfrac{\partial D_j}{\partial P^\beta} & \dfrac{\partial D_j}{\partial \alpha_j^\beta} \\ \dfrac{\partial D_i}{\partial P^\beta} & \dfrac{\partial D_i}{\partial \alpha_j^\beta} \end{bmatrix} \begin{bmatrix} \partial P^{\beta+1} \\ \partial \alpha_j^{\beta+1} \end{bmatrix} = \begin{bmatrix} -D_j^\beta \\ -D_i^\beta \end{bmatrix}$$

and then obtain the pressure $P^{\beta+1}$ and the vapor volume fraction $\alpha_j^{\beta+1}$.

(6) The thermodynamic vapor density ρ_j is obtained from the equation of state $\rho_j = \rho_j(P, h_j)$ and liquid ρ_i from $\rho_i = \rho_i(P, h_i)$.

(7) The macroscopic vapor density is $\rho_j' = \alpha_j \rho_j$, and that of liquid is $\rho_i' = \alpha_i \rho_i$.

(8) Calculate the new vapor velocities $u_j^{t+\Delta t}$ and $v_j^{t+\Delta t}$ and the new liquid velocities $u_i^{t+\Delta t}$ and $v_i^{t+\Delta t}$ from the momentum equations.

(9) Check the convergence, i.e. to see if $(D_j)^\beta < \varepsilon_1$ and $(D_i)^\beta < \varepsilon_2$ where ε_1 and ε_2 are some specified small numbers.

(10) If step 9 is satisfactory, then the calculation is finished; otherwise, it should go back to step 3 for new iteration until it converges or stops at some desired iteration step, e.g. iteration = 100.

(11) After step 10 is done, the energy equations will be used to calculate the enthalpies h_j and h_i.

The detailed derivations of the matrix coefficients appearing in step 5 are given in Ref. [7.4].

7.7 Description of Test Problems

This section describes each of the three test problems chosen for simulation.

7.7.1 Batch Settling of a Two-phase Mixture

The first test problem consists of a single vertical column 0.1 m high and 0.01 m wide filled with a steam-water mixture of uniform vapor fraction $\alpha_j = 0.40$ at a pressure $P = 1.013 \times 10^5$ Pa and temperature $T = 373°$K. Ten finite difference cells were used, each 0.01 m high. The top and the bottom are closed so that no material enters or leaves the system. At time 0^+, gravity is switched on, and liquid starts to settle to the bottom while vapor rises to the top. This is a thought problem that tests the rate of phase separation and the ability to predict counter-current flow. A fixed value of the interface drag function K equal to 1,000 was used with no mass transfer ($\Gamma_e = \Gamma_c = 0$).

7.7.2 One-dimensional Fluidized Bed

The experimental apparatus consisted in part of a vertical tubular glass column through which air is passed at known flow rates [7.18]. Initially, a bed of solid, spherical particles rests on a fine mesh screen at the bottom of the tube. A flow-straightening section to minimize circumferential motion (swirling) of the gas is located at the entry section below the screen. As gas is passed through the particles, they are fluidized. Flow conditions, such as air flow rate, initial bed depth, and particle size and density were varied, and the behavior of the bed was recorded. For the flow conditions studied, the behavior of the bed is primarily a periodic slug flow or aggregate fluidization. An oscillatory period begins with the bed lifting off the screen as a uniform mass. As it rises, it becomes more diffuse, and particles begin to drop off the bottom collecting on the screen forming the next slug. The upper boundary of the rising slug remains relatively flat. When the interparticle distance becomes great enough and the local fluid velocity is not sufficient to support the particles, the remaining slug falls, joining the lower slug as it is lifting off the screen.

Several experiments were carried out, but only one is simulated here. Motionless glass beads of 3 mm diameter (density $= 2.2 \times 10^3$ kg/m^3) initially fill the bottom 160 mm of the bed at a pressure of 1.013×10^5 Pa, a temperature of 298°K, and a porosity, $\alpha_j = 0.4$. The air is treated as an ideal gas. Twenty cells are used for a total system height of 1.6 m. Each cell is 80 mm wide and 80 mm high. At time 0^+, the pure gas is introduced at 3 m/s and the pressure is reduced at the top to 7.72×10^4 Pa (atmosphere pressure in Los Alamos, New Mexico, USA). The solids velocity is set to zero at the bottom and top so that no glass beads leave the system. This simulation corresponds to the second run in Rexroth and Starkovich's Table I [7.18] for which a time-averaged voidage oscillation period of 0.89 s was obtained. We used the same drag function K as Rexroth and Starkovich [7.18].

7.7.3 Simulation of Two-phase Jet Impinged on Vertical Plate

A horizontal two-phase, steam-water jet impinging on a vertical flat plate experiment [7.6] was also simulated. The experiment consisted of a round 10 mm tube discharging against a wall 5 mm away. The discharge conditions were recorded as: pressure $= 3.4 \times 10^6$ Pa, temperature $= 510.8°$K, vapor volume fraction, $\alpha_j = 0.67$, and mass flow rate $= 3.055 \times 10^4\,\mathrm{kg/(m^2s)}$. At time $t = 0^+$, the high-pressure jet containing the mixture of steam and water enters into a stagnant atmosphere and impinges on the vertical plate. On the plate, there are five stations recording the stagnation pressure. This experiment is simulated in a three-dimensional Cartesian coordinates with 10 cells in each of the x and y directions ($\Delta x = \Delta y = 5$ mm, and 5 in the z direction normal to the wall $\Delta z = 1$ mm). Four cells were used in the jet. The total mass flow was set equal to the experimental value.

7.7.4 Results of Computations and Discussion

The purpose of the first test problem, the batch-settling of a two-phase mixture, is to determine the effect of the interfacial drag on phase separation. From the computational experiment, it was found that the drag function K has a lower limit of approximately $10\,\mathrm{kg/(m^3s)}$. When using such a small value of K, the computation tends to become unstable. This instability is true for all five of the models. However, the instability is due to the numerical scheme used for solving the governing equations. Therefore, for a more stable finite difference method, the value of K may be less than 10. For large values of K, i.e. K greater than 1,000, all models produce basically the same results. Figure 7.2 shows the transient void fraction for the top and bottom cells by using the basic set. With constant $K = 1,000$, it takes approximately 0.21 s for the bottom cell to be filled with water and 0.3 s for the top cell to be filled with steam. For larger values of K, the separation rate decreases as expected as shown in Fig. 7.2. The transient vapor volume fractions for the top and bottom cells are presented in Fig. 7.3 for

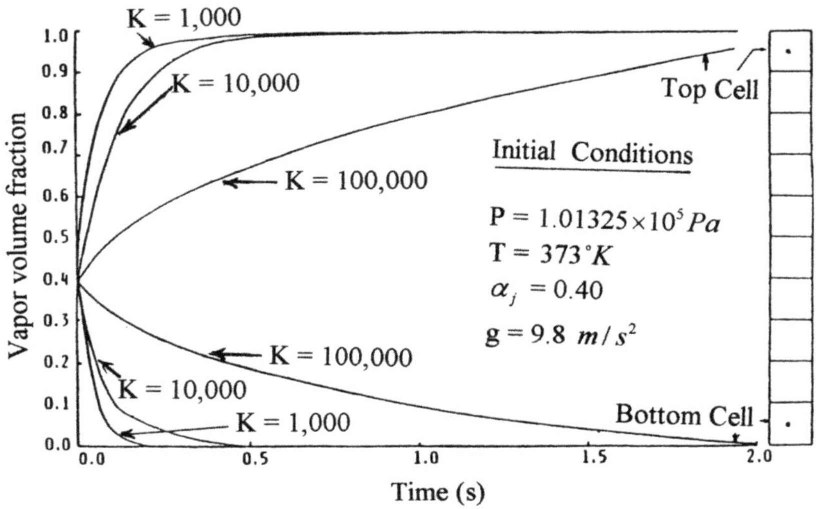

Figure 7.2. Transient vapor volume fraction for top and bottom cells using the basic set Eqs. (7.1)–(7.3a).

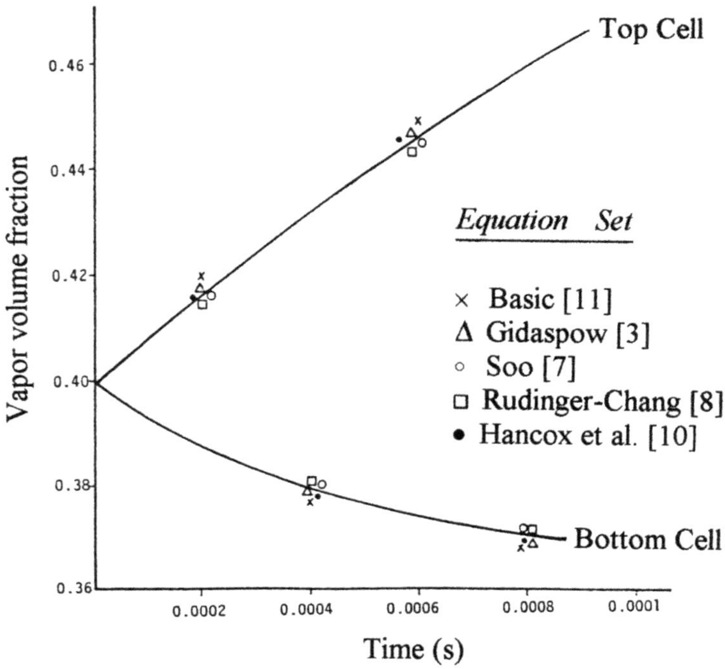

Figure 7.3. Transient vapor volume fraction for top and bottom cells in the early time using all five models.

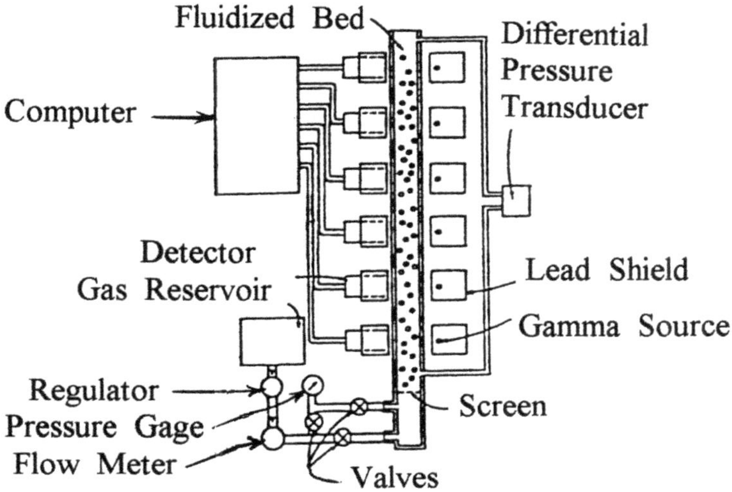

Figure 7.4. Apparatus for interfiled drag experiment (adapted from Reference [7.18]).

very early time of all the models. As indicated there, all of the five models give basically the same results.

Figure 7.4 is a schematic of the experimental apparatus for the inter-field drag experiment, whereas Fig. 7.5 defines the modeling mesh and inflow, outflow and internal boundary velocities. An artificial reduced flow area is imposed on the cell below the bed of glass beads so that a high air velocity can be obtained and consequently a non-negative particle velocity will be maintained. The maximum particle volume fraction oscillation period of 1.025 s and the slug height of 0.64 m obtained by using the basic model are shown in Fig. 7.6. The experimental data give a time-averaged oscillation period of 0.89 s and a height of 0.8 m. This discrepancy can be reduced by using a better drag function K. It was found that all models except Soo's give essentially the same oscillatory behavior but have a maximum slug height of about 25 percent deviation from the basic set results. This deviation can be seen in Fig. 7.7 which presents the transient particle volume fraction at the location of 0.2 m above the screen, i.e. the base of the glass bead bed initial height, for early time. It can be noticed that Soo's model produces results which deviate very much from the other models. It is not clear at this stage which term causes such a deviation but is believed to be the inertial coupling terms [7.7]. Since these results are preliminary (as are

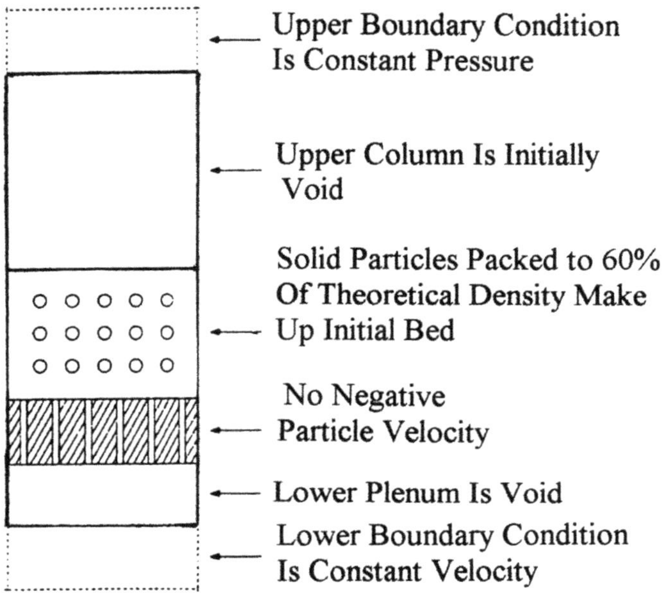

Figure 7.5. Calculational mesh set-up (adapted from Reference [7.18]).

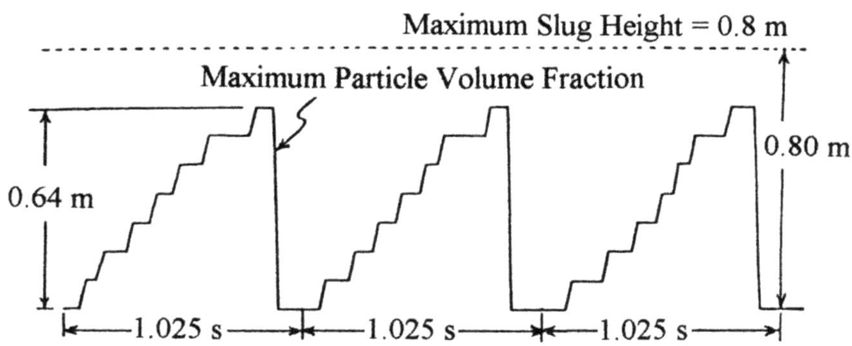

Figure 7.6. The particle oscillation period and the slug height.

all of them), a further investigation will be carried out in the near future to clarify such points. Hence, the results are subject to change.

Figure 7.8 shows the experimental setup and initial operating conditions for the two-phase jet impingement problem. The calculated results of the pressure distribution on the vertical plate by using the basic model are

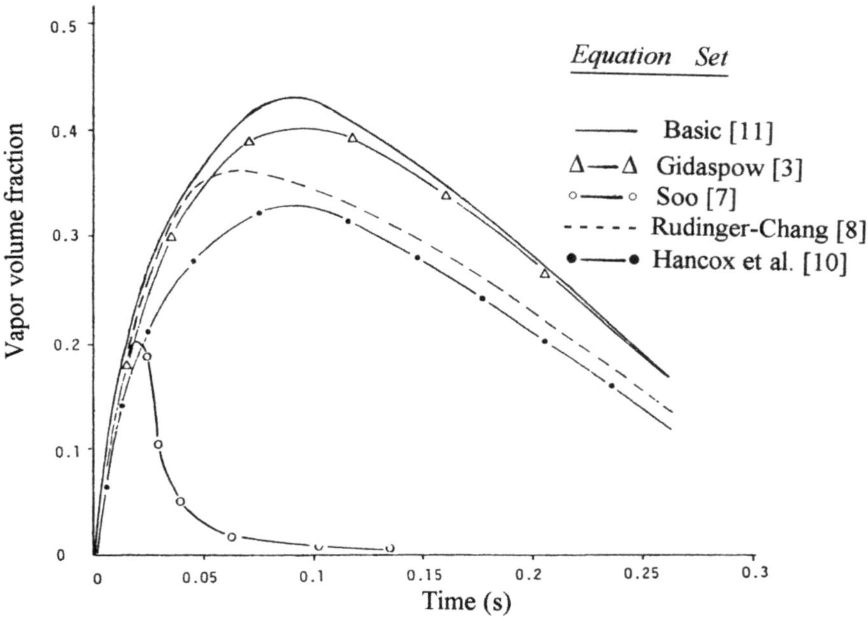

Figure 7.7. Transient particle volume fraction at 0.2 m above the screen.

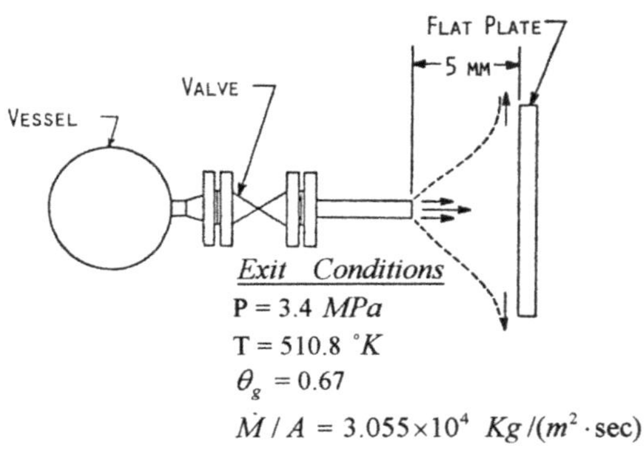

Figure 7.8. Two-phase jet impingement experiment (adapted from Reference [7.6]).

given in Fig. 7.9 for different values of interfacial drag function K and evaporation function Γ_e. The best simulations are obtained when $K = 2.0 \times 10^{12}$ and $\Gamma_e = 100$. The transient and steady state results of the pressure distribution for all models except the basic set are shown in Fig. 7.10. At

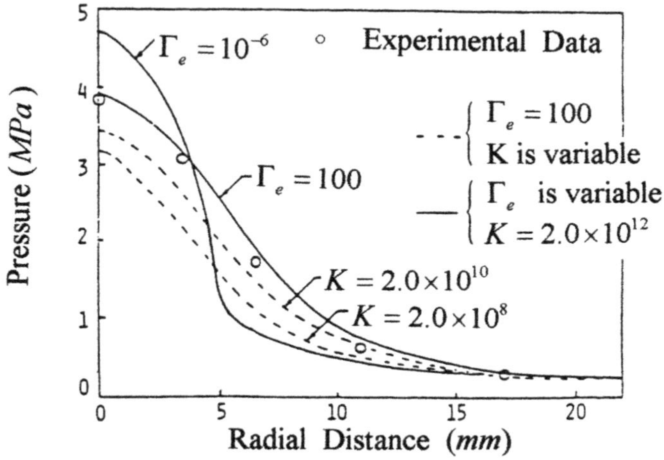

Figure 7.9. Basic model results as compared with experimental data.

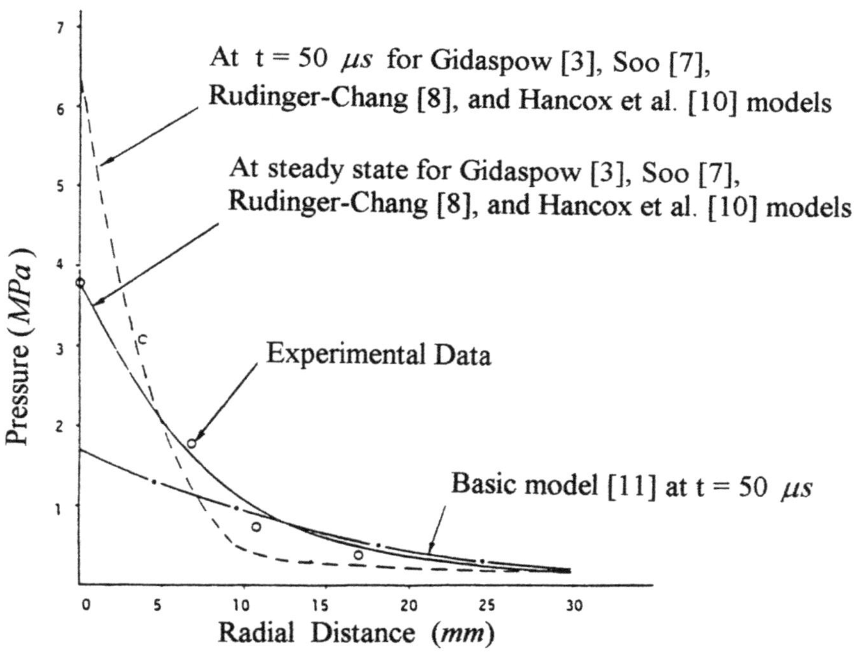

Figure 7.10. Computed results as compared with experimental data for all models.

time $t = 50\,\mu s$ and steady state, the models of Gidaspow, Soo, Rutinger-Chang, and Hancox *et al.* compute the same results.

The steady state pressure distributions are very close to the measurements but not as good as the basic set as shown in Fig. 7.9. It is interesting to note that at $t = 50\,\mu s$, the results of the basic set are quite different from the other four models, with the basic set lying below the steady state pressure and the other four sets overshooting it.

7.8 Conclusions

It is concluded that of the five models studied, four of them (basic [7.11], Gidaspow [7.3], Rudinger-Chang [7.8], and Hancox *et al.* [7.10]) are in close agreement with each other and the data for the low-pressure (0.1 MPa) fluidized-bed experiment [7.18]. Soo's model [7.7] disagrees with the other four models and damps oscillations out, probably due to the presence of the inertial coupling terms. It is further concluded that all of the five models studied, four of them (Gidaspow [7.3], Soo [7.7], Rudinger-Chang [7.8], and Hancox *et al.* [7.10]) gave essentially the same results and are in very close agreement with the data for the high-pressure (3.4 MPa) jet impingement experiment [7.6]. All five models gave essentially the same results for the batch settling simulation.

A first-order finite difference scheme such as used herein to solve the governing equations introduces second-order numerical viscous stresses into the momentum equations when we discretize the differential equations. This will certainly change the stability for the numerical scheme of the two-phase flow equation sets. For example, Eqs. (7.1)–(7.3a) may become stable after discretization, depending on the grid size. It is also found that the models of Gidaspow [7.3] and Hancox *et al.* [7.10] are computationally slightly more unstable than the other three models because of the presence of the discretized time derivatives on the right-hand side of the momentum equations in the numerical schemes described in this chapter.

For shaped charge problems, the Hancox *et al.* model [7.10] is recommended since it assumes that the interfacial phase pressure is different from the average phase pressure. Also, the single sphere drag function is included in the dynamic pressure coefficient, i.e. C_P. The standard sphere drag coefficient is useful for computing the late time liner jet breaking into particles.

References

7.1 Stuhmiller, JH, The influence of interfacial pressure forces on the character of two-phase flow model equations, Int. J. Multiphase flow, 3, pp. 551–560 (1977).

7.2 Banerjee, S, Ferch, RL, Mathers, WG, and McDonald, BH, The dynamics of two-phase flow in a duct, Sixth International Heat Transfer Conference, Toronto, Canada, August 7–11 (1978); General Papers, 1, pp. 345–350, Paper No. FB-13, Hemisphere Pub. Corp., Washington (1978).

7.3 Gidaspow, D, A set of hyperbolic incompressible two-phase flow equations for two components, Heat Transfer Research and Application AIChE Symposium Series, 74(174), pp. 186–190, JC Chen, ed., American Institute of Chemical Engineers (1978).

7.4 Lee, WH, A pressure iteration scheme for two-phase flow modeling, Multiphase Transport Fundamentals, Reactor Safety, Applications, TN Veziroglu, ed., 1, pp. 407–431, Hemisphere Pub. Corp., Washington (1980).

7.5 Soo, SL, Multiphase Fluid Dynamics, pp. 367–372, Gower Technical, Beijing, Distributed by Ashgate Pub. Co., Brookfield, VT (1990).

7.6 Schweickert, H, Untersuchung der Vorgage in einem mehrfach unterteilten containment biem Bruch einer Kuhlmitteleitung Wassergekuhlter Reaktoren, Technischer Bericht BFR 50-32-C12-1, Kraftwerk Union, Germany (1976).

7.7 Soo, SL, Equations of multiphase-multidomain mechanics, Multiphase Transport Fundamentals, Reactor Safety, Applications, TN Veziroglu, ed., 1, pp. 291–305, Hemisphere Pub. Corp., Washington (1980).

7.8 Rudinger, G, and Chang, A, Analysis of non-steady two-phase flow, Phys. of Fluids, 7, pp. 1747–1754 (1964).

7.9 Lyczkowski, RW, Transient propagation behavior of two-phase flow equations, Heat Transfer: Research and Application, AIChE Symposium Series, 74(174), pp. 165–174, American Institute of Chemical Engineers, New York (1978).

7.10 Hancox, WT, Ferch, RL, Liu, WS, and Nieman, RE, On one-dimensional models for transient gas-liquid flow in ducts, Int. J. Multiphase Flow, 6, pp. 25–40 (1980).

7.11 Lyczkowski, RW, Gidaspow, D, Solbrig, CW, and Hughes, ED, Characteristics and stability analyses of transient one-dimensional two-phase flow equations and their finite difference approximations, Nuc. Sci. and Eng., 66(3), pp. 378–396, (1978).

7.12 TRAC-P1: An advanced best estimate computer program for PWR LOCA analysis, Los Alamos Scientific Laboratory Report NUREG/Cr-0063, LA-7279-MS, I, Los Alamos, NM (June 1978).

7.13 Arastoopour, H, and Gidaspow, D, Vertical pneumatic conveying using four hydrodynamic models, Ind. Eng. and Fund., 18, pp. 123–130 (1979).

7.14 Shih, YT, and Arastoopour, H, Hydrodynamic analysis of horizontal solids transport, Ind. Eng. and Fund., 21, pp. 37–43 (1980).

7.15 Lee, WH, and Shah, VL, Numerical simulation for two-phase jet problem, Numerical Methods in Laminar and Turbulent Flows, Proc. 2nd Int. Conf.

Venice, Pineridge Press, U. K., C Taylor and BA Schrefler, eds., pp. 841–852 (1981).

7.16 Lee, WH, and Lyczkowski, RW, Development of a state of the art reactor containment program, 17th National Heat Transfer Conference, Salt Lake City, Utah, Preprints of AIChE Papers, pp. 130–138, American Institute of Chemical Engineers, New York (1977).

7.17 Solbrig, CW, McFadden, JH, Lyczkowski, RW, and Hughes, ED, Heat transfer and friction correlations required to describe steam-water behavior in nuclear safety studies, Heat Transfer Research and Applications, AIChE Symposium Series 74(174), pp. 100–128, JC Chen, ed, American Institute of Chemical Engineers, New York (1978).

7.18 Rexroth, PE, and Starkovich, VS, Measurement and analysis in multiphase flow systems, Multiphase Transport Fundamentals, Reactor Safety, Applications, TN Veziroglu, ed., 5, pp. 2469–2481, Hemisphere Pub. Corp., Washington (1980).

7.19 Lyczkowski, RW, Gidaspow, D, and Solbrig, CW, Multiphase flow models for nuclear, fossil, and biomass energy production, Advances in Transport Processes, 2, pp. 198–340, AS Majumdar and RA Mashelkar, eds., Wiley Eastern, New Delhi (1982).

CHAPTER 8

EQUATION OF STATE, CONSTITUTIVE RELATIONSHIP AND HIGH EXPLOSIVE

8.1 Introduction to the Equation of State

Most of the equation of state (EOS) uses pressure as a function of density and internal energy. Occasionally, there is a need for calculating temperature as a function of density and internal energy. A simple EOS model uses the u_s (shock speed) and u_p (particle speed) relationship or $Mei - Grüneisen$ [8.2] EOS. The most complicated formula of the EOS will be the one proposed by Osborne [8.1]. This model includes the second order polynomial function of density and internal energy.

In the calculation of the burned product pressure of a high explosive, the JWL EOS is quite suitable. The JWL EOS for high explosive is described in Section 8.9.

8.2 The *Mie – Grüneisen* EOS and the Simple u_s, u_p Model

The $Mei - Grüneisen$ [8.2] EOS is

$$P = P_H + \rho\gamma(\varepsilon - \varepsilon_H)\,. \tag{8.1}$$

In Eq. (8.1), P is the pressure, ρ the density, γ the *Grüneisen* parameter, ε the internal energy, P_H and ε_H are the pressure and internal energy on the principal Hugoniot curve. The pressure P and the internal energy ε are function of density and temperature, while P_H and ε_H are function of density only. The *Grüneisen* parameter is defined as

$$\gamma = \frac{1}{\rho}\left(\frac{\partial P}{\partial \varepsilon}\right)_\rho\,. \tag{8.2}$$

A good approximation for γ is

$$\rho\gamma = \rho_0\gamma_0\,, \tag{8.3}$$

or

$$\gamma = \frac{\rho_0\gamma_0}{\rho}\,, \tag{8.4}$$

where ρ_0 and γ_0 are at zero-Kelvin states. Based on the experimental data, the shock wave velocity u_s and the particle velocity u_P can be related to by

$$u_s = C_0 + s u_P\,, \tag{8.5}$$

where C_0 and s are parameters dependent on material. For example, some of the material used for oil well perforator problems as described in Chapter 12 are given in Table 8.1.

Using Eq. (8.5) and the Hugoniot equation (8.1), one can derive the Hugoniot pressure

$$P_H = \frac{\rho_0 C_0 \eta}{(1 - s\eta)^2}\,, \tag{8.6}$$

where

$$\eta = 1 - \frac{\rho_0}{\rho}\,. \tag{8.7}$$

Also, from the Hugoniot relationship the internal energy ε_H can be written as

$$\varepsilon_H = \frac{1}{2}\frac{P_H \mu}{\rho}\,, \tag{8.8}$$

where

$$\mu = \frac{\rho_0}{\rho} - 1\,. \tag{8.9}$$

During the hydrodynamic code calculation, for example, at step 14 in Section 2.5, where the density ρ and internal energy ε are available from the code computation, one will use Eq. (8.6) for computing P_H, Eq. (8.8) for ε_H, and Eq. (8.1) for P. The material dependent constants ρ_0, C_0, s, γ_0 are obtained from Table 8.1. More Hugoniot data are provided in Appendix C.

Table 8.1. Parameters for the *Mei – Grüneisen* EOS.

Material	$\rho_0(\text{g/cm}^3)$	$C_0(\text{km/s})$	s	γ_0
Stainless Steel 304	7.9	4.57	1.49	2.17
Copper OFHC 1/2 Hard	8.93	3.94	1.49	1.96
Water	0.9979	2.393	1.333	0.50
Granite	2.672	3.712	1.086	0.90
Bronze	8.733	3.814	1.452	2.12

8.3 The Osborne Model

The Osborne model [8.1] is a quadratic form polynomial that calculates the pressure as a function of density ratio and internal energy. The expression is

$$P(\text{MBar}) = \frac{A_1\mu + A_2\mu\,|\mu| + (B_0 + B_1\mu + B_2\mu^2)\varepsilon + (C_0 + C_1\mu)\varepsilon^2}{\varepsilon + D_0}\,, \tag{8.10}$$

where

$$\mu = \frac{\rho}{\rho_0} - 1\,. \tag{8.11}$$

In Eq. (8.10), ρ is the density (g/cm^3), ρ_0 is the initial density (g/cm^3), ε is the specific internal energy (Mbar-cm^3/g), A_1, A_2, B_0, B_1, B_2, C_0, C_1 and D_0 are coefficients that are material dependent. The coefficients for some material are given in Table 8.2.

8.4 The Tillotson Equation of State

The Tillotson [8.3] equation of state is

$$P = P_c = \left[a + \frac{b}{\frac{I}{I_0\eta^2} + 1}\right] I\rho + A_1\mu + B_1\mu^2\,, \quad \text{for } I < I_s\,, \tag{8.12}$$

$$P = P_E = aI\rho + \left[\frac{bI\rho}{\frac{I}{I_0\eta^2} + 1} + A_1\mu e^{-\beta\left(\frac{\rho_0}{\rho} - 1\right)}\right] e^{-\alpha\left(\frac{\rho_0}{\rho} - 1\right)2}, \quad \text{for } I > I_s'\,, \tag{8.13}$$

Table 8.2. Osborne coefficients.

	Tungsten	Aluminum	Steel	Uranium	Copper
A_1	21.67419	1.1867466	4.9578323	2.4562457	4.9578323
A_2	14.93338	0.762995	3.6883726	4.6163216	3.6883726
B_0	10.195827	3.4447654	7.4727361	4.3432909	7.4727361
B_1	12.263234	1.5452573	11.519148	0.76214541	11.519148
B_2	9.3051515	0.96429632	5.5251138	6.4410793	5.5251138
C_0	0.33388437	0.43381656	0.39492613	0.31988993	0.39492613
C_1	0.48248861	0.54873462	0.52883412	0.46744784	0.52883412
D_0	7.0	1.5	3.6	2.2	0.6000001
ρ_0	19.17	2.806	7.9	18.983	8.899

and

$$P = \frac{(I - I_s)P_E + (I_s' - I)P_c}{I_s' - I_s}, \quad \text{for } I_s < I < I_s'. \tag{8.14}$$

For example, the parameters for the oil shale are: $a = 0.5$; $b = 1.0$; $A_1 = 0.28$; $B_1 = 0.11$; $\alpha = 5.0$; $\beta = 5.0$; $I_0 = 0.11$; $I_s = 0.032$; $I_s' = 0.16$; and $\rho_0 = 2.3\,(\text{g/cm}^3)$. The oil shale data are used in Chapter 12 for the oil well perforator problems.

8.5 Introduction to the Constitutive Relationship

Various expressions for the shear modulus, G, and the yield strength, Y, are available. Here, we simply present options made available for use in either Eulerian or Lagrangian code.

(a) Hydrodynamic flow

$$G = Y = S^{ij} = 0. \tag{8.15}$$

(b) Elastic perfect plastic flow

$$G = G_0 = \text{constant}. \tag{8.16}$$

$$Y = Y_0 = \text{constant}. \tag{8.17}$$

(c) Strain-hardened elastic-plastic model

(1) Quadratic EOS.
(2) Steinberg-Guinan model.
(3) Combination of (1) and (2) for multi-material problems.

8.6 Quadratic Model

In this model, the dynamic yield strength, Y, and the melting energy, E_m, are expressed in terms of pressure and internal energy as below.

$$Y = (Y_0 + \alpha P)\left(1 - \frac{E}{E_m}\right), \tag{8.18}$$

and

$$E_m = E_{m0} + E_{m1}(1 - v) + E_{m2}(1 - v)^2, \tag{8.19}$$

where Y_0 = yield strength; α = coefficient for pressure dependence; E_{m0}, E_{m1}, E_{m2} = coefficients for melting function; and v = normalized specific volume, i.e. $v = \rho_0/\rho$.

8.7 Steinberg-Guinan Model

This model is adopted from Steinberg and Guinan [8.4], who used the following constitutive relations for high-strain rate.

$$G = G_0 \left\{ 1 + b\frac{P}{\eta^{1/3}} + h\left[\frac{E - E_0(x)}{3R'} - 300\right]\right\} \exp\left[-\frac{fE}{E_m(x) - E}\right], \tag{8.20}$$

$$Y = Y_0(1+\beta e)^n \left\{ 1 + qb\frac{P}{\eta^{1/3}} + h\left[\frac{E - E_0(x)}{3R'} - 300\right]\right\}$$
$$\times \exp\left[-\frac{gE}{E_m(x) - E}\right], \tag{8.21}$$

$$(1+\beta e)^n \leq Y_{\max}, \tag{8.22}$$

$$E_m(x) = E_0(x) + 3R'T_m(x), \tag{8.23}$$

$$T_m(x) = \frac{T_{m0}\exp(2ax)}{(1-x)^\alpha}, \tag{8.24}$$

$$\alpha = 2\left(\gamma_0 - a - \frac{1}{3}\right), \tag{8.25}$$

$$E_0(x) = \int_0^x P(x)\,dx - 3R'\,TAD, \tag{8.26}$$

and

$$TAD = \frac{300\exp(ax)}{(1-x)^{(\gamma_0 - a)}}, \tag{8.27}$$

where P = pressure (Mbar),

G_0 = shear modulus (Mbar) at reference state ($T = 300°\mathrm{K}$, $P = 0$, $e = 0$),

$$b = \frac{1}{G_0}\frac{\partial G}{\partial T} = \frac{G_P}{G_0}, \tag{8.28}$$

$$h = \frac{1}{G_0}\frac{\partial G}{\partial P} = \frac{G_T}{G_0}, \tag{8.29}$$

$$\rho = \text{density}(\text{g/cc}),$$

$$\eta = \rho/\rho_0, \tag{8.30}$$

E = energy (Mbar-cm^3/cm$_0^3$),
A = atomic weight (g/mole),
R = gas constant = 8.314×10^{-5} Mbar-cm^3/(mole – °K),

$$R' = R\rho_0/A\,, \tag{8.31}$$

f, g = shaping parameters in the melting region,
β, n = work hardening parameters,
q = ratio of pressure dependence of yield strength to that of the shear modulus,
e = equivalent plastic strain,
$Y_{\max}$ = maximum value of yield strength $Y(e)$ at $T = 300$°K and $P = 0$,
T_{m0} = melting temperature at $\rho = \rho_0$(°K),
T_m = melting temperature at $P = 0$(°K),

$$x = 1 - \frac{1}{\eta} = 1 - \frac{v}{v_0}\,, \tag{8.32}$$

$$\gamma = \text{thermodynamic gamma} = \frac{C_p}{C_v}\,, \tag{8.33}$$

a = correction of first-order volume correction to γ.

For the Eulerian code , the integral expression in Eq. (8.26) is replaced by a polynomial form:

$$\int_0^x P(x)dx = y^2\frac{T_1}{T_2}\,, \tag{8.34}$$

where

$$y = \eta - 1 = \frac{\rho}{\rho_0}\,, \tag{8.35}$$

$$T_1 = \alpha_0 + [\alpha_1 + (\alpha_2 + \alpha_3 y)\,y]\,y\,, \tag{8.36}$$

and

$$T_2 = \beta_0 + [\beta_1 + (\beta_2 + \beta_3 y)\,y]\,y\,. \tag{8.37}$$

For example, for copper, the coefficients are: $\alpha_0 = 0.68462$; $\alpha_1 = -0.00868$; $\alpha_2 = 0.26429$; $\alpha_3 = 0.0068119$; $\beta_0 = 1.0$; $\beta_1 = 0.00903$; $\beta_2 = 0.170902$; and $\beta_3 = 0.0417657$.

Equations (8.20) and (8.21) are in the same form except for the work-hardening term in Eq. (8.21). The temperature dependence is characterized by

$$G_0 h\left[\frac{E-E_0(x)}{3R'}-300\right]=G_T(T-300)\,. \tag{8.38}$$

The coefficients of the Steinberg-Guinan model of some material are given in Table 8.3.

8.8 Steinberg's New Model

In 1991, Steinberg [8.5] reports the material properties for 37 elemental metals and alloys, 34 of which have a rate-independent constitutive model and 7 of which have a rate-dependent model. In addition, there are 9 plastic materials plus graphite, LiF and WC. Most of the material in this section is adopted from Steinberg's report [8.5].

The rate-independent model is described in Ref. [8.6], and other applications of the model are given in Refs. [8.7] and [8.8]. The principal equations are as follows:

$$Y=Y_0 f(\varepsilon_p)G(P,T)/G_0\,, \tag{8.39}$$

where

$$Y_0 f(\varepsilon_p)=Y_0\left[1+\beta(\varepsilon_p+\varepsilon_i)\right]^n\le Y_{\max}\,, \tag{8.40}$$

and

$$G(P,T)=G_0\left[1+\frac{AP}{\eta^{1/3}}-B(T-300)\right]. \tag{8.41}$$

Here

$$A=\frac{1}{G_0}\frac{dG}{dP}\,, \tag{8.42}$$

and

$$B=\frac{1}{G_0}\frac{dG}{dT}\,. \tag{8.43}$$

Melting is based on a modified Lindemann law and is described in Ref. [8.6]. The melting temperature T_m is given by

$$T_m=T_{m0}\exp\left[2a\left(1-\frac{1}{\eta}\right)\right]\eta^{2(\gamma_0-a-\frac{1}{3})}\,. \tag{8.44}$$

Table 8.3. The Coefficients of the Steinberg-Guinan model for Tungsten, Aluminum, Copper, and Uranium.

Material	Tungsten	Aluminum	Copper	Uranium
G_0	1.6	0.276	0.477	0.844
Y_0	0.022	0.0029	0.0012	0.0012
β	7.7	125.0	36	16000
n	0.13	0.1	0.45	0.26
$Y_{\max}$	0.04	0.0068	0.0064	0.0168
b	1.375	7.971	3.1446541	4.739
h	−0.0001375	−0.0067159	−0.000377358	−0.0008056
q	1.0	1.0	1.0	1.0
f	0.001	0.001	0.001	0.001
g	0.001	0.001	0.001	0.001
R' (Mbar/°K)	0.000008671	0.000008326	0.00001164	0.00000663
T_{m0}	4520	1220.0	1790	1710
$\gamma_0 - a$	0.27	0.49	0.52	0.92
a	1.4	1.7	1.5	1.5

When T equals or exceeds T_m, Y and G are set to zero. Because the EOS is energy-based, it is easier to test for melt on the basis of energy. The melt energy E_m is defined as

$$E_m = E_c + c_p T_m \,, \tag{8.45}$$

where E_c is given by

$$E_c = E(\eta) - 300 c_p \exp\left[a\left(1 - \frac{1}{\eta}\right)\right] \eta^{\gamma_0 - a} \,, \tag{8.46}$$

and $E(\eta)$ is the integral of PdV on the zero Kelvin isotherm.

A simplified model is

$$Y = Y_0(1 + \beta\varepsilon_p)^n \left[1 + \frac{CP}{\eta^{-1/3}} - m(T - 300)\right] , \tag{8.47}$$

and

$$G = G_0\left[1 + \frac{CP}{\eta^{-1/3}} - m(T - 300)\right] . \tag{8.48}$$

The parameters for some material are given in Table 8.4.

The formula of the constitutive model with the data for metals being subjected to large strain, high strain rate, and high temperature are given by Johnson and Cook [8.9]. The basic model is well suited for computations because it uses the information that is readily available in most of the

Table 8.4. Coefficients used in Eqs. (8.40) and (8.41) for some materials.

Material	G_0 (GPa)	Y_0 (GPa)	β	n	C	m	T_{melt} (ev)
Stainless Steel 304	0.77	0.0034	43.	0.35	2.26	0.000455	
Copper OFHC 1/2 Hard	0.477	0.0012	36.	0.45	2.83	0.000377	
Bronze	0.409	0.00123	39.	0.27	3.03	0.000411	
Aluminum 6061-T6	0.33	0.324	125.	0.10	6.52	0.000616	0.08
Gold	0.425	0.02	49.	0.39	3.75	0.000311	0.092

Table 8.5. The Coefficients of Eq. (8.49) (Johnson-Cook Model) for some materials.

Metal	θ	Y_0 (GPa)	β	n	C	m	T_{melt} (ev)
Copper		0.0897	3.26	0.31	0.025	1.09	0.119
Steel 4340		0.793	0.643	0.26	0.014	1.03	0.157
Tungsten		1.507	0.117	0.12	0.016	1.0	0.151
Granite Westerly	300.	0.048	1.0	1.0	0.0	0.5	

computer codes. The von Mises flow stress is given by

$$\sigma_{eq} = Y_0(1 + \beta\varepsilon_p^n)\left[1 + C\ln\left(\frac{\dot{\varepsilon}_p}{\dot{\varepsilon}_0}\right)\right](1 - \theta^m)\,, \tag{8.49}$$

where

$$\dot{\varepsilon}_0 = 1.0/s\,,$$

$$\theta = \frac{T - T_0}{T_m - T_0}\,, \tag{8.50}$$

and

$$T_0 = 300°\text{K}$$

The coefficients of Eq. (8.49) for some material are given in Table 8.5.

In Eq. (8.49), the definitions of the variables are

σ_{eq} : von Mises flow stress
ε_p : equivalent plastic strain
$\dot{\varepsilon}_p$: equivalent plastic strain rate
θ : reduced temperature
T_m : melt temperature.

8.9 High Explosive

8.9.1 Introduction

A programmed burn model of high explosive (HE) is adequate for the Lagrangian and Eulerian codes calculations. Huygens' principle and the Chapman-Jouguet theory are used in defining the detonation velocity and location where the high-explosive energy is released. Precalculated burn information is implemented into the code before the hydrodynamic actions take place.

In HE programmed burn, the basic assumption is that the detonation wave front travels in all directions at the Chapman-Jouguet detonation velocity. Information concerning the energy released from HE such as burn time (BT) and burn interval (BI) is precalculated and stored in the code at time $t = 0$. During the run, when the problem time T^n at cycle n becomes greater than the BT value of a HE but less than (BT + BI), a fraction of the specific energy for the particular HE is deposited in the cell. This fraction is given by $(T^n - \text{BT})/\text{BI}$. On the next cycle, at T^{n+1}, if T^{n+1} is still less than (BT + BI), another fraction $(T^{n+1} - T^n)/\text{BI}$ of specific energy is deposited into the cell. This continues until all the energy is deposited into the cell and $T^n > (\text{BT} + \text{BI})$. Then the cell is completely burned.

8.9.2 JWL Equation of State

A number of different equations of state have been developed to describe the pressure due to the HE burn. The Jones-Wilkins-Lee (JWL) EOS [8.10] is accepted as one that accurately describes the expansion process for nearly ideal explosives:

$$P = A\left[1 - \left(\frac{\omega}{R_1 V}\right)\right] e^{-R_1 V} + B\left[1 - \left(\frac{\omega}{R_2 V}\right)\right] e^{-R_2 V} + \frac{\omega E}{V}, \qquad (8.51)$$

and

$$P_s = Ae^{-R_1 V} + Be^{-R_2 V} + CV^{-(\omega+1)}, \qquad (8.52)$$

where

$P =$ pressure
$P_s =$ pressure along the expansion isentrope, i.e. pressure as a function of volume at constant entropy
$A, B, C =$ linear coefficient in units of P and P_s
$R_1, R_2, \omega =$ nonlinear, unitless coefficients
$V = v/v_0 =$ detonation product volume (v)/initial high-explosive volume (v_0)
$E =$ detonation energy per unit volume

The coefficients are obtained by comparing EOS calculations with experimental expansion data. Table 8.6 lists the JWL parameters for a number

Table 8.6. JWL equation of state parameters.

HE	ρ_0^a (Mg/m^3)	D^b (km/s)	E_0^c (GPa)	A (GPa)	B (GPa)	R_1	R_2	ω
Comp B	1.717	7.98	8.50	524.2	7.678	4.20	1.10	0.34
Comp C-4	1.601	8.19	9.00	609.8	12.95	4.50	1.40	0.25
Cyclotol 77/23	1.754	8.25	9.20	603.4	9.924	4.30	1.10	0.35
H-6	1.760	7.47	10.3	758.1	8.513	4.90	1.10	0.20
HMX	1.891	9.11	10.5	778.3	7.071	4.20	1.00	0.30
LX-04-1	1.865	8.47	9.5	849.8	15.277	4.65	1.30	0.35
LX-07	1.865	8.64	10.0	871.0	13.896	4.60	1.15	0.30
LX-09	1.838	8.84	10.5	868.4	18.711	4.60	1.25	0.25
LX-10	1.875	8.82	10.4	880.2	17.437	4.60	1.20	0.30
LX-11	1.875	8.32	9.0	779.1	10.668	4.50	1.15	0.30
LX-13	1.540	7.35	6.6	2714.0	17.930	7.00	1.60	0.35
LX-14-0	1.835	8.80	10.2	826.1	17.240	4.55	1.32	0.36
LX-17-0	1.900	7.60	6.90	446.0	13.399	3.85	1.30	0.46
Nitro-methane	1.128	6.28	5.10	209.2	5.689	4.40	1.20	0.30
Octol	1.821	8.48	9.60	748.6	13.380	4.50	1.20	0.38
PBX-9010	1.787	8.39	9.00	581.4	6.801	4.10	1.00	0.35
PBX-9011	1.777	8.50	8.90	634.7	7.998	4.20	1.00	0.30
PBX-9404-3	1.840	8.80	10.12	852.4	20.493	4.60	1.35	0.25
PBX-9407	1.600	7.91	8.60	573.19	14.639	4.60	1.40	0.32
PBX-9501^d	1.840	8.80	10.12	852.4	18.02	4.60	1.30	0.38
PBX-9502^d	1.895	7.62	7.07	460.3	9.544	4.00	1.70	0.48
Pentolite	1.670	7.47	8.00	491.1	9.061	4.40	1.10	0.30
PETN	1.770	8.30	10.10	617.0	16.926	4.40	1.20	0.25
Tetryl	1.730	7.91	8.20	586.8	10.671	4.40	1.20	0.28
TNT	1.630	6.93	7.00	371.2	3.231	4.15	0.95	0.30
Deta Sheet C^d	1.48	7.00	3.69	349.0	4.524	4.10	1.20	0.30

of explosives, along with the C-J detonation parameters used in obtaining this information.

ρ_0^a = undetonated explosive density.
D^b = detonation velocity.
E_0^c = detonation energy.
$^d Data$ = data contributed by J. Jacobson, Group M-4, Los Alamos National Laboratory.

8.9.3 Small Variation of JWL-EOS

In the Eulerian code, we use a slightly different form of JWL-EOS which is

$$P(\text{Mbar}) = A\left[1-\left(\frac{\omega}{R_1 V}\right)\right]e^{-R_1 V} + B\left[1-\left(\frac{\omega}{R_2 V}\right)\right]e^{-R_2 V} + \frac{\omega(E-E_1)}{V}. \tag{8.53}$$

For Octol 75/25, the coefficients are: $A = 7.486$ Mbar; $B = 0.1338$ Mbar; $R_1 = 4.5$; $R_2 = 1.2$; $\omega = 0.38$; $E_1 = 0.272$ Mbar-cm^3/cm^3; V is the normalized specific volume (see Section 8.9.2); and E the detonation energy (Mbar-cm^3/cm^3). For the Chapman-Jouguet parameters: $\rho_0 = 1.821\,\text{g/cm}^3$, detonation velocity; $D = 0.849\,\text{cm}/\mu\text{s}$; $E_0 = 0.098$ Mbar-cm^3/cm^3; and $E_{chemical} = E_0 + E_1 = 0.37$ Mbar-cm^3/cm^3.

In Eq. (8.53), the $E_1 = 0.272$ Mbar-cm^3/cm^3 is calculated by setting $P = 0$, $E = 0$, and $V = 1$ in Eq. (8.53), and using other parameters from the JWL-EOS in Table 8.6. In the code calculation the high explosive energy released for the hydrodynamic energy equation is the term $E_{chemical}$. In calculating E_1, it is very convenient to use the following unit conversion factors for pressure and energy: 1 Mbar = 100 GPa; 1 Mbar-cm^3/cm^3 = 10^{11} J/m^3.

References

8.1 Jacobson, J, Private communication, Los Alamos National Laboratory, Las Alamos, NM (1993).
8.2 McQueen, R, Marsh, S, Taylor, J, Fritz, J, and Carter, W, The equation of state of solids from shock wave measurements, High-velocity impact, edited by R Kinslow, Academic Press, New York (1970).
8.3 Tillotson, J, Metallic equation of state for hypervelocity impact, General Atomic Report GA-3216, July (1962).
8.4 Steinberg, DJ, and Guinan, MW, Constitutive relations for the Kospall code, Lawrence National Laboratory Report UCID-16326 (December 1973).

8.5 Steinberg, DJ, Equation of state and strength properties of selected materials, UCRL-MA-106439, Lawrence Livermore National Laboratory, February (1991).
8.6 Steinberg, DJ, Cochran, SG, and Guinan, MW, A constitutive model for metals applicable at high-stain rate, J. Appl. Physics, 51, 1498 (1980).
8.7 Steinberg, DJ, Constitutive model used in computer simulation of time-resolved, shock-wave data, Int. J. Impact Engrg., 5, 603 (1987).
8.8 Steinberg, DJ, and Lund, CM, A constitutive model for strain rates from 10^{-4} to $10^{6}\,s^{-1}$, J. Appl. Physics, 65, 1528 (1989).
8.9 Johnson, GR, and Cook, WH, A constitutive model and data for metals subjected to large strains, high strain rates and high temperature, Proceedings of 7th Int. Symposium on Ballistics, The Hague, The Netherland, April (1983).
8.10 Lee, E, Horning, H, and Kury, J, Adiabatic expansion of high explosive detonation products, Lawrence Livermore National Laboratory Report UCRL-50422 (1968).

CHAPTER 9

SHAPED CHARGE PROBLEMS

9.1 Introduction

Shaped charges are used to penetrate hard targets such as tanks, concrete walls, or oil well pipes. A typical shaped charge consists of a metal shell liner (cone, hemisphere, or disk) and high explosive contained inside a metal cylindrical case. Usually, the high explosive is detonated at a single point. Occasionally, there are ring or surface detonations for special design purposes. The most important factors for designing a good shaped charge are the geometry, the grain size and the thickness of the liner, and the amount and uniformity of the high explosive. For oil well perforators, the geometry and materials used for the casing are also very important.

9.2 Shaped Charge Calculations by Lagrangian Method without Slip

The first problem calculates the jet formation of a shaped charge with a hemispherical titanium liner. Figure 9.1 shows the initial dimensions of the test problem with the hollow hemispherical titanium liner of thickness 0.43 cm and outside diameter of 8 cm. The bared PBX-9404 high explosive with an outside diameter of 10 cm is detonated by a single point initiated at point A. There are 33 zones in both the Z and R directions with 45 sectors (each sector is 2°). Therefore, we have $L = 1, 2, 3, \ldots, 34$ and $K = 1, 2, 3, \ldots, 46$ for the logical meshes. The titanium is divided into 8 uniform zones, and the PBX-9404 is divided into 25 zones. The calculated meshes at time $t = 0.0, 10.0$, and $20.0\,\mu s$ are shown in Fig. 9.2. The dimensions of the titanium liner are shown in Fig. 9.3 at times $25\,\mu s$ and $30\,\mu s$. However, since the present calculation does not use the sliding line treatment, there is a long thin zone near the interface region between the liner and the high explosive.

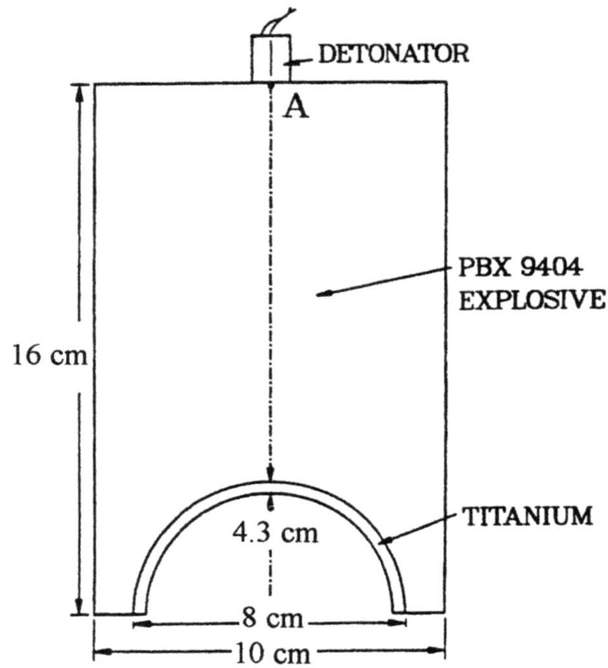

Figure 9.1. Schematic of the hemispherical titanium shaped charge.

9.3 Calculation of Shaped Charge Problem with Slip by Lagrangian Method

The shaped charge used for the slip study is described in Fig. 9.4. The aluminum conic liner has an outside diameter of 12 cm and a thickness of 0.7023 cm, and it is divided into six zones. The shaped charge is detonated with a EX-12 detonator which initiates the detonation-sheet explosive. The detonation wave in the detonation sheet propagates radially outward until it detonates the LX-14 explosive that is divided into 40 zones. The foam wave shaper prevents the detonation wave from pre-igniting the LX-14 in the interior region. This wave-shaping method changes the angle of the incidence of the detonation wave on the liner, providing for an extremely high collapse and jet tip velocity. Figure 9.5 shows the zoning of the current simulation including the aluminum liner and the LX-14 only. Due to the axisymmetry, we only model half of the problem. There are 40 zones in the radial direction and 46 zones in the axial direction. The high explosive has a ring detonation located at $R = 6$ cm and $Z = 11.529$ cm.

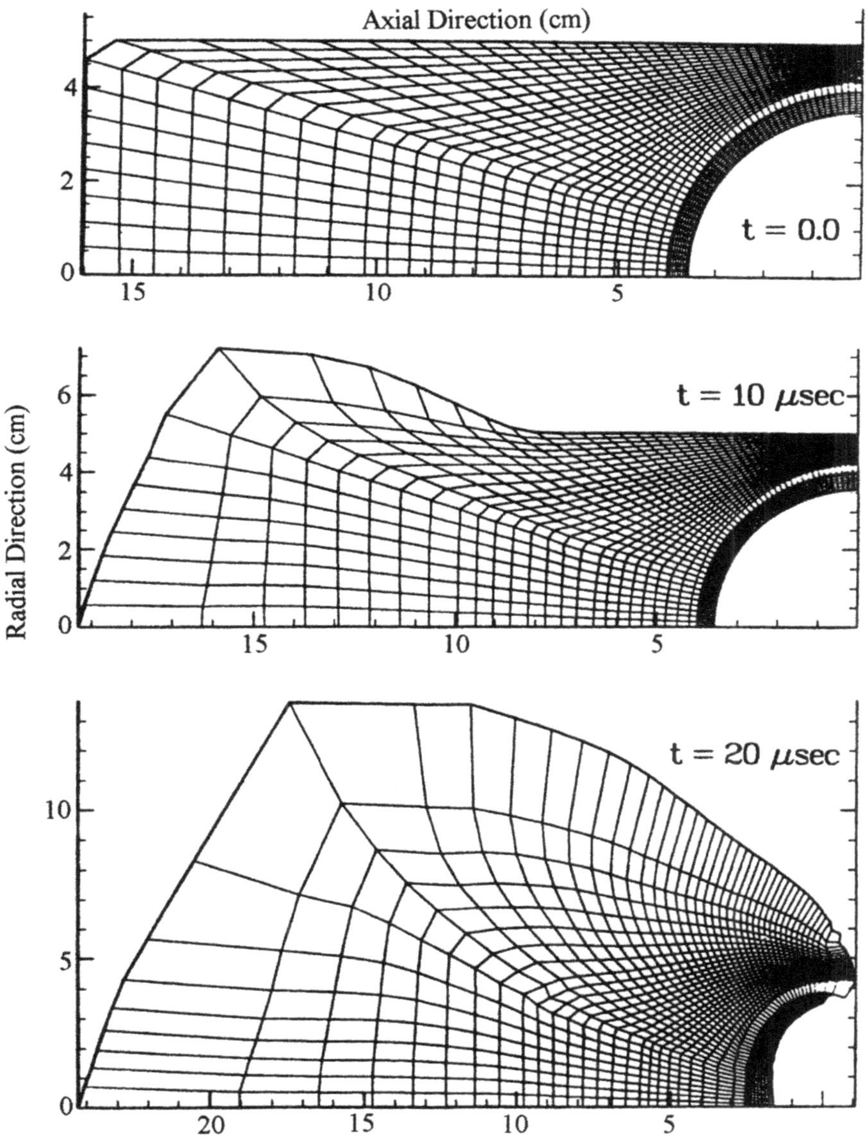

Figure 9.2. Lagrangian mesh of the explosive and the titanium liner at time 0.0, 10 and 20 μs.

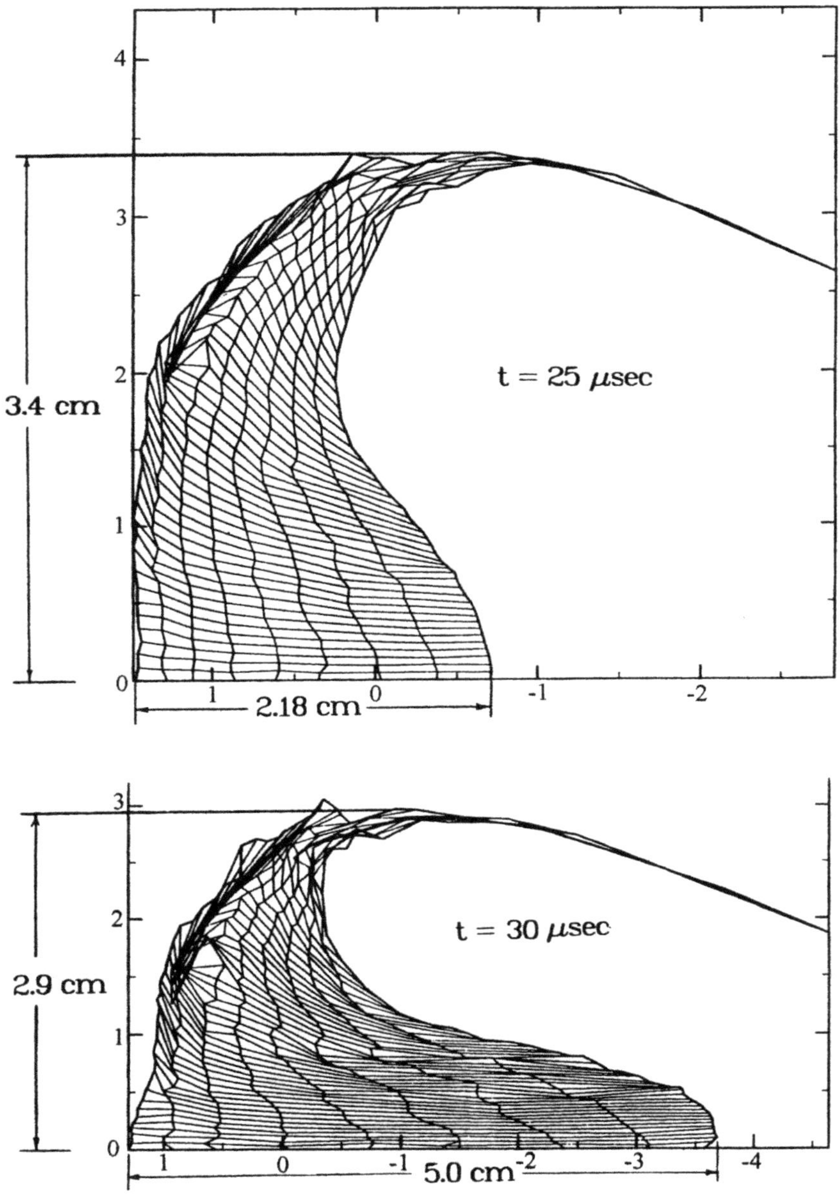

Figure 9.3. The dimension of the titanium liner at time 25 and 30 μs.

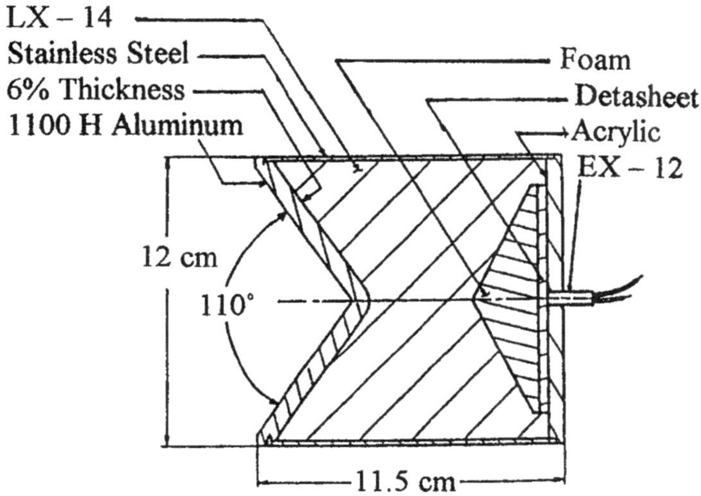

Figure 9.4. Shaped charge design configuration for code calibration and verification.

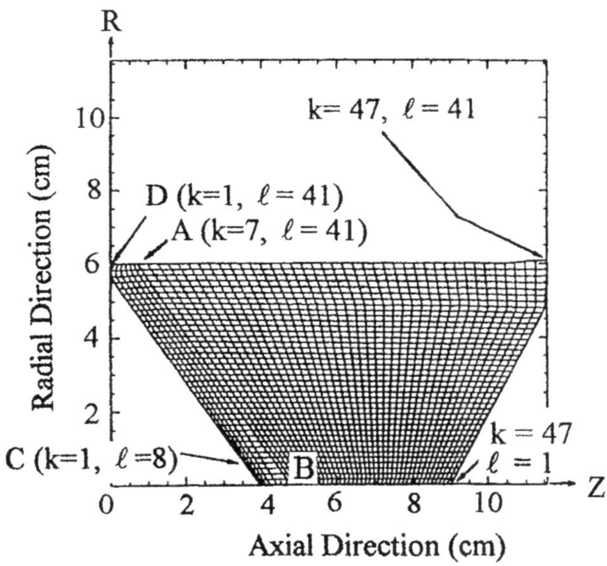

Figure 9.5. The 2D Lagrangian grid at time 1 μs with 40 zones axially in the high explosive, 6 zones axially in the aluminum liner and 40 zones in the radial direction. The slip surface is from $A(k = 7, \ell = 41)$ to $B(k = 7, \ell = 1)$ which is also the material interface.

Figure 9.6 shows the sequential calculations at time 20, 30, and 40 μs with and without slip treatment. When one uses a 2D Lagrangian code without slip for material interfaces to simulate the shaped charge problems, the zones located at the outside ring surface, i.e. $\overline{DA}$ as shown in Fig. 9.5 for this problem, always present some difficulties for calculation. Part of the problem is due to the turbulent nature of burned product gas of the high explosive near the interface. Most of the existing 2D Lagrangian code for calculating the metal-explosive interaction does not include the turbulent physics or separation due to turbulence. The other problem is due to the lack of slip at the material interface that results in high shearing stress inside the metal zones. This fact is observed at locations D $(k = 1, \ell = 4)$ and C $(k = 1, \ell = 8)$ at time 40 μs on the left column of Fig. 9.6. The final jet velocity and the shape of the calculations with slip option are much closer to the experimental data.

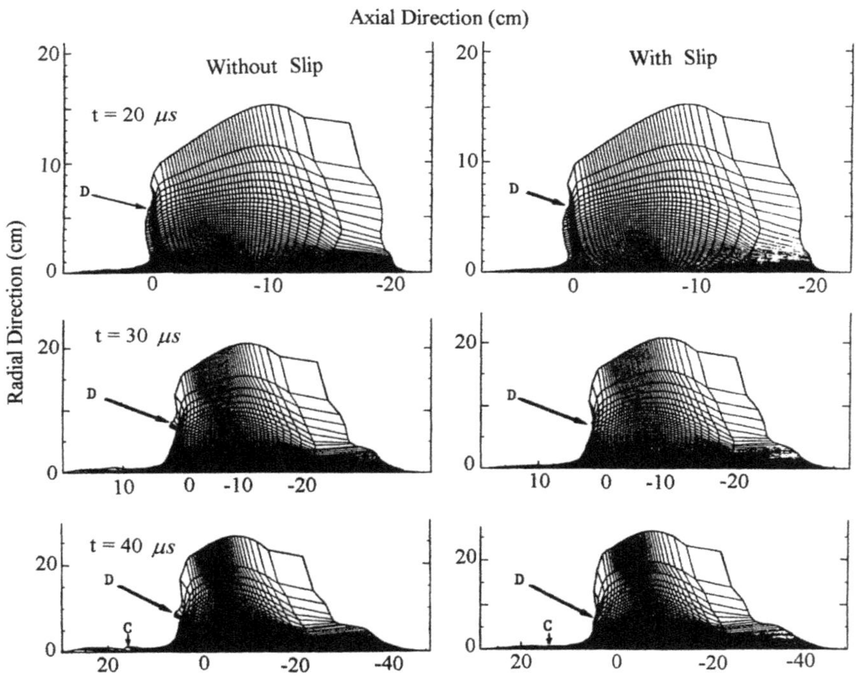

Figure 9.6. The grid formation of the aluminum liner and the high explosive at time 20, 30 and 40 μs with slip treatment (right column) and without slip (left column). At time 40 μs, the liner surface at $D(k = 1, \ell = 41)$ and $C(k = 1, \ell = 8)$ are much closer to experimental data for the calculations with slip.

9.4 Shaped Charge Calculations using Eulerian Method

It is more accurate to calculate the shaped charge jet formations by using the Eulerian code. In the following sections, the calculations of 10 different shaped charges are presented, showing the formations of the jet and other pertinent physics.

9.4.1 Viper Shaped Charge

The Viper shaped charge was designed by the Warhead Division of General Dynamics in the mid-1970s as a light, shoulder-launched, anti-armor weapon for the U.S. Army, but was never fielded. Mason and Hanger continued to produce 3,000–4,000 Viper warheads per year as a standard charge for ballistic testing. The Viper shaped charge employs a 77 g, 44° conical copper liner 0.12 cm thick at the hemispherical apex and 0.10 cm thick elsewhere. The high explosive charge consists of 427 g of LX-14 pressed over the liner after which it is lightly machined at room temperature to a final 65 mm diameter [9.1]. The booster initiator is made of PBX-9407. The upper plot of Fig. 9.7 shows the initial setup of the Viper shaped charge. The white square is the PBX-9407 booster, the yellow region is the main charge made of LX-14, the copper liner is in red and the purple region is the air. The computer simulations are confined in $-16.0\,\text{cm} \le z \le 16.0\,\text{cm}$ and $0.0\,\text{cm} \le r \le 5.0\,\text{cm}$ with 61 and 297 zones in the r and z directions respectively. Very fine zones are assigned to the rectangular region defined in $0.0\,\text{cm} \le r \le 0.5\,\text{cm}$ and $-8.0\,\text{cm} \le z \le 0.0\,\text{cm}$.

The lower plot of Fig. 9.7 shows the jet formation at time $t = 15.0\,\mu\text{s}$ after the HE detonated. The liner jets at time $t = 31.4\,\mu\text{s}$ and $t = 35.0\,\mu\text{s}$ are shown in Fig. 9.8. The X-ray radiograph of the Viper jet at time $t = 31.4\,\mu\text{s}$ is shown in the upper plot of Fig. 9.9 with the corresponding jet formation from the code calculations shown in the lower plot of Fig. 9.9. The discrepancy of the jet length and the jet tip velocity between the calculation and the experiment is about 2%. Figure 9.10 shows the calculation of the Viper jet at time $t = 10\,\mu\text{s}$ by using 2D Eulerian code which utilizes adaptive mesh refinement (AMR) and SLIC [9.3] methods for material interface tracking. The jet formation at time $t = 20.0\,\mu\text{s}$ is shown in Fig. 9.11. At time $t = 31.4\,\mu\text{s}$, the jet length of the calculations (the upper plot of Fig. 9.12) is very close to the experimental results. However, the calculated shape of the jet tip section is different from the experimental radiograph.

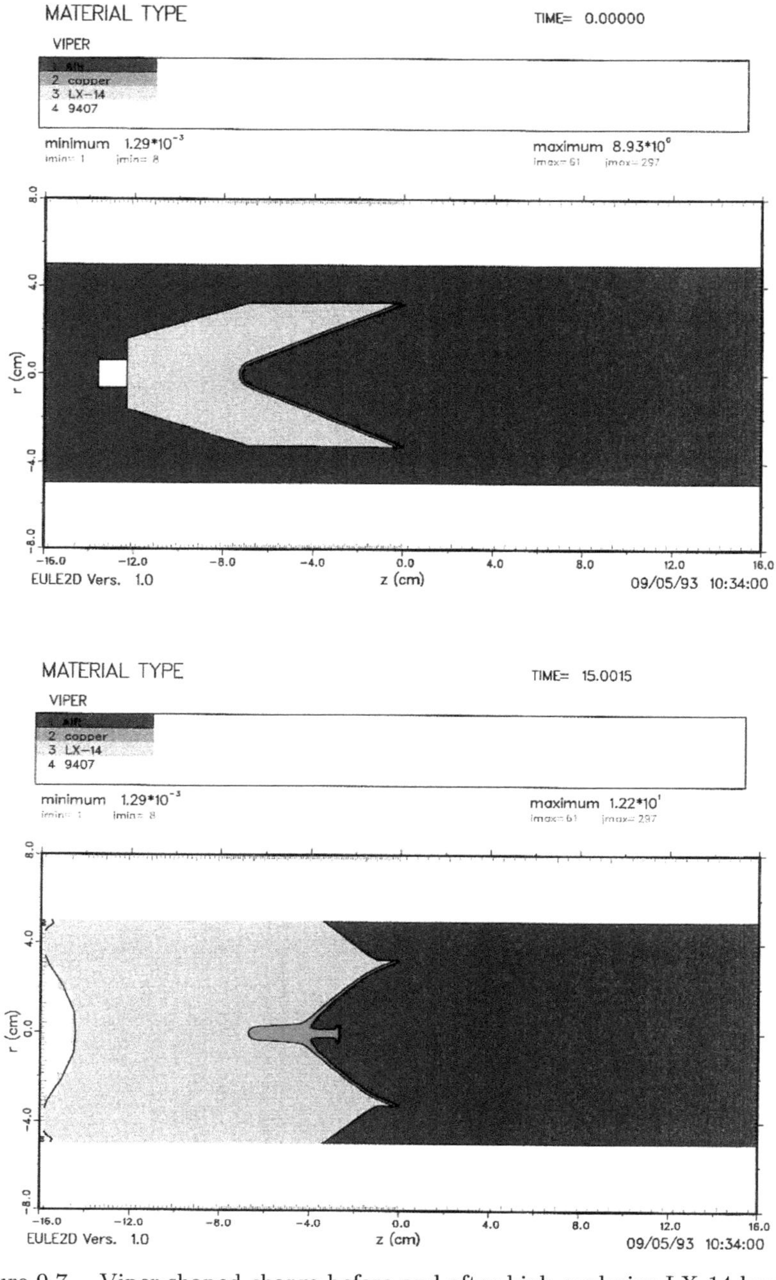

Figure 9.7. Viper shaped charge before and after high explosive LX-14 burned.

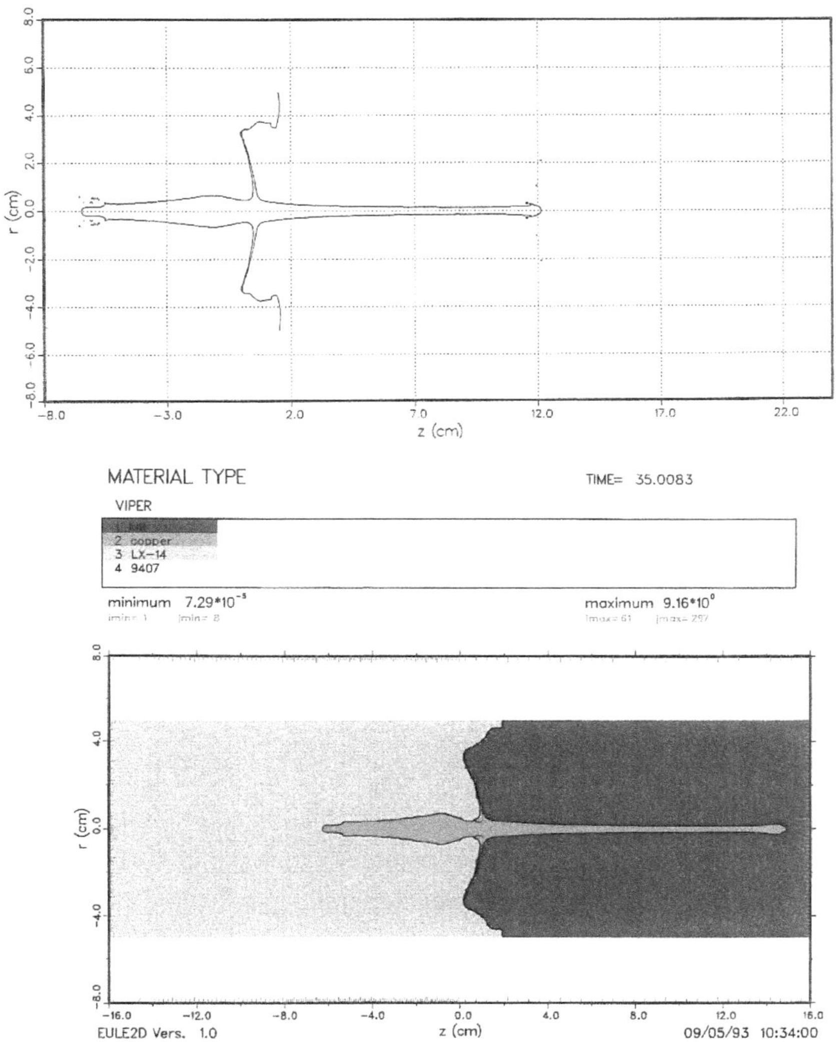

Figure 9.8. Viper shaped charge at time 31.4 μs (same time as in Fig. 9.9) and 35 μs.

9.4.2 Tantalum Hemi-spherical Shaped Charge

A good shaped charge should have high material density and dynamic ductility so that the stretched jet can endure a maximum length before it breaks up. This is why tantalum is an attractive material for making the shaped

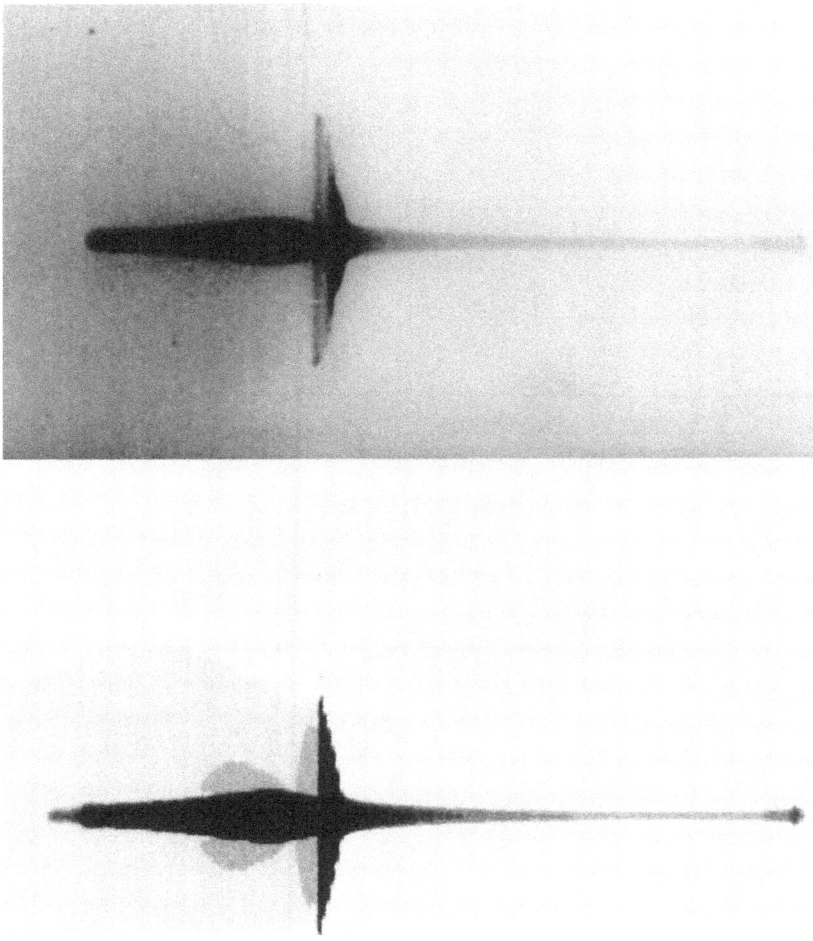

Figure 9.9. Viper shaped charge X-ray radiograph at time 31.4 μs (top) and the Eule 2D calculation (bottom). The discrepancy of the length and the tip velocity is about 2%.

charge liner. The upper plot of Fig. 9.13 shows the initial configuration of the hemi-spherical tantalum shaped charge which has a liner of 0.15 cm thick with inside diameter 10.70 cm and outside diameter 11.0 cm. The bare cylinder of diameter 13.4 cm is the main charge of PBX-9501 which was detonated at $r = 0.0$ cm and $z = -24.5$ cm. The shape of the liner jet and its density at time 60 μs are shown in the lower plot of Fig. 9.13. The high density of about 17.5 g/cm^3 is located near $z = 2.0$ cm while the low density

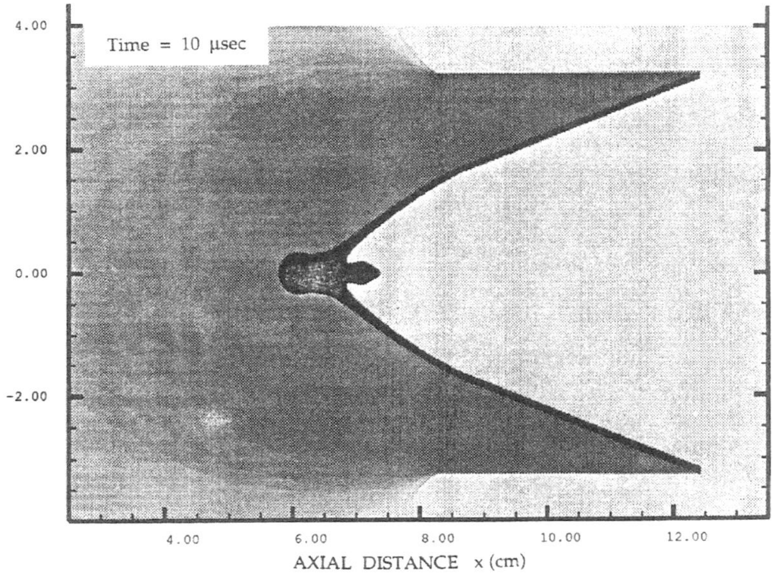

Figure 9.10. The Viper jet formation at time 10 μs as computed by SLIC method.

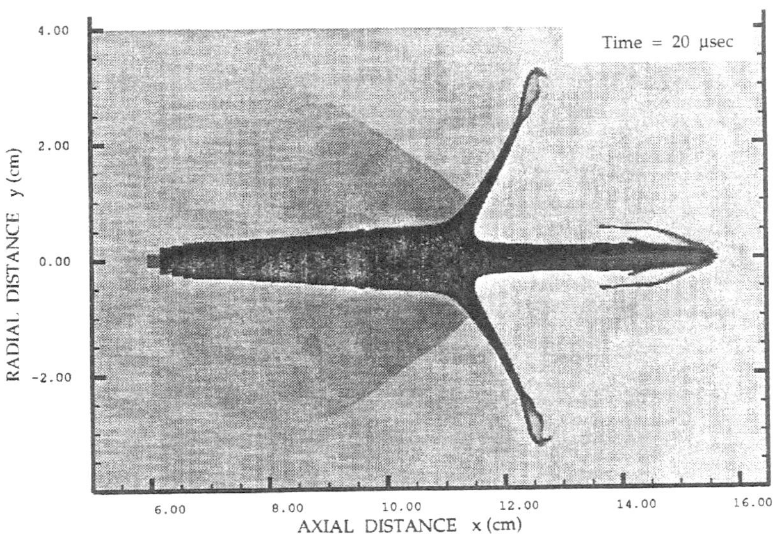

Figure 9.11. The formation of the copper jet at time 20 μs.

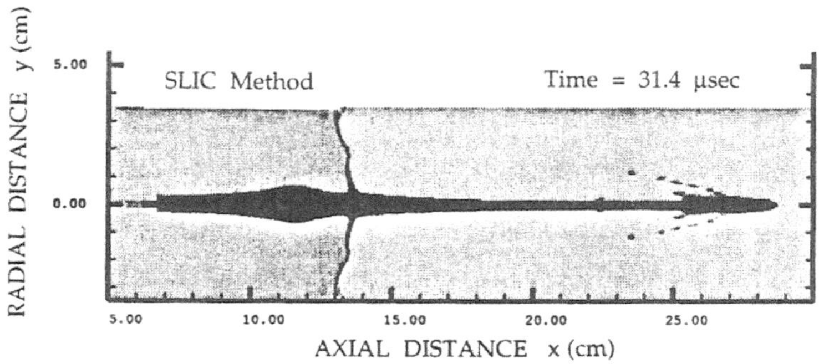

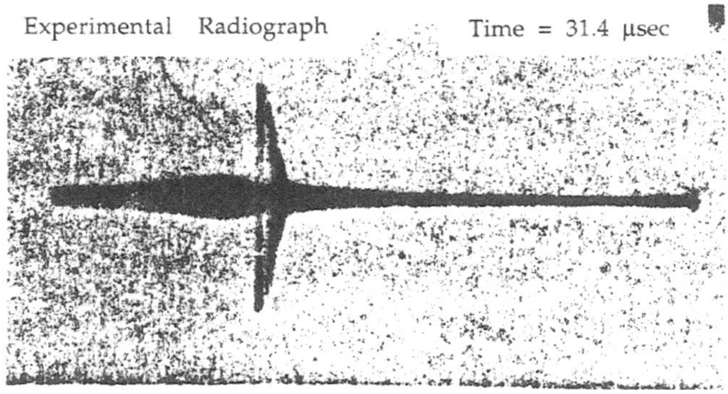

Figure 9.12. Viper jet formation at time 31.4 μs as computed by SLIC method (top) comparing with the experimental radiograph at the same time and scale (bottom).

of 15.7 g/cm^3 is at $z = 4.0$ cm. The formation of the tantalum jet particles at time $t = 550$ μs is shown in the upper plot of Fig. 9.14 for both experimental data and code calculations. The velocity distribution of the same jet particle can be seen in the lower plot of Fig. 9.14 for both experiments and calculations. The particle velocity and the jet length are surprisingly good for the code calculations.

9.4.3 Bi-conical Copper Shaped Charge

This bi-conical copper liner shaped charge is similar to the family of HELLFIRE [9.2]. As shown in the upper plot of Fig. 9.15 the bi-conical liner is made of copper with a hemi-spherical apex. The blue block is the LX-10 booster, the green block is the lead wave shaper, the yellow block is the LX-14 main charge, and the red block is the steel case. The lower plot is the jet

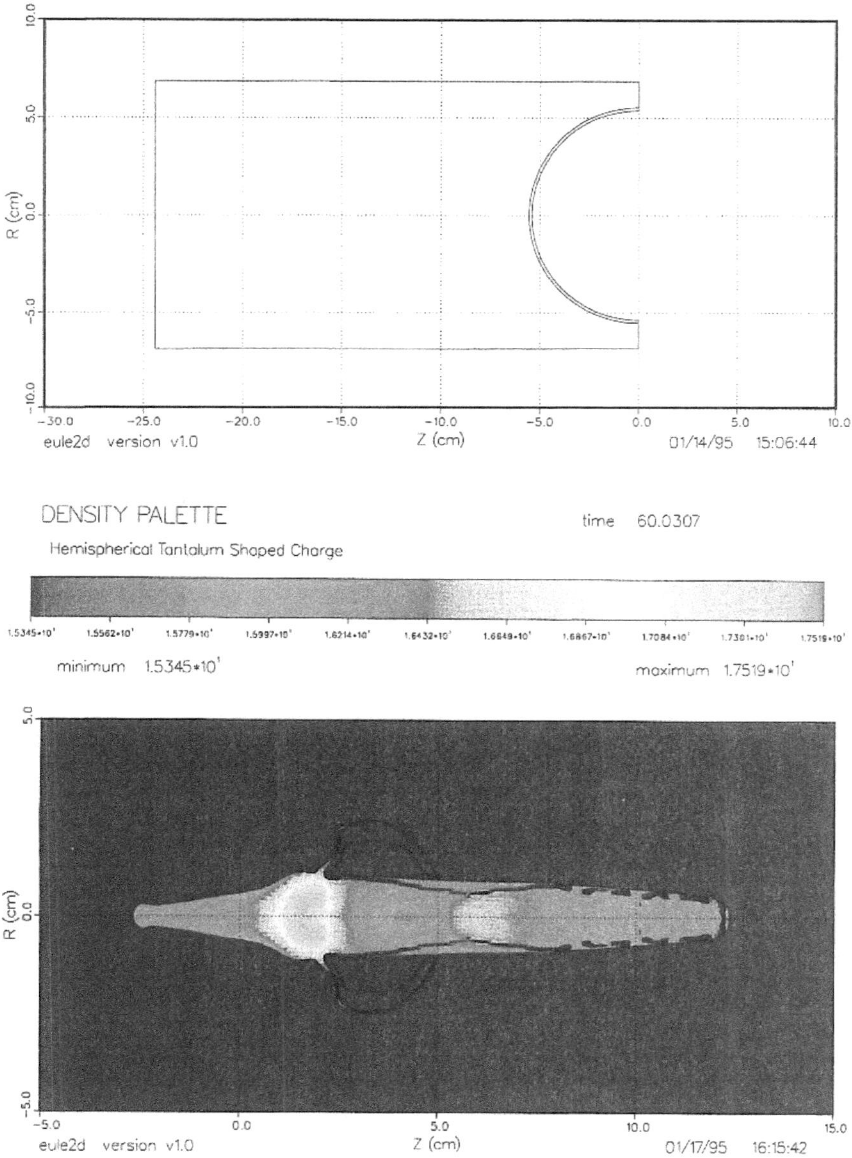

Figure 9.13. Tantalum hemi-spherical shaped charge before the high explosive PBX-9501 burned (top) and the formation of the tantalum jet at time 60 μs (bottom).

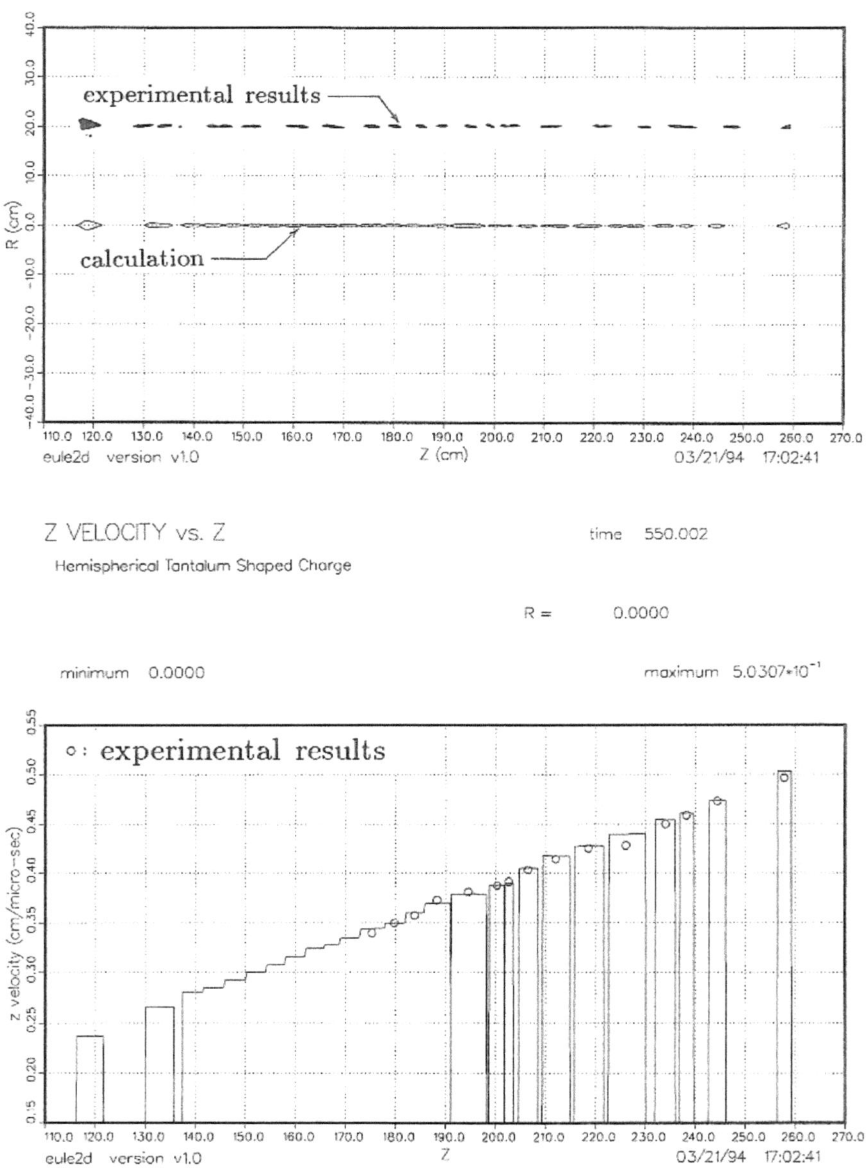

Figure 9.14. The formations of the tantalum jet particles at time 550 μs for the calculation and the experimental results are shown in the top plot. The velocity distributions for the same jet particles are shown in the bottom plot.

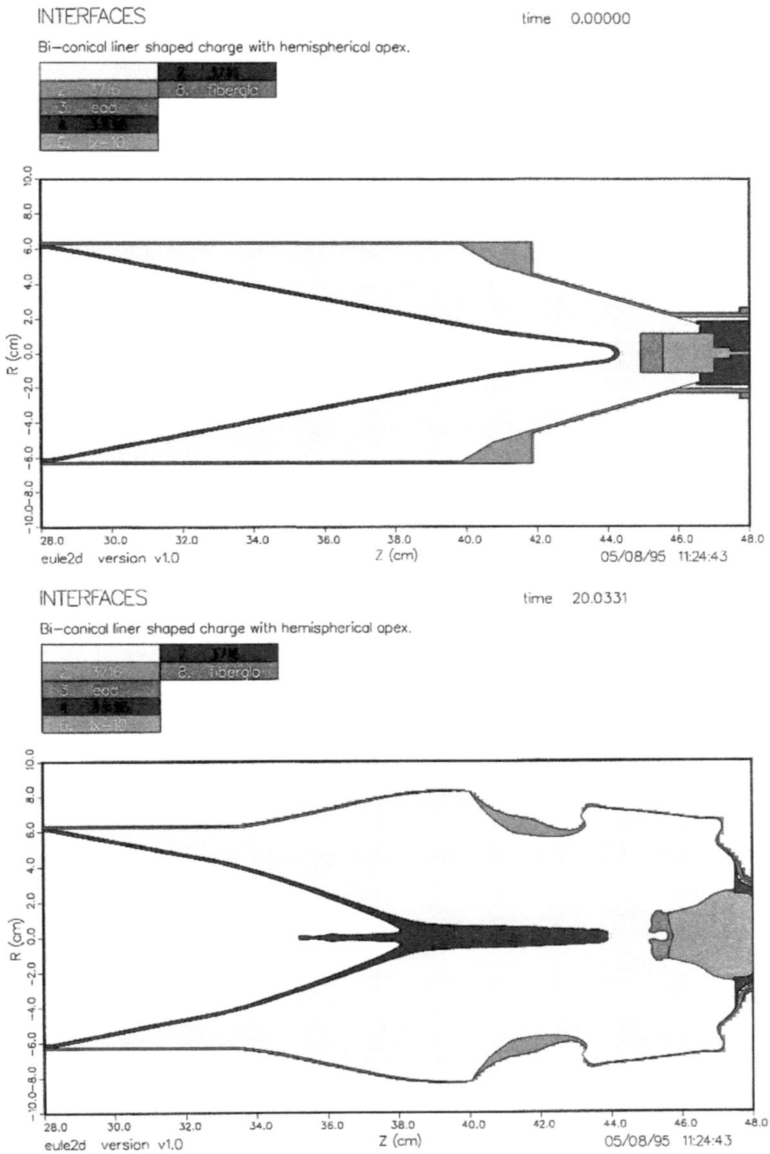

Figure 9.15. The initial set-up for the bi-conical copper liner with hemi-spherical apex is shown in the top plot. The main charge is LX-14 with booster explosive LX-10 and wave shaper lead. The formation of the jet at time 20 μs is shown in the bottom plot.

formation at time $t = 20.0\,\mu s$. Figure 9.16 shows the jet formation (upper plot) and the pressure contour at time $t = 30.0\,\mu s$ (lower plot). Along the liner the highest pressure is located at $z = 34.0$ cm while the lowest one is at $z = 44.0$ cm.

9.4.4 Oil Well Perforator P-C

This oil well perforator P-C uses a copper liner consisting of a spherical apex section with a hole along the axis of symmetry and the tangent truncated conical section. The case is made of stainless steel where the high explosive is composed of RDX. Figure 9.17 shows the initial setup (upper plot) and the jet formation at time $t = 24.0\,\mu s$ (lower plot). The jet formation at time $t = 36.0\,\mu s$ (upper plot) and $46.0\,\mu s$ (lower plot) are shown in Fig. 9.18. On both plots, one can see the hollow section along the z-axis (center of the jet cylinder) near the jet tip positions, i.e. $z = 9.0$ cm for the upper plot and $z = 14.0$ cm for the lower plot. Some of the burned high explosive gas is pushed out by the liner jet.

9.4.5 Copper Hemi-spherical Liner with PBX-W-113 (Energetic Explosive)

Figure 9.19 shows the copper hemi-spherical liner that has an outside radius of 8.89 cm (center at $r = 0.0$ cm and $z = 0.0$ cm) and thickness of 0.47489 cm. The outside aluminum cylinder case is 0.635 cm thick with a diameter of 21.59 cm, i.e. $\overline{GF}$, and length of 21.635 cm, i.e. $\overline{EF}$. The high explosive PBX-W-113 is detonated at the whole circular area along the line $\overline{AB}$. Figures 9.20–9.22 show the copper jet formation by using the Particle-in-Cell (PIC) 2D Eulerian code at time $t = 40.0$, 70.0 and $100.0\,\mu s$ respectively. The cumulative mass versus velocity at time $t = 100\,\mu s$ is shown in Fig. 9.23.

The same problem is calculated by using a multi-fluid 2D Eulerian code (different from PIC code) which uses SLIC [9.3] scheme for material interface tracking. The jet formations at time 40.0, 70.0, and $100.0\,\mu s$ are shown in Fig. 9.24. The results are very similar to the PIC code calculations.

9.4.6 Calculations of the Hemi-spherical Copper Shaped Charge

This hemi-spherical copper liner shaped charge also was calculated by Harrison [9.4]. The initial dimension of this problem with the hollow

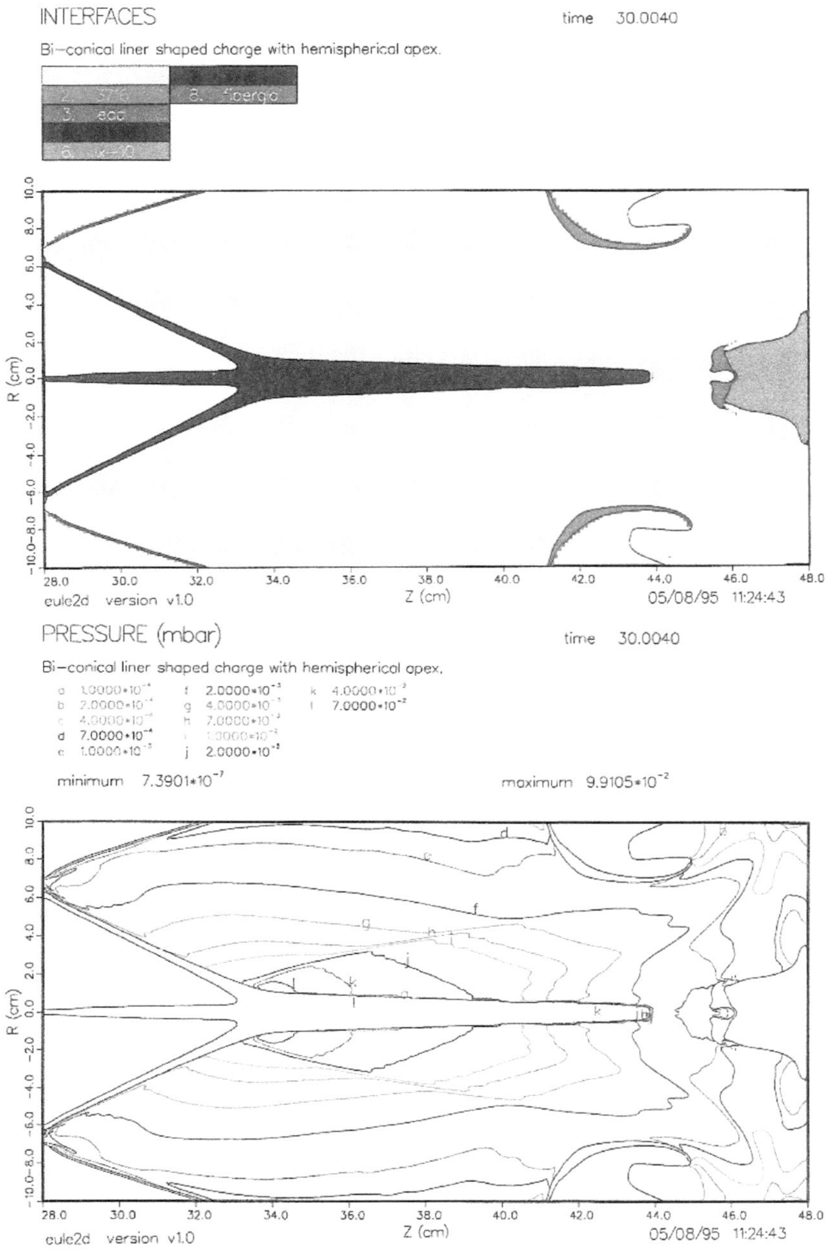

Figure 9.16. The material interface at time 30 μs is shown in the upper plot, while the pressure contour is shown in the lower plot.

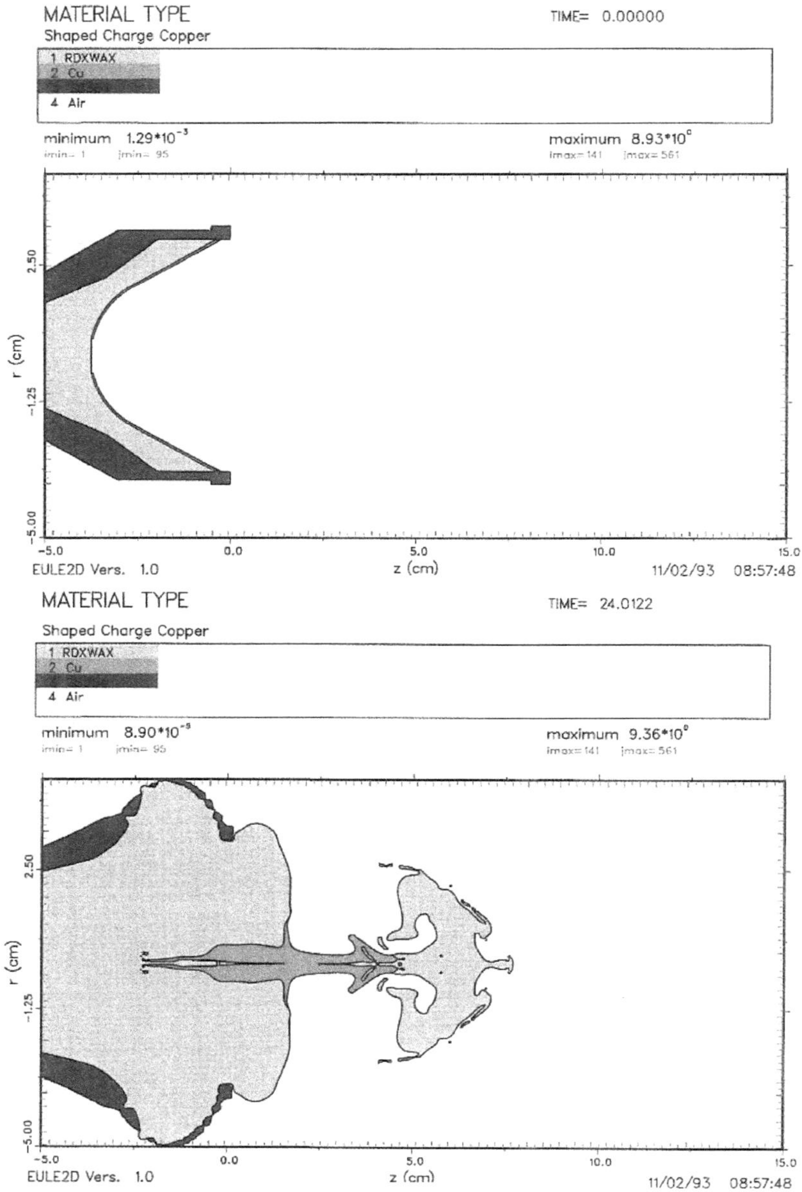

Figure 9.17. Copper liner consisting of the spherical apex section with a hole along the axis of symmetry and the tangent truncated conical section is shown in the upper plot, while the jet formation at time 24 μs is shown in the lower plot.

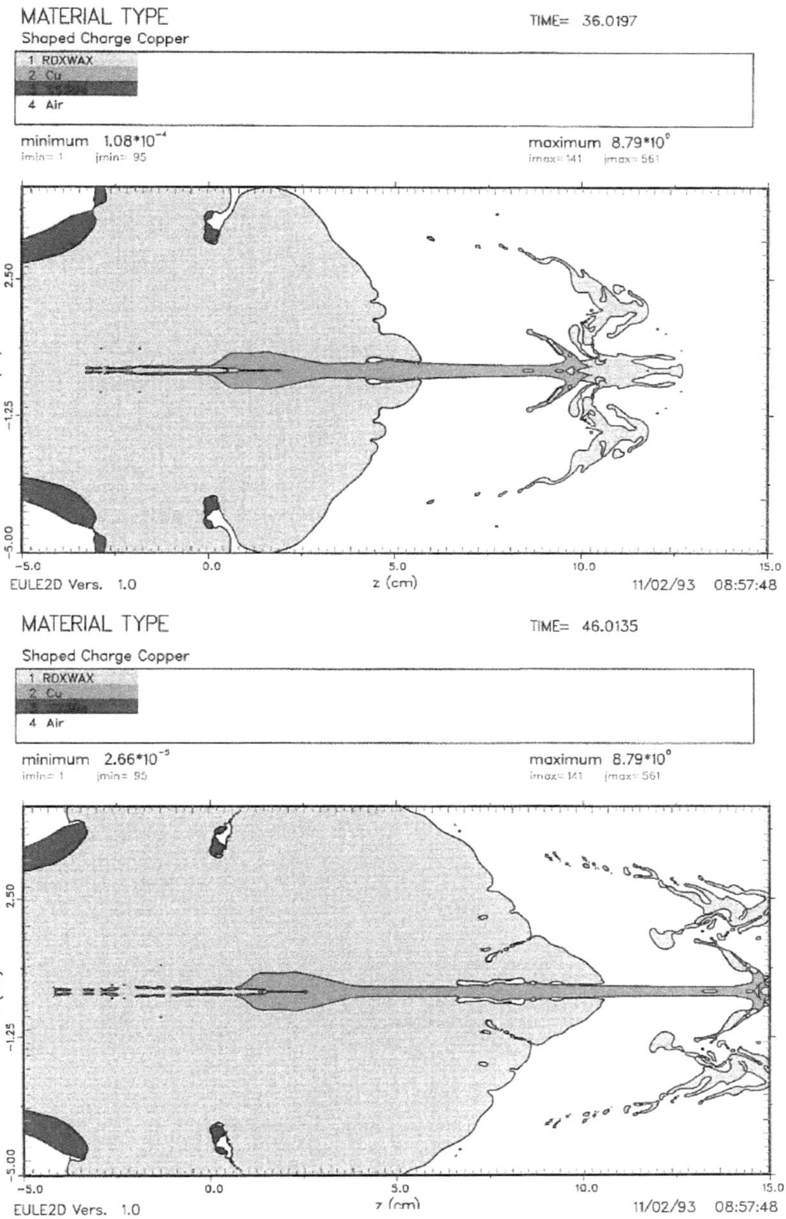

Figure 9.18. The jet formations of the shaped charge defined in Fig. 9.17 at time 36 μs (top) and 46 μs (bottom).

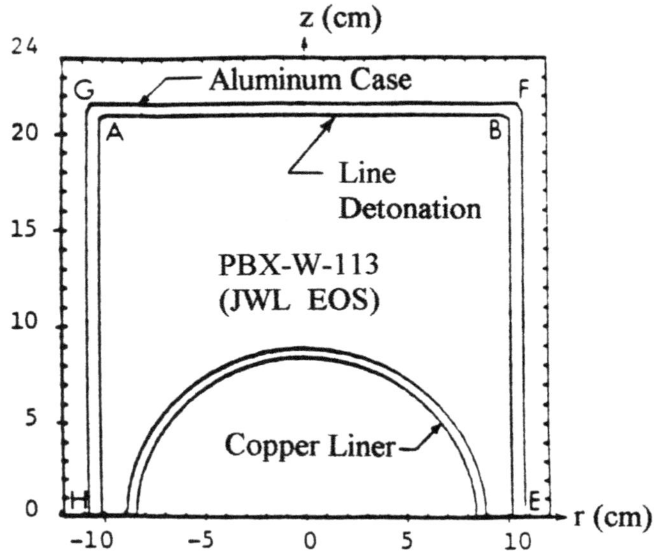

Figure 9.19. The initial setup of the copper hemi-spherical shaped charge with energetic explosive PBX-W-113 is shown.

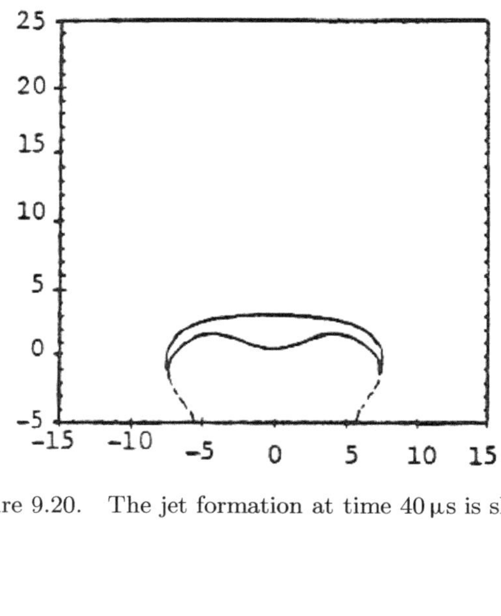

Figure 9.20. The jet formation at time 40 μs is shown.

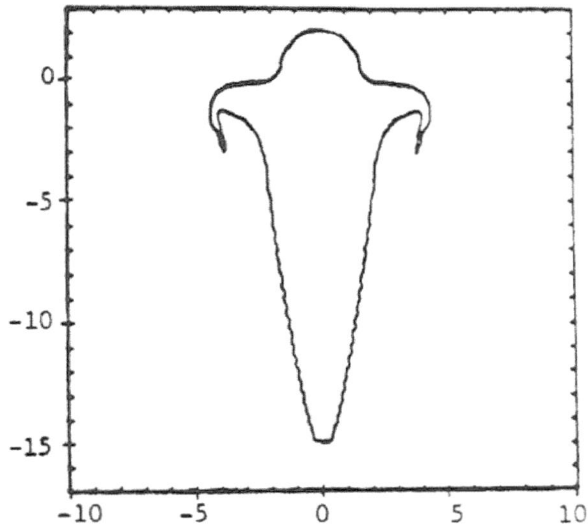

Figure 9.21. The jet formation at time 70 μs.

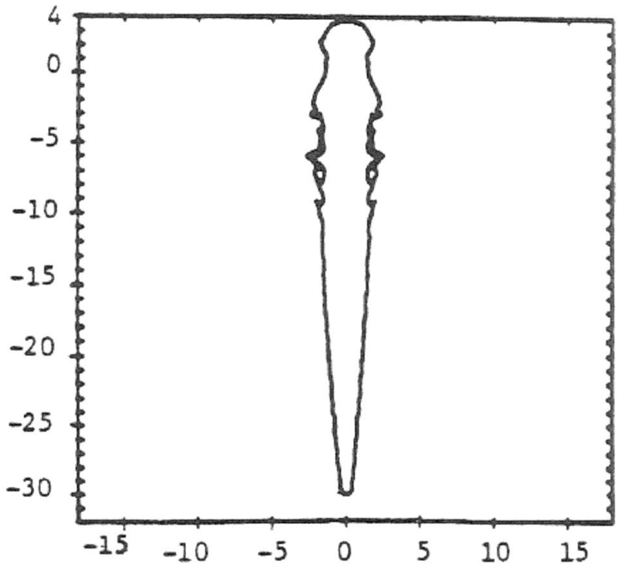

Figure 9.22. The jet formation at time 100 μs.

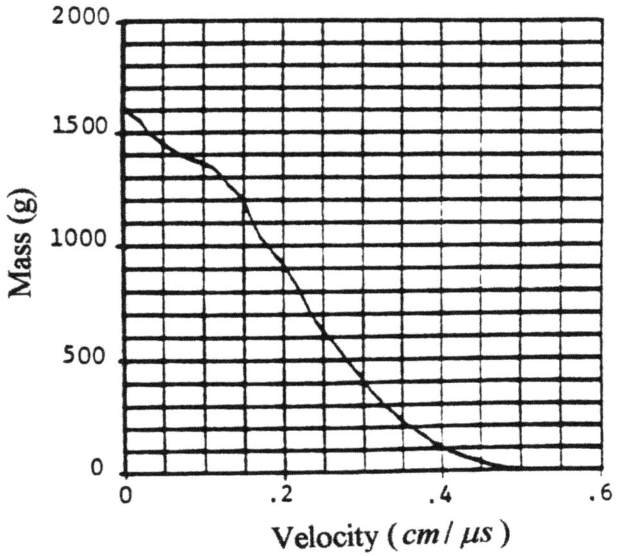

Figure 9.23. The cumulative mass versus velocity plot is shown for time 100 µs.

hemi-spherical copper liner 0.206 cm thick and 6.35 cm outside diameter are shown in Fig. 9.25. The bare 75/25 OCTOL high explosive with a 6.985 cm outside diameter was detonated by a single point initiated at point A. We use a 2D Lagrangian code to set up the input and run the problem up to 15 µs when the liner becomes about 0.52 cm thick. The Eulerian code calculations start at 15 µs and stop at 90 µs using a window of 7.5 cm (r-direction) X 30 cm (z-direction). The grid size is a 0.15 cm square mesh equivalent to a grid number of 50 X 200. For 75/25 OCTOL, the following JWL equation of state is used

$$P(\text{Mbar}) = A\left[1-\left(\frac{\omega}{R_1V}\right)\right]e^{-R_1V} + B\left[1-\left(\frac{\omega}{R_2V}\right)\right]e^{-R_2V} + \frac{\omega(E-E_1)}{V} \tag{9.1}$$

where $A = 7.486\,\text{Mbar}$; $B = 0.1338\,\text{Mbar}$; $R_1 = 4.5$; $R_2 = 1.2$; $\omega = 0.38$; $E_1 = 0.272\,\text{Mbar cm}^3/\text{cm}^3$; V is the specific volume; and E the detonation energy. For the Chapman-Jouguet parameters: $\rho_0 = 1.821\,\text{g/cm}^3$; detonation velocity $D = 0.849\,\text{cm/µs}$; $E_0 = 0.098\,\text{Mbar cm}^3/\text{cm}^3$; and $E_{chemical} = E_0 + E_1 = 0.37\,\text{Mbar cm}^3/\text{cm}^3$. A quadratic equation of state

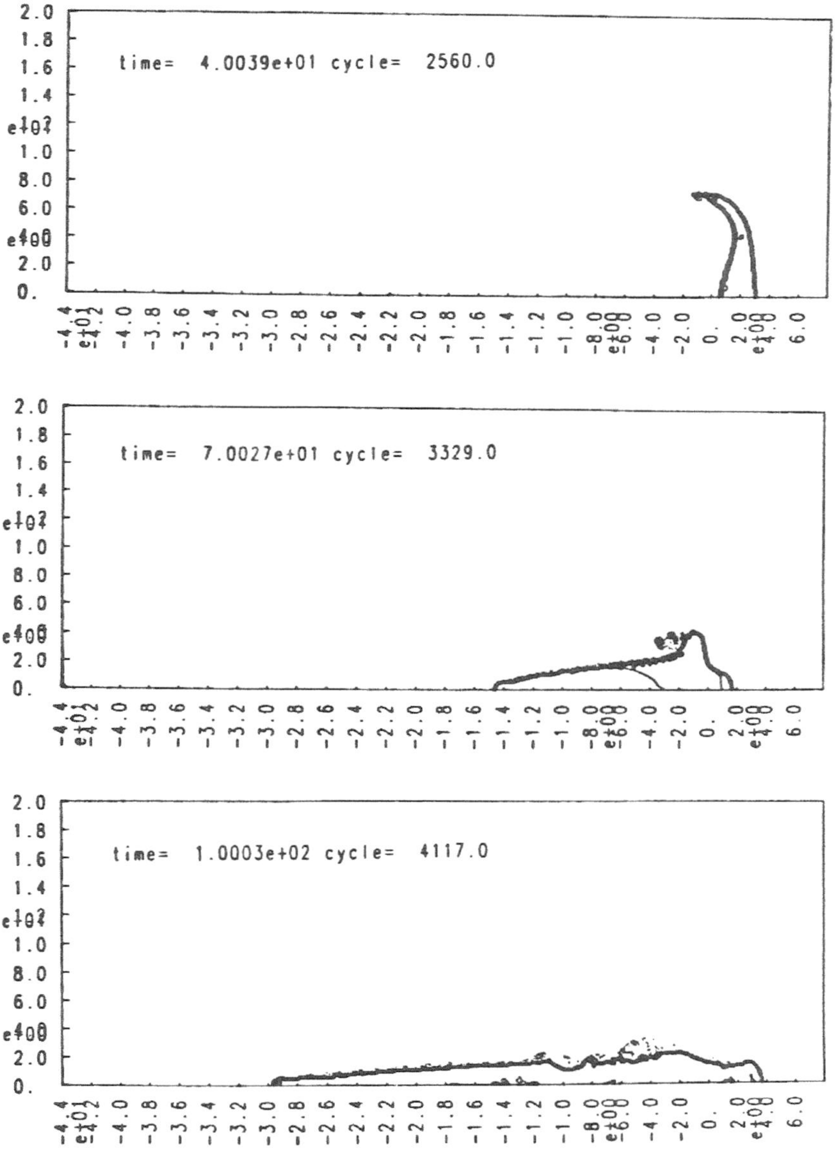

Figure 9.24. The jet formations of the same problem defined in Fig. 9.19 are shown as computed by a multi-fluid 2D Eulerian code using SLIC scheme for material interfacial tracking. The results are very similar to those by using PIC code.

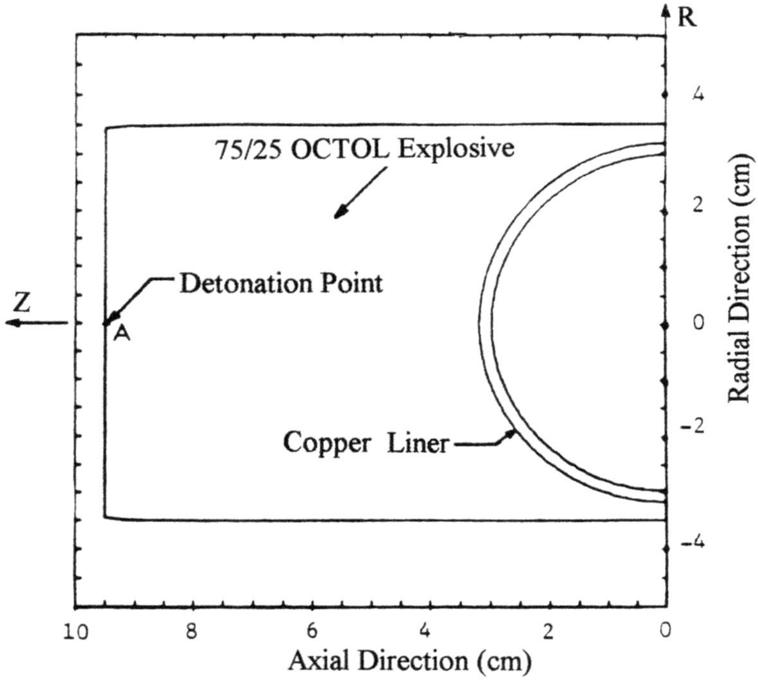

Figure 9.25. The initial dimensions of the copper liner shaped charge with 75/25 OCTOL as high explosive.

is used for copper liner as

$$P(\text{Mbar}) = \frac{A_1\mu + A_2\mu\,|\mu| + (B_0 + B_1\mu + B_2\mu^2)\varepsilon + (C_0 + C_1\mu)\varepsilon^2}{\varepsilon + D_0} \tag{9.2}$$

where $\mu = \rho/\rho_0 - 1$; $\varepsilon = \rho_0 I$ (Mbar cm^3/cm^3); $A_1 = 4.9578323$; $A_2 = 3.6883726$; $B_0 = 7.4727361$; $B_1 = 11.519148$; $B_2 = 5.5251138$; $C_0 = 0.39492613$; $C_1 = 0.52883412$; $D_0 = 0.6000001$; and $\rho_0 = 8.899\,\text{g/cm}^3$. The necessary parameters for copper, as described in Ref. [8.4], are: spallation pressure $= -10.$; $G_0 = 0.477$; $Y_0 = 0.0012$; $\beta = 36$; $n = 0.45$; $Y_{\max} = 0.0064$; $b = 3.14465$; $h = -0.000377$; $q = 1.0$; $f = 0.001$; $g = 0.001$; $R' = 0.00001164$; $T_{m0} = 1790$; $\gamma_0 - a = 0.52$; $a = 1.5$; $\alpha_0 = 0.68462$; $\alpha_1 = -0.00868$; $\alpha_2 = 0.26429$; $\alpha_3 = 0.0068119$; $\beta_0 = 1.0$; $\beta_1 = 0.00903$; $\beta_2 = 0.170902$; and $\beta_3 = 0.0417657$.

The initial grid setup for the 2D Lagrangian code is shown in Fig. 9.26. There are 31 zones in the HE region and 11 in the liner along the z-axis and 90 sectors (1° in one sector) for the first quadrant. At 15 μs, the liner has become about 0.52 cm thicker near the z-axis, as shown in Fig. 9.27, and the HE region has slipped way below the liner wring. The geometry of the collapsing copper liner and jet formations at the times of 20, 30, 40 and 50 μs are shown in Fig. 9.28. The velocity distribution along the z axis for the liner are provided in Fig. 9.29, which shows that the jet slug section is moving in the positive z direction, and that many velocity fluctuations exist inside the jet between the slug and tip. The observed cumulative mass versus jet velocity is plotted in Fig. 9.30, along with the computer simulations of HOIL code [9.4] and the PIC code calculations. Experimental data show that the tip velocity is 0.422 cm/μs, compared with the PIC code of 0.43 cm/μs.

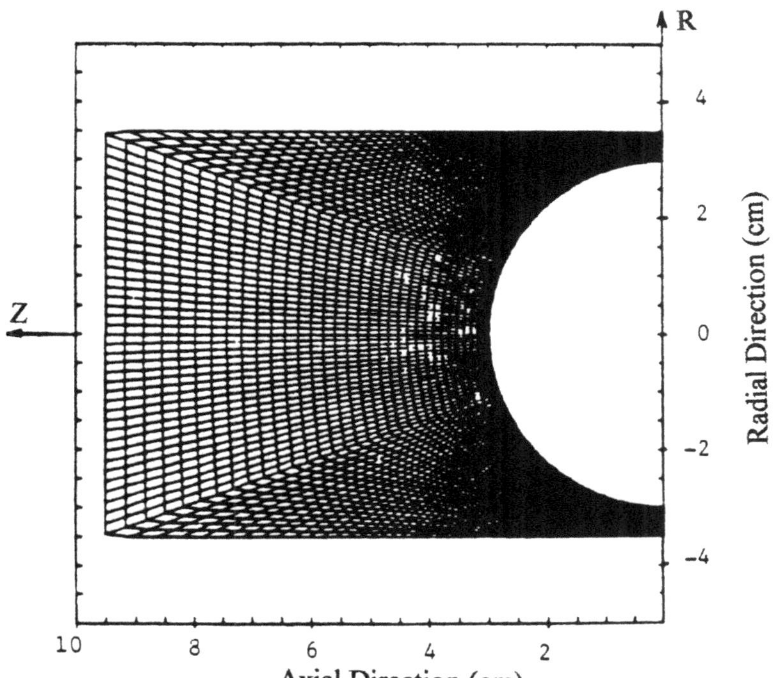

Figure 9.26. The initial grid setup for the 2D Lagrangian code has 31 zones in HE, 11 in the liner and 90 sectors in the first quadrant.

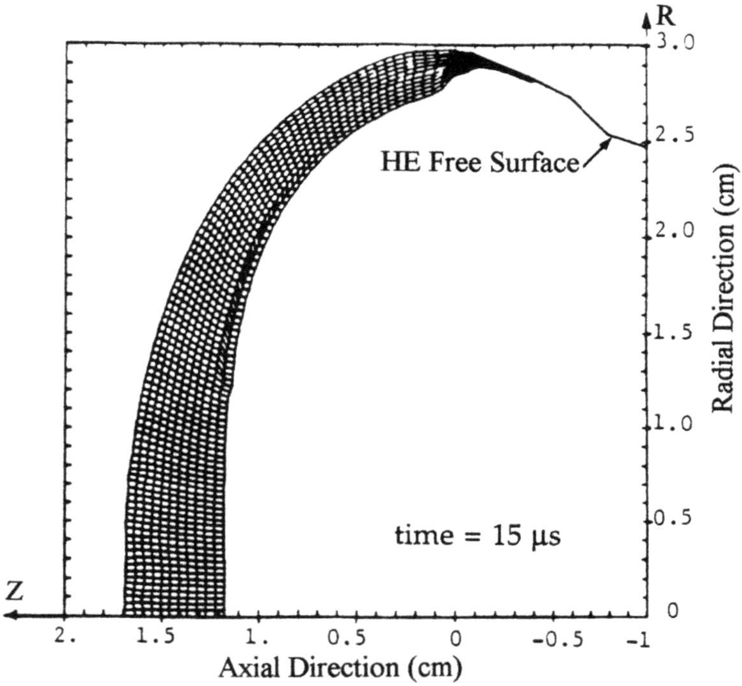

Figure 9.27. The copper liner has a thickness of approximately 0.52 cm along the z-axis at time 15 μs.

9.4.7 Bi-conical Copper Shaped Charge with PBX-9404

The copper liner has the initial dimension of spherical apex centered at $x = -19.77136$ cm and $y = 0.0$ cm as shown in the upper plot of Fig. 9.31. The inside wall of the apex region starts at $x = -20.25142$ cm and $y = 0.0$ cm and stops at $x = -19.88779$ cm and $y = 0.4662653$, while the outside wall starts at $x = -20.42668$ cm and $y = 0.0$ cm and stops at $x = -19.94097$ cm and $y = 0.6329906$ cm. The position of the first inside conical wall begins at $x = -19.88779$ cm and $y = 0.4662653$ cm and stops at $x = -14.98828$ cm and $y = 1.690606$ cm. The second conical inside wall starts at $x = -14.98828$ cm and $y = 1.690606$ cm and stops at $x = -0.9907803$ cm and $y = 7.06374$ cm. The horizontal inside wall starts at $x = -0.9907803$ cm and $y = 7.06374$ cm and stops at $x = 0.0$ cm and $y = 7.06374$ cm. The first conical outside wall starts at $x = -19.94097$ cm and $y = 0.6329906$ cm and stops at $x = -14.77613$ cm and $y = 2.016906$ cm.

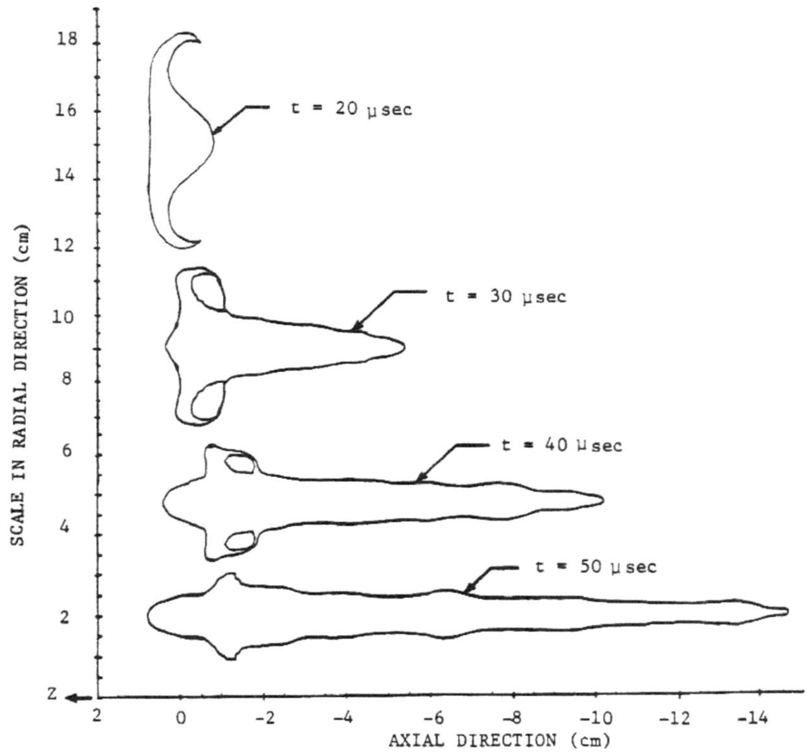

Figure 9.28. The jet formation for times 20, 30, 40 and 50 μs.

Then, the second conical outside wall stops at $x = -1.191956$ cm and $y = 7.23138$ cm. Then, the outside wall extends vertically to the point at $x = -1.191956$ cm and $y = 7.30758$ cm and finally stops at $x = 0.0$ cm and $y = 7.30758$ cm.

The contour of the lead wave shaper is defined by four points, i.e. ($x = -21.94506$ cm and $y = 0.0$ cm), ($x = 21.94306$ cm and $y = 1.41859$ cm), ($x = -21.16328$ cm and $y = 1.41859$ cm), and ($x = -21.16328$ cm and $y = 0.0$ cm). The outside wall of the PBX-9404 is defined by five points, i.e. ($x = -23.14448$ cm and $y = 0.0$ cm), ($x = -23.14448$ cm and $y = 1.7145$ cm), ($x = -13.58088$ cm and $y = 7.23646$ cm), ($x = -13.58088$ cm and $y = 7.30758$ cm), and ($x = -1.191956$ cm and $y = 7.30758$ cm). The outside wall of the aluminum case is defined by 10 points, i.e. ($x = -25.0$ cm and $y = 0.0$ cm), ($x = -25.0$ cm and $y = 2.30124$ cm), ($x = -22.38294$ cm and $y = 2.30124$ cm), ($x = -16.50696$ cm and $y = 5.693742$ cm),

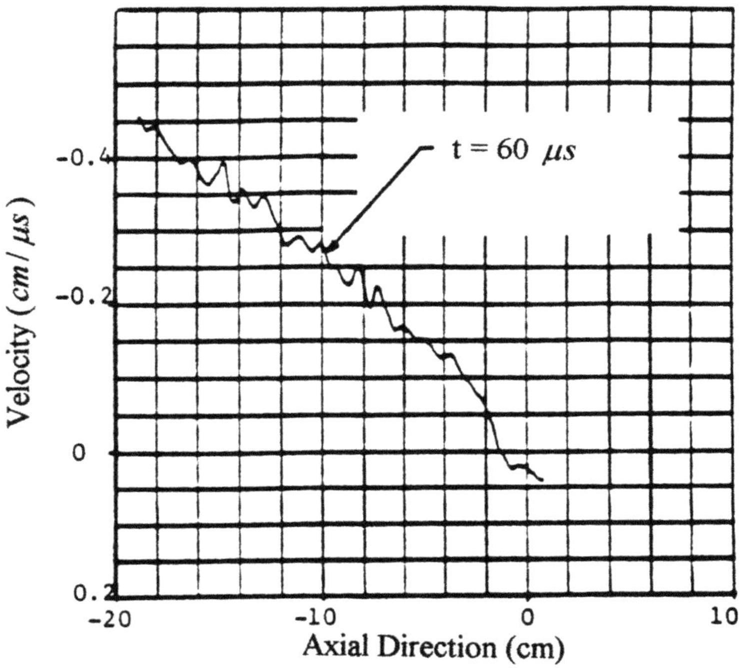

Figure 9.29. The velocity distributions at time 60 μs are shown for the liner along the z-axis.

($x = -16.50696$ cm and $y = 7.30758$ cm), ($x = -15.04392$ cm and $y = 7.43458$ cm), ($x = 1.9$ cm and $y = 7.43458$ cm), ($x = 1.9$ cm and $y = 2.75143$ cm), ($x = 5.0$ cm and $y = 2.75143$ cm), and ($x = 5.0$ cm and $y = 2.60143$ cm). The rest of the aluminum inside wall is defined by the following 4 points, i.e. ($x = 1.65$ cm and $y = 7.06374$ cm), ($x = 1.65$ cm and $y = 2.75143$ cm), ($x = 1.9$ cm and $y = 2.60143$ cm), and ($x = 5.0$ cm and $y = 2.60143$ cm). The high explosive is detonated at $x = -23.14448$ cm and $y = 0.0$ cm. The jet formation at time $t = 32.0$ μs is shown in the lower plot of Fig. 9.31. The cumulative mass versus velocity plot at time 78 μs is shown on the top plot of Fig. 9.32, while the jet formation at time $t = 152$ μs is shown on the lower plot.

9.4.8 Calculations of the BRL Precision Copper Shaped Charge

This problem consists of a shaped charge loaded with COMP B. It has a conical 42° copper liner that is 0.20574 cm thick. Both experimental and

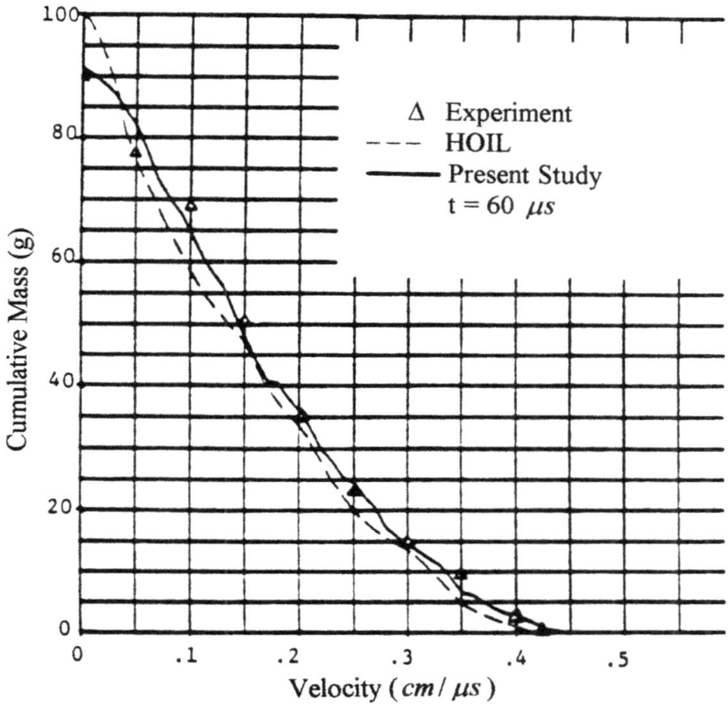

Figure 9.30. The cumulative mass vs. velocity plots are given for the experiment, HOIL and the PIC code calculations.

computational charges were confined with aluminum bodies and had cone diameters of 8.382 cm as shown in Fig. 9.33. The explosive region was detonated at the single point A for the modeling calculations. We used the JWL EOS for the high explosive.

The experimental measurements recorded were the collapse angles along the inside and the outside surfaces of the liner wall, β^{in} and β^{out}, respectively, and the distance L from the stagnation point to the rear of the slug. The radiographs of the collapse process were taken at delay times 25, 31, and 37 μs after the initiation of the charge. Due to the detonator/booster assembly in the experiment, there is a difference of 9.5 μs between the data and the calculations. Table 9.1 shows the comparison among the data, HEMP [9.5], and present code calculations. The HEMP code calculations and the detailed geometry of the shaped charge are reported in Ref. [9.6]. The cumulative mass versus jet velocity, which is plotted in Fig. 9.34,

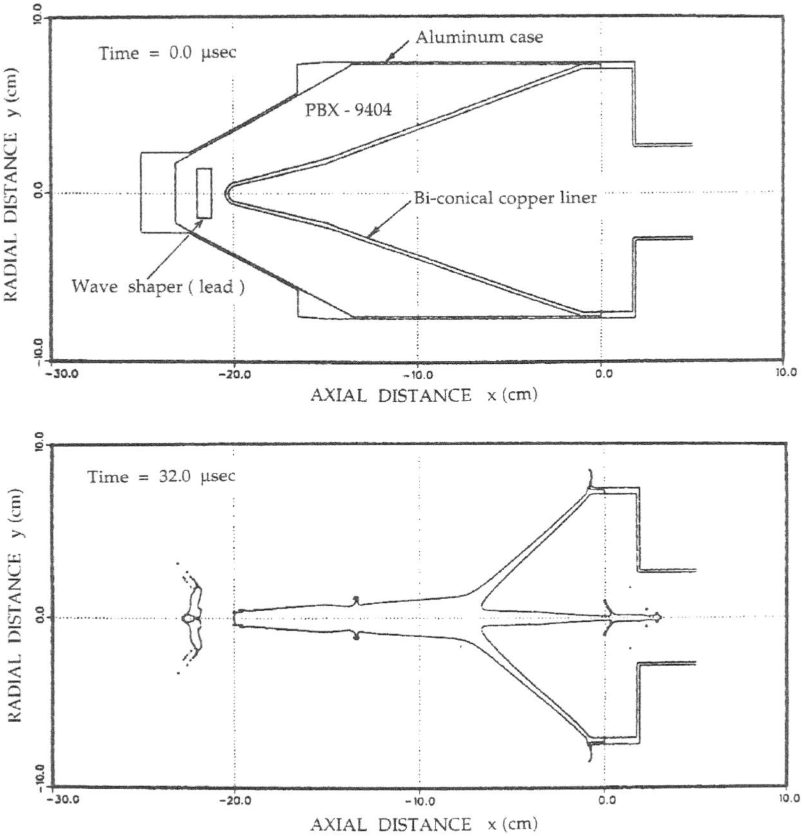

Figure 9.31. The initial setup for the bi-conical copper shaped charge that has PBX-9404 as the explosive and a disk of lead as the wave shaper is shown in the upper plot. The jet formation at time 32 μs is given at the lower plot.

compares well with the experimental data. The computed tip velocity is 0.73 cm/μs as compared to the experimental value of 0.77 cm/μs. The code simulation also shows that the jet starts to break at 106 μs (see Fig. 9.35) as compared to the data of 106.4 μs [9.6].

9.4.9 Shaped Charges of Tungsten-copper Alloy

In this section, two shaped charges, one with a hemispherical liner and the other with an elliptical liner, are calculated. The high explosive used in these charge was pressed and bonded with explosive PBX-9501. The JWL

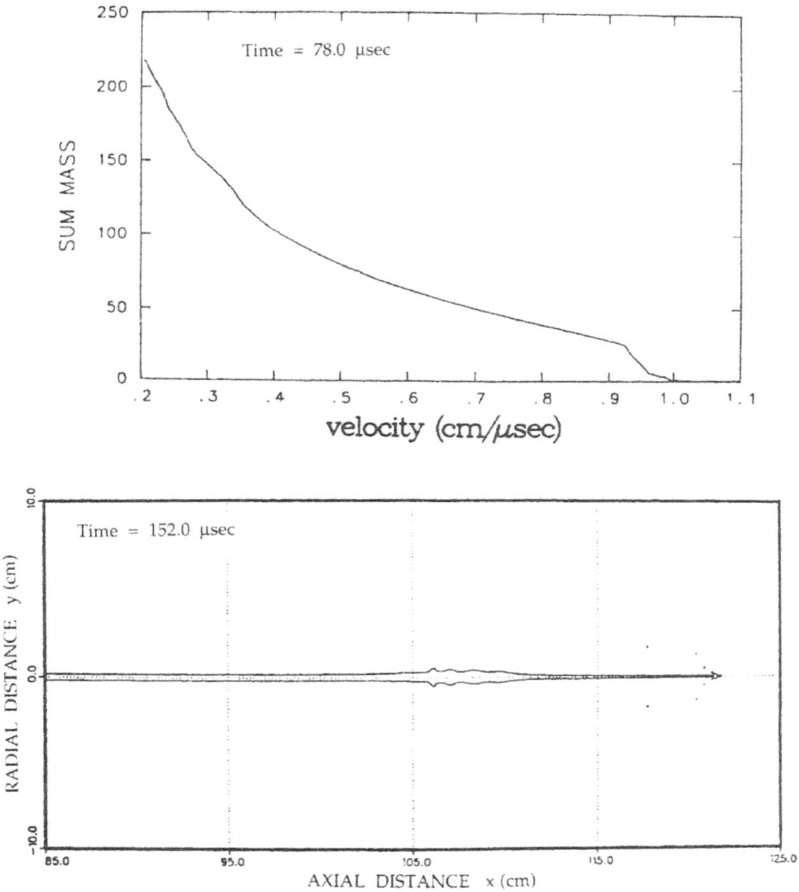

Figure 9.32. The cumulative mass vs. velocity plot at time 78 μs is shown on the top figure, while the jet formation at time 152 μs is shown on the bottom one. The penetration of this jet into the steel block is about 76 cm.

EOS was used for the explosive with parameters; $\rho_0 = 1.84$; $E_0 = 0.05543$; D $= 0.88$; $\omega = 0.25$; $A = 8.5445$; $B = 0.20493$; $R_1 = 4.6$; and $R_2 = 1.35$. The liners were made of annealed, fine-grain, tungsten-copper sheet which had been cast, deep drawn, annealed at 600°C for about 15 min, and machined on tape-controlled milling machines. The constants for the liner equation of state, i.e. Eq. (8.10), are $A_1 = 2.4562457$; $A_2 = 4.6163216$; $B_0 = 4.3432909$; $B_1 = 0.76214541$; $B_2 = 6.4410793$; $C_0 = 0.31988993$; $C_1 = 0.46744784$; $D_0 = 2.2$; and $\rho_0 = 18.983$. The parameters used for

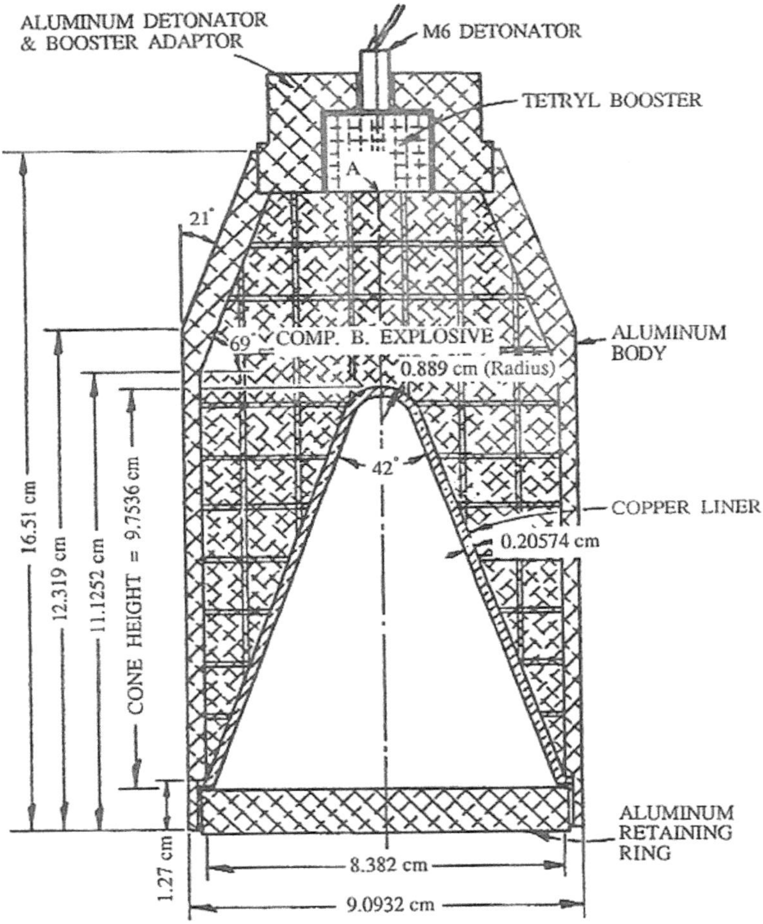

Figure 9.33. Sketch of the 8.382 cm BRL precision shaped charge.

the constitutive relations, i.e. Eqs. (8.20) and (8.21), are $G_0 = 0.844$; $Y_0 = 0.0012$; $\beta = 1.6 \times 10^4$; $n = 0.26$; $Y_{\max} = 0.0168$; $b = 4.739$; $h = -0.0008056$; $q = 1.0$; $f = 0.001$; $g = 0.001$; $R' = 0.00000663$; $Tm_0 = 1710$; $\gamma - a = 0.92$; and $a = 1.5$. Figure 9.36 shows the initial setup for the hemispherical shaped charge which has a bore cylindrical explosive of outside radius of 5.0 cm and height of 16 cm. The liner has a thickness of 0.1 cm and an outside radius of 4.0 cm. The calculated liner jet (solid line) and the radiograph at $t = 40\,\mu\text{s}$

Table 9.1. Collapse sequence for a 42° Comp. B charge X-ray observations versus code calculations.

	Elapsed Time (μs)	Collapsed Angle (deg)	L^{a} (mm)
(1) X-rays	25.0	$\beta^{in} = 35,\ \beta^{out} = 32$	20
HEMP	15.5	$\beta^{in} = 35,\ \beta^{out} = 32$	26
Present study	15.5	$\beta^{in} = 35,\ \beta^{out} = 32$	21
(2) X-rays	31.0	$\beta^{in} = 42,\ \beta^{out} = 39$	44
HEMP	21.5	$\beta^{in} = 42,\ \beta^{out} = 36$	50
Present study	21.5	$\beta^{in} = 42,\ \beta^{out} = 39$	45
(3) X-rays	37.0	$\beta^{in} = 53,\ \beta^{out} = 49$	58
HEMP	27.5	$\beta^{in} = 54,\ \beta^{out} = 47$	76
Present study	27.5	$\beta^{in} = 53,\ \beta^{out} = 49$	68
X-ray time lagging = 9.5 μs			

[a] L is the distance from the stagnation point to the rear of the slug.

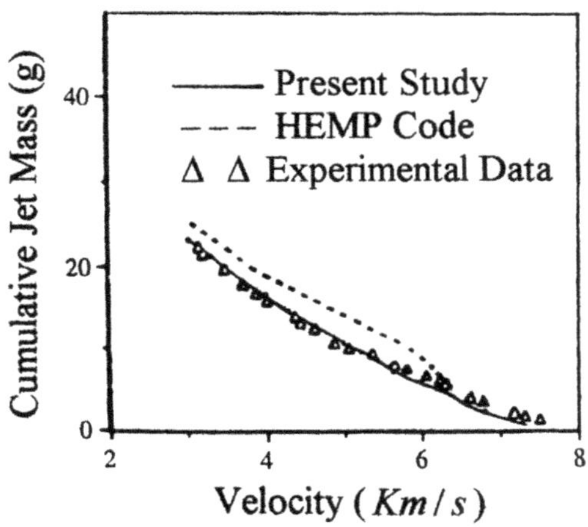

Figure 9.34. Cumulative mass versus velocity plots for present study, HEMP and data.

(dash line) are shown in Fig. 9.37. The same information for $t = 60\,\mu s$ is shown in Fig. 9.38. The calculated lengths and widths of the liner are in excellent agreement with the experimental data.

The next problem is a 0.1-cm-thick, elliptical-liner-shaped charge which has a short outside radius (along the radial direction) of 6.0 cm and a long

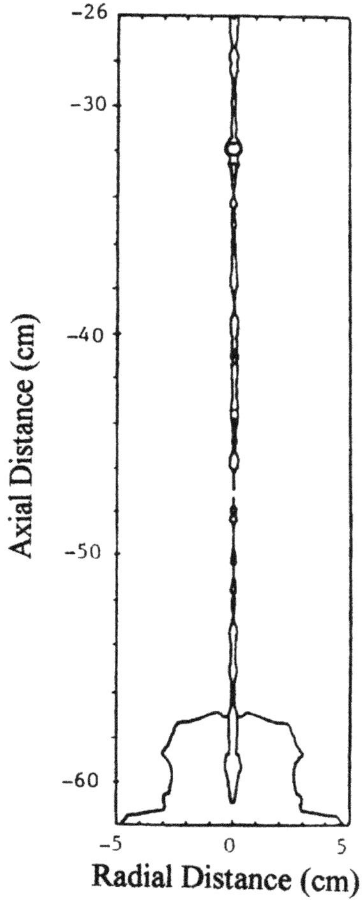

Figure 9.35. The jet starts to break at time 106 μs (only 24.2 g of jet are shown).

outside radius of 10.0 cm (along the axial direction) as shown in Fig. 9.39. The computed liner jet and the radiograph at $t = 74.0\,\mu\text{s}$ are shown in Fig. 9.40. The tip shape and the length of the liner jet as calculated by present method are in good agreement with the data. Figure 9.41 shows the cumulative mass versus velocity plots as calculated by present code (solid line) and as obtained by Harvey [9.7] (dash line) using the SLIC [9.3] method.

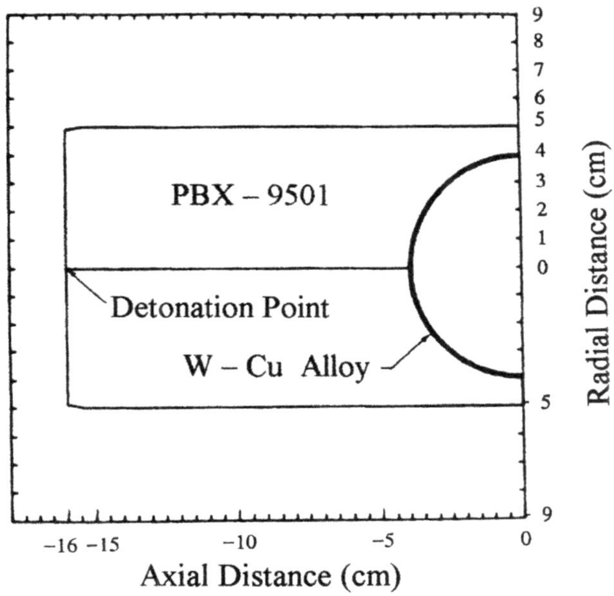

Figure 9.36. The initial conditions of the W-Cu hemispherical shaped charge with PBX-9501 as the high explosive.

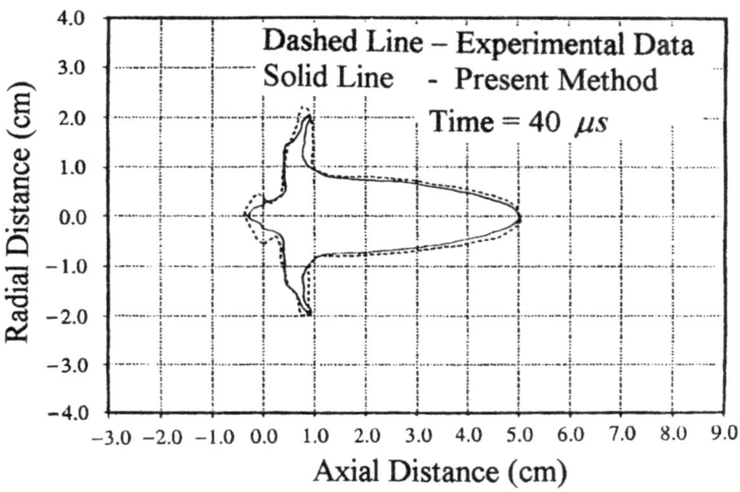

Figure 9.37. Calculated jet formation (solid line) and the radiograph data (dashed line) at time 40 μs.

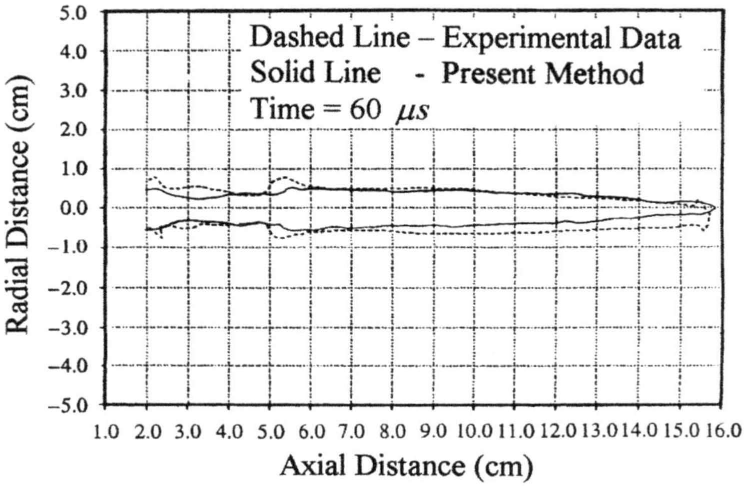

Figure 9.38. Calculated jet formation (solid line) and the radiograph data (dashed line) at time 60 μs.

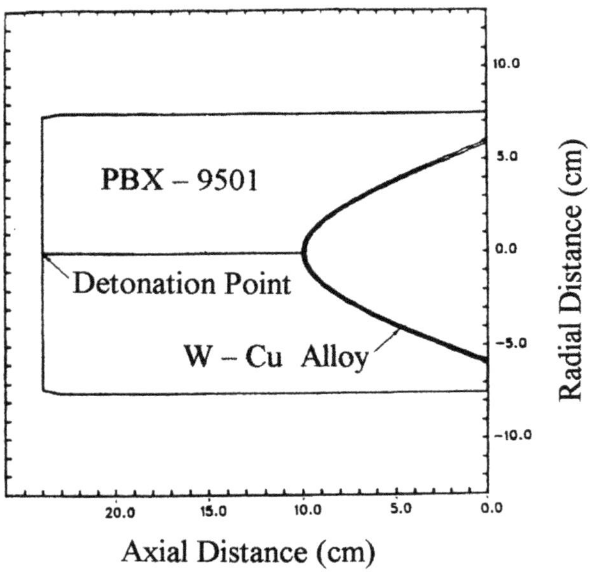

Figure 9.39. The initial conditions of the W-Cu elliptic shaped charge with PBX-9501 as the high explosive.

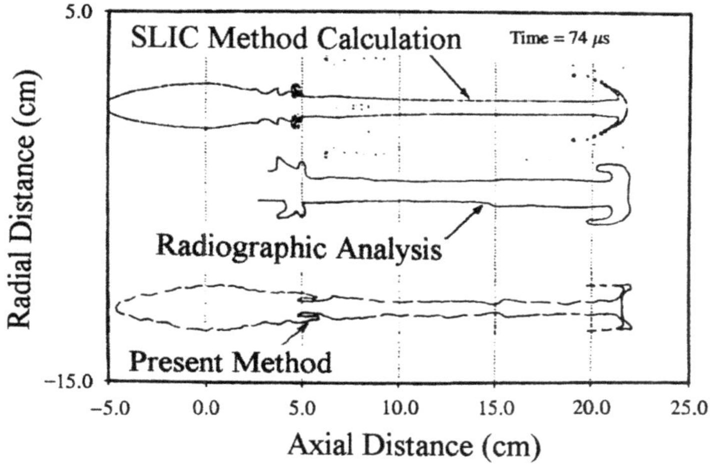

Figure 9.40. Calculated jet formation (dashed line) and the radiograph data (solid line) at time 74 μs.

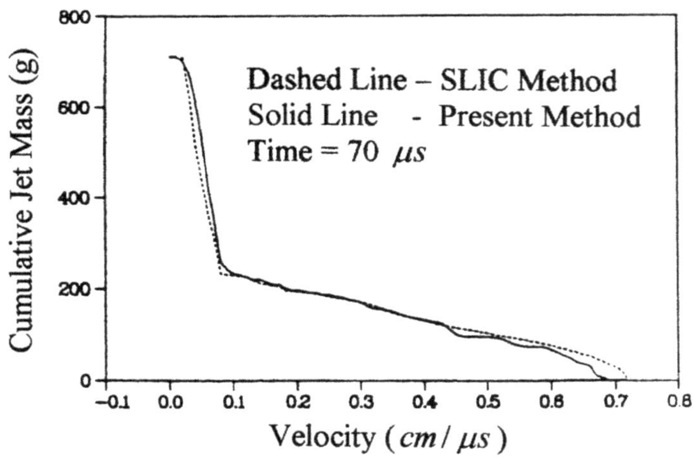

Figure 9.41. The cumulative mass versus velocity plots for the elliptic liner problem as computed using PIC (solid line) and SLIC (dashed line) methods at time 70 μs.

References

9.1 Baum, DW, Shaw, LL, Simonson, SC, and Winer, KA, Liner collapse and early jet formation in a shaped charge, Ballistics'93, 14th Int. Symp., 26–29 September 1993, Quebec City, Canada, 2, pp. 13–22 (1993).

9.2 Tactical missile warheads, edited by Carleone, J, 155, progress in astronautics and aeronautics, pp. 33 (1993).
9.3 Noh, WF, and Woodward, P, SLIC (simple line interface calculation), Report UCRL-77651, Lawrence Livermore Laboratory (1976).
9.4 Harrison, JT, A two stage, hydrodynamic, numerical technique, "HOIL" and an analysis of the results of a hemispherical, shaped charge liner collapse, 8th Int. Symp. on Ballistics, Orlando, Florida, October 23–25 (1984).
9.5 Wilkins, ML, Computer simulation of dynamic phenomena, Springer, Berlin, Germany (1999).
9.6 Simon, J, The effect of explosive detonation characteristics on shaped charge performance, BRL-MR-2414, Ballistic Research Laboratory (1974).
9.7 Harvey, W, Private communication, Los Alamos National Laboratory (1997).

CHAPTER 10

EXPLOSIVE FORMED PROJECTILE

10.1 Introduction

The explosive formed projectile (EFP) family includes a self-forging fragment, the Misznay-Schardin charge, and the P-charge. Like the shaped charge, EFP is made by a metal disk and high explosive with the single initiation point filled inside a cylindrical case. About 50–60 μs after the high explosive has detonated, the EFP disk forms a flared rod that has a velocity of 2–3 km/s. The rod will hit the tank from the top (either tank cap or body) where the metal is thin. EFP is very efficient in punching a large hole to a relatively soft target.

10.2 Calculation of Explosive Formed Projectile with Combination of Lagrangian and Eulerian Codes

This self-forging fragment has a copper liner, an aluminum, a steel case, and an Octol 75/25 high explosive. This problem is calculated using the Eulerian code and a two-dimensional Lagrangian code with the high explosive initiated at $z = 6.8834$ cm on the z-axis. For the equation of state, we used JWL EOS for the high explosive and the quadratic formula (Eq. (8.10)) for aluminum, steel, and copper. Figure 10.1(a) shows the initial setup grid for the Lagrangian code; there are 31 zones in the z direction and 90 sectors between the coordinates. Figure 10.1(b) shows the liner grid at time 12 μs, and Figs. 10.1(c) and (d) show velocity vectors for the high explosive at time 12 μs. Figure 10.2 shows the particle plots of the fragment at time 20, 40, 60, 80, and 100 μs. Figure 10.3 presents the comparison of the calculation and the experimental data for the fragment at time 164 μs. For calculations, we tried several models for the yield strength and shear modulus (including elastic-perfect plastic) and found that the constitutive models detailed by Eqs. (8.20) and (8.21) produced the best results.

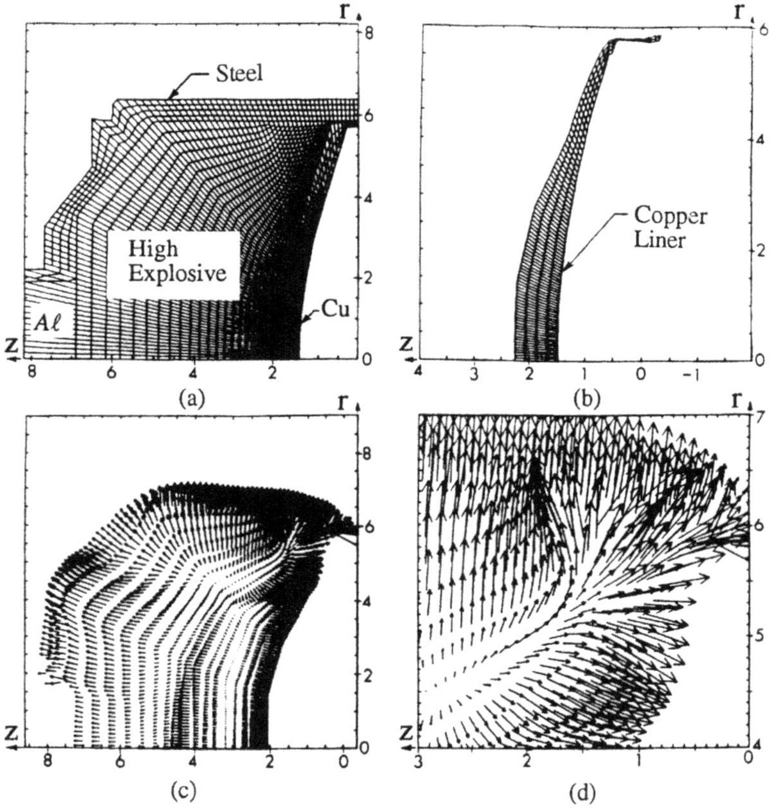

Figure 10.1. The Lagrangian meshes and the velocity vector (the coordinate units are in cm). (a) The initial Lagrangian meshes with 1 zone in aluminum, 4 zones in steel, 22 zones in high explosive, and 4 zones in copper liner. (b) The liner meshes at time 12 μs. (c) The velocity vector plot for high explosive. (d) A blowup of the velocity vectors near the upper right corner.

10.3 Copper EFP with Foam

The dimensions of the copper liner are defined inside the region of $0.0 \leq r \leq 14.224\,\text{cm}$ with

$$Z_1 = 18.2 - \sqrt{331.24 - r^2} \tag{10.1}$$

and

$$Z_2 = 0.4306 + 0.0118769 \cdot r + 0.0267 \cdot r^2 \tag{10.2}$$

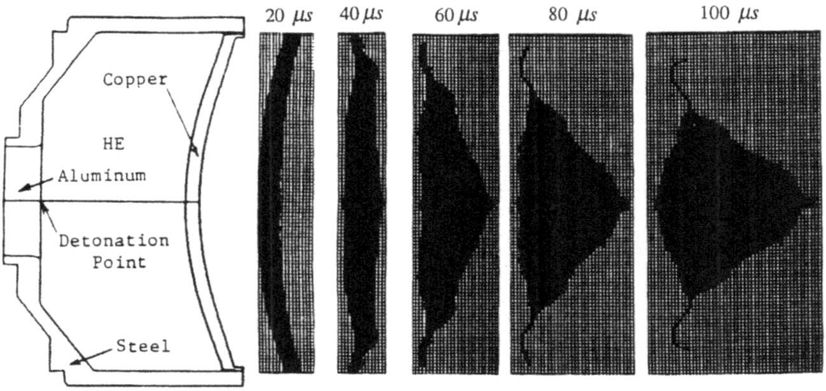

Figure 10.2. Computer simulations of the fragment at 20, 40, 60, 80 and 100 μs.

	Maximum velocity (Km/s)	Minimum velocity (Km/s)
Data	2.22	1.77
Present code	2.2	1.7

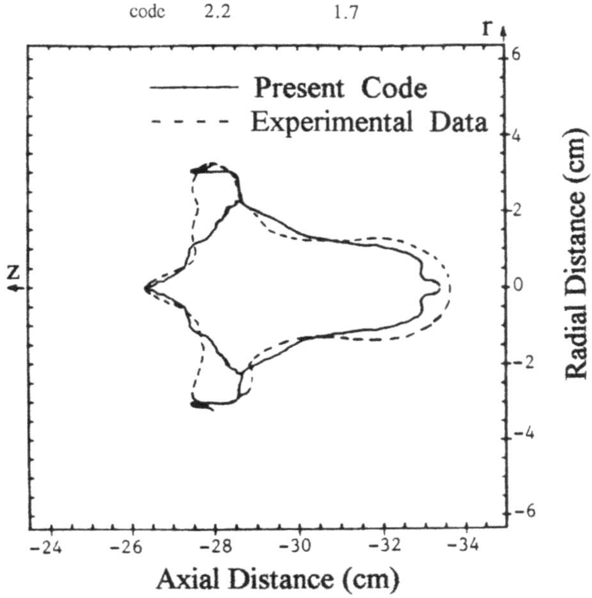

Figure 10.3. Computer simulation and the radiograph of the fragment at time 164 μs.

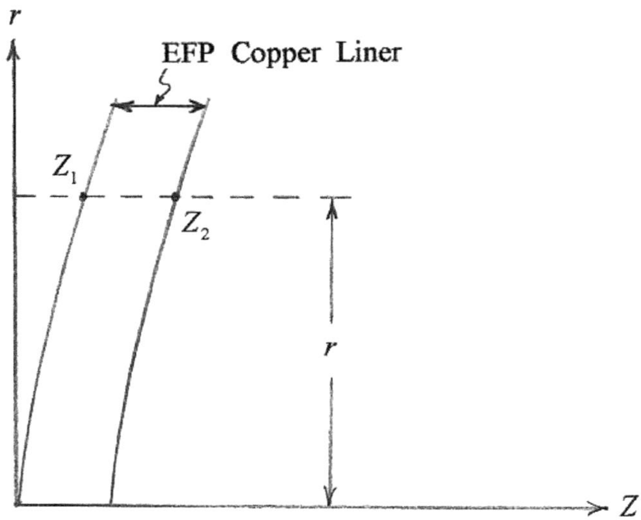

Figure 10.4. The dimensions of the EFP copper liner are described by Z_1 and Z_2 as given in Eqs. (10.1) and (10.2).

The locations of r, Z_1, and Z_2 are described in Fig. 10.4 with all units in cm. The steel case has a length of 11.4935 cm, a thickness of 1.3335 cm, an outside diameter of 15.494 cm and an inside diameter of 14.224 cm. The high explosive OCTOL is detonated with one point located at $r = 0.0$ cm and $Z = -11.4935$ cm. This copper EFP has a diameter of 14.224 cm with the liner sandwiched between the high explosive and the low density foam (polyurethane with density between 0.1 g/cm^3 to 0.2 g/cm^3) as shown in the upper plot of Fig. 10.5.

The pressure contour at time $t = 10.0\,\mu$s is shown in the lower plot of Fig. 10.5 with the maximum pressure located at the detonation shock front. The velocity vectors at time $t = 10.0\,\mu$s are shown in the upper plot of Fig. 10.6. Again, the maximum velocity vectors coincide with the shock front. The pressure contour at time $t = 25.0\,\mu$s is given in the lower plot of Fig. 10.6. Near the central region, the interaction of the reflection wave from the stainless steel case and the copper liner produces an interesting picture.

At time $t = 410.0\,\mu$s, the calculated results of the EFP are shown in the top plot of Fig. 10.7 along with the experimental results shown in the bottom plot of Fig. 10.7. The bottom plot of Fig. 10.7 shows the internal and external contours determined by the least square fitting an analytical model to the film density profile. The tomography reconstruction in the

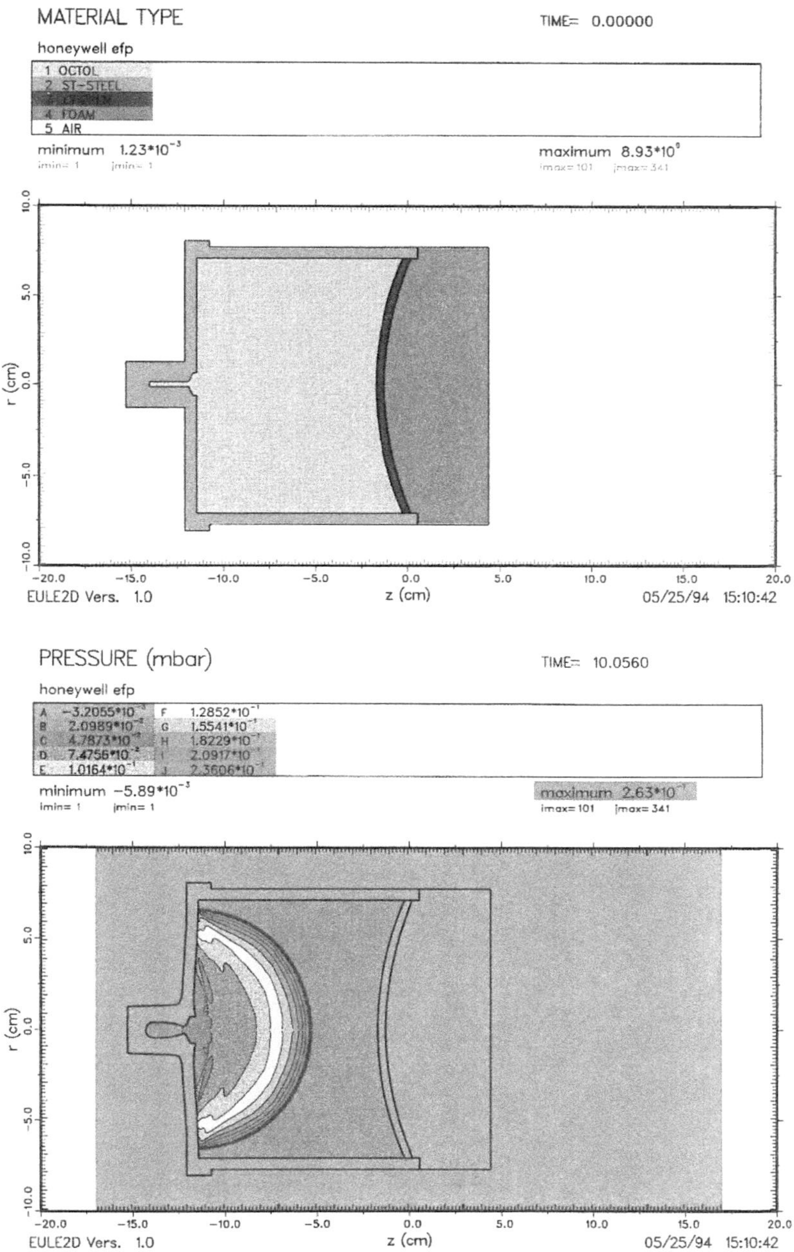

Figure 10.5. Honeywell copper EFP initial setup (upper) and pressure contour at time 10 μs (bottom).

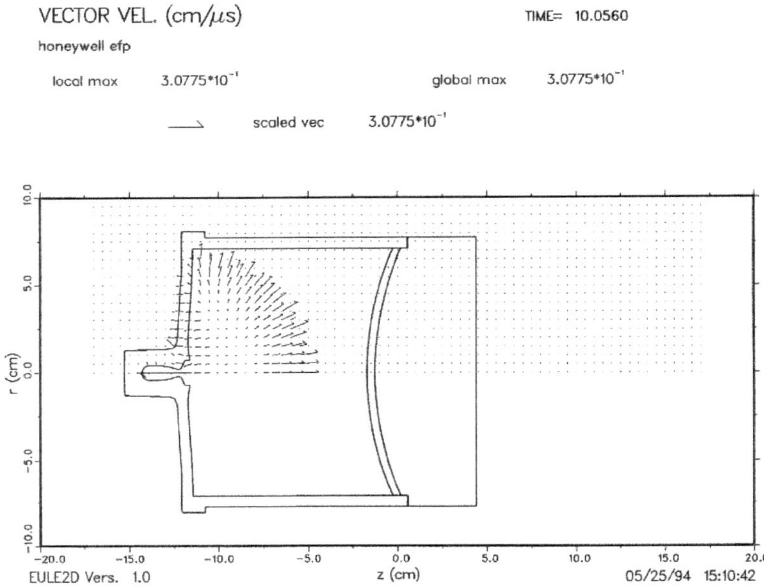

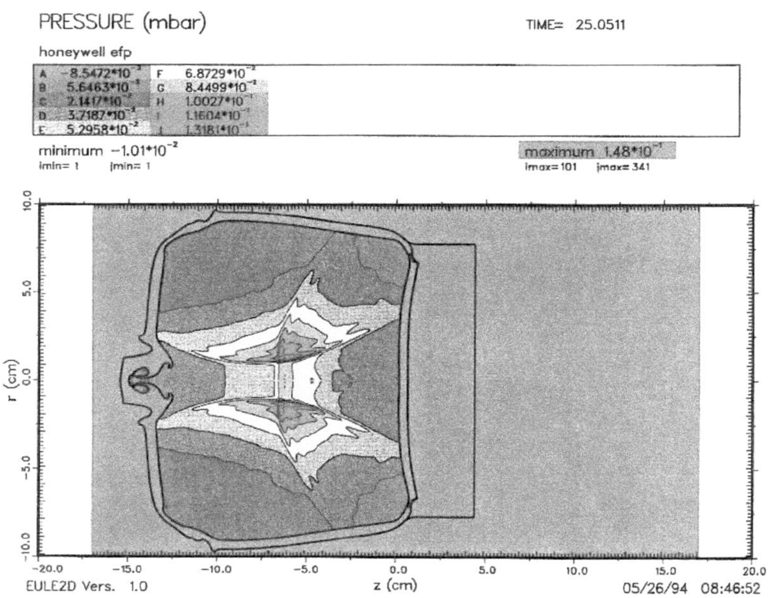

Figure 10.6. Honeywell copper EFP velocity vector at time 10 μs (upper) and pressure contour at time 25 μs (bottom).

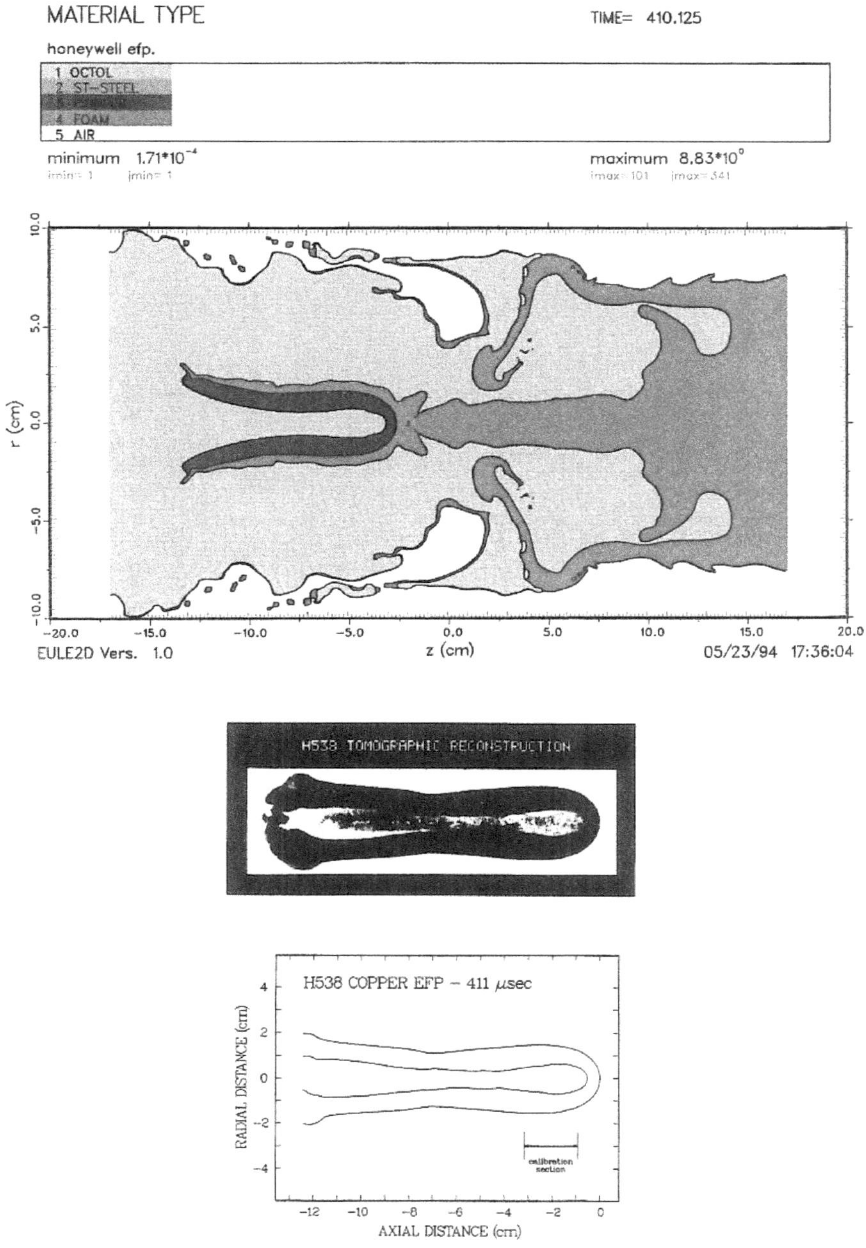

Figure 10.7. Honeywell copper EFP at time 410 μs. The discrepancy of the liner length between the experiment and code calculation is about 4%.

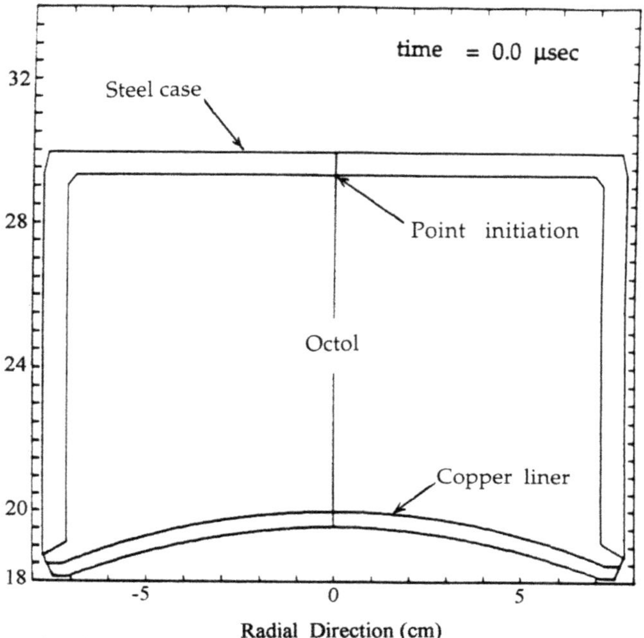

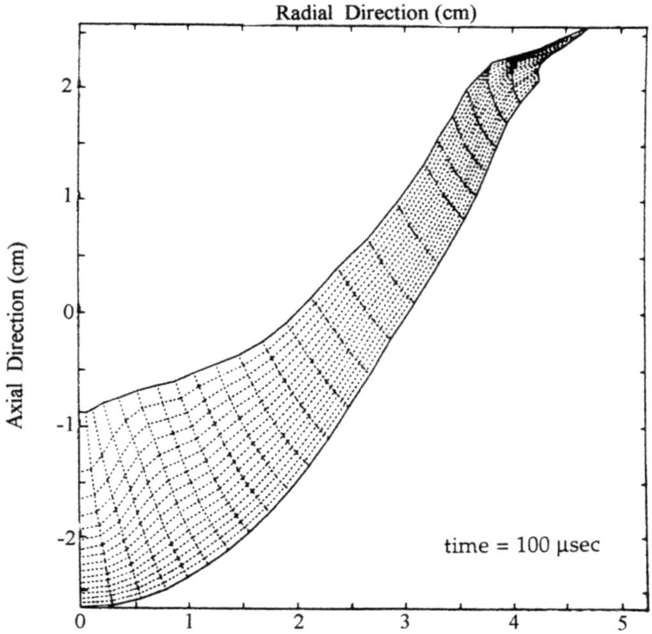

Figure 10.8. Initial setup for Honeywell EFP (upper) and the formation of the liner at time 100 μs (bottom) as computed by the 2D Lagrangian code.

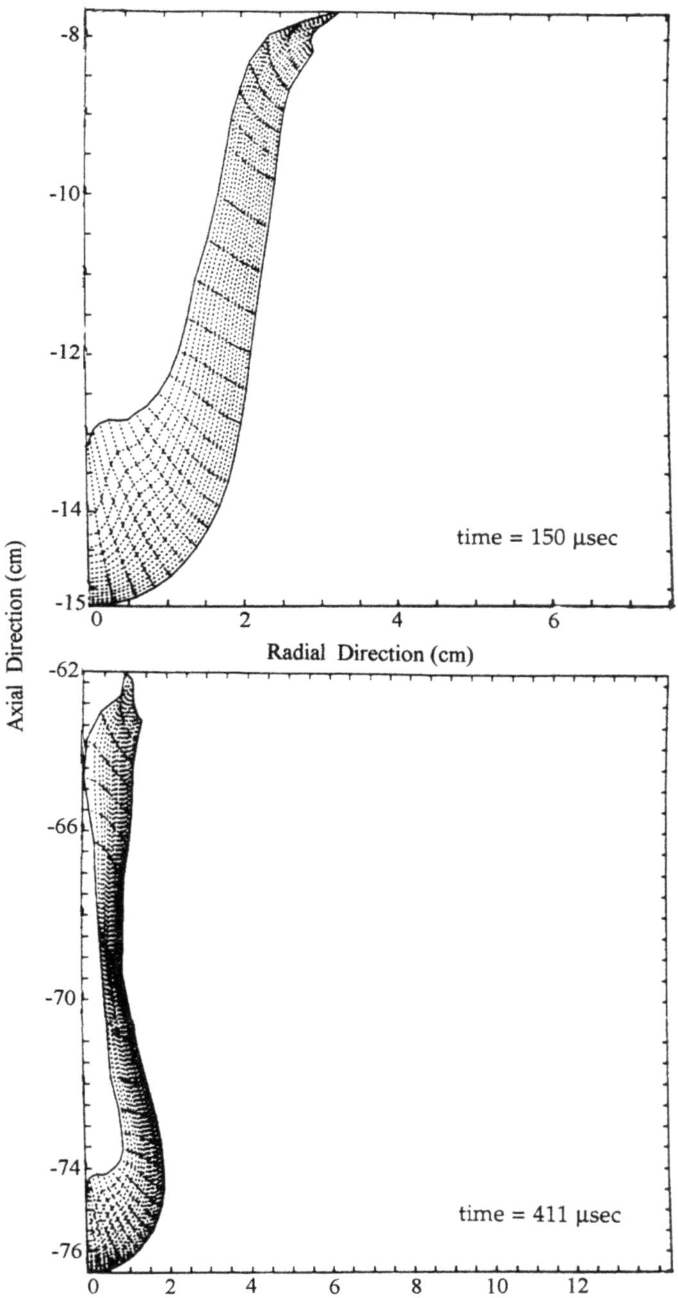

Figure 10.9. The EFP formation at time 150 μs (upper) and time 411 μs (lower).

center plot of Fig. 10.7 represents a gray-scale map of the EFP's material densities in a cross-section view.

10.4 Copper EFP without Foam

The dimensions of the copper liner are the same as those described in Sec. 10.3 except that there is no foam attached to the liner. Here, the formation of the EFP by using only the 2D Lagrangian code is computed. The initial setup is shown in the upper plot of Fig. 10.8 with the EFP formation at time $t = 100\,\mu s$ at the lower plot. Again, the EFP formations of time $t = 150\,\mu s$ and $411\,\mu s$ are shown in the upper and lower plots of Fig. 10.9. The shape of the EFP formation at time $t = 411\,\mu s$ is very close to the experimental results.

10.5 Non-axisymmetric Tantalum EFP Warhead

The non-axisymmetric tantalum EFP warhead consists of an open-sided cylindrical case of steel, a charge of OCTOL explosive, and a rectangular shaped liner of tantalum, approximately 80 by 120 mm and nominally 2 mm thick as shown in Fig. 10.10. The computer simulation of this problem

Figure 10.10. The tantalum EFP warhead. It consists of an open-sided cylindrical case of steel, a charge of OCTAL explosive, and a rectangular shaped liner of tantalum, approximately 80 by 120 mm and nominally 2 mm thick.

is performed using a 3D Eulerian code with the same numerical method described in Chapter 12. The gray-scale plots of the material densities in a planar cut through the major axis of the warhead are shown in Fig. 10.11. The high explosive is detonated at the center of the curved face opposite the liner. After 5 to 10 μs, the spherically diverging detonation wave impacts the liner, launching it in a trajectory perpendicular to the cylindrical axis of the warhead. The shock wave front and the shapes of the EFP liner at times 0.0, 5.0, 10.0, 15.0, 20.0, and 30.0 μs are shown in Fig. 10.11. The synthetic radiograph of the computer simulation at time 22.0 μs is shown in Fig. 10.12. The Eulerian 3D code calculations of the EFP formation at time 0.0, 50.0, and 100.0 μs are shown in Fig. 10.13. Again, Fig. 10.14 shows the formation of the EFP at times 200.0, 400.0, and 750.0 μs. The liner material along both the major and minor axes folds backward and stretches to form a projectile which attains equilibrium shape in about 500.0 μs. Figure 10.15 shows the synthetic framing camera record from the code calculation for density at time $t = 750.0$ μs.

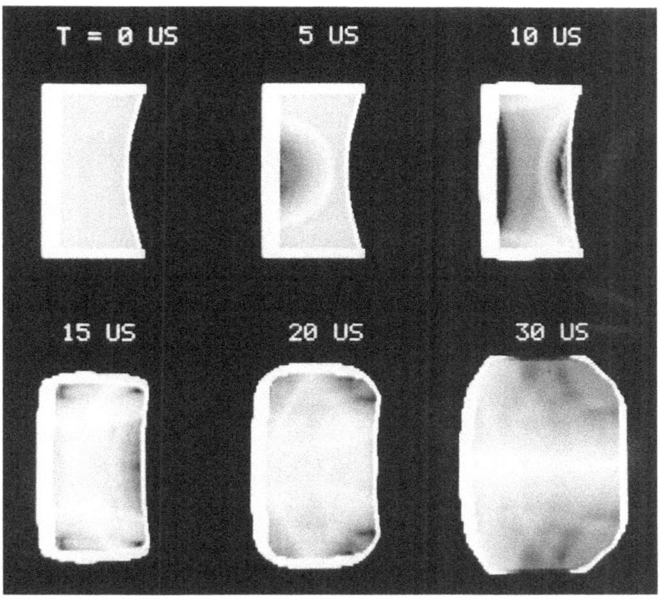

Figure 10.11. The gray-scale plots of the material densities in a planar cut through the major axis of the warhead. The high explosive is detonated at the center of the curved face opposite the liner. After 5 to 10 μs, the spherically diverging detonation wave impacts the liner, launching it in a trajectory perpendicular to the cylindrical axis of the warhead.

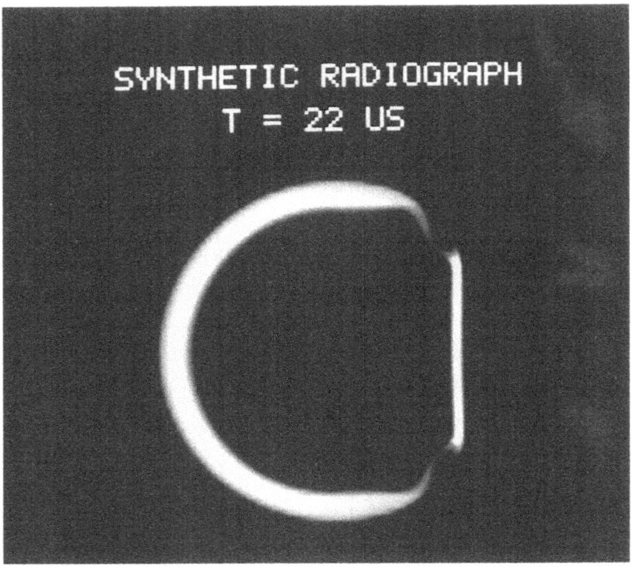

Figure 10.12. The synthetic radiograph of the code calculation at time 22 μs.

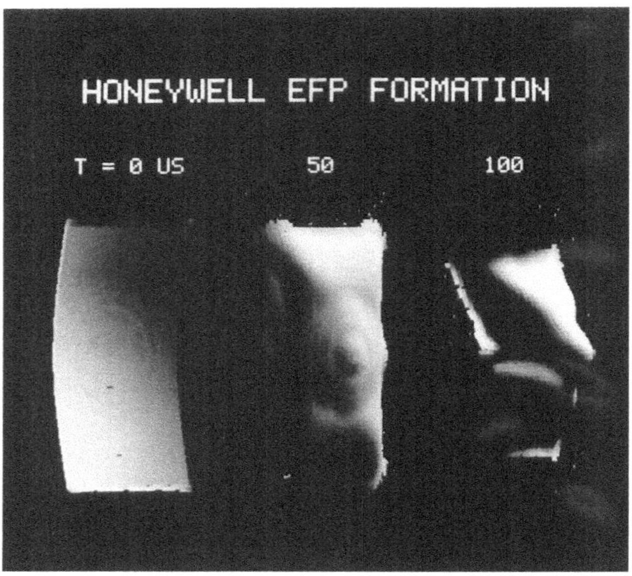

Figure 10.13. The Eulerian 3D code calculations at time 0, 50 and 100 μs.

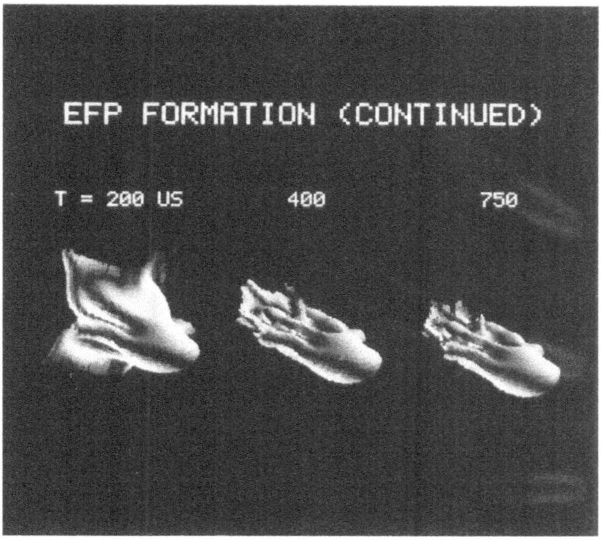

Figure 10.14. The Eulerian 3D calculations at time 200, 400 and 700 μs. The liner material along both the major and minor axes folds backward and stretches to form a projectile, which attains an equilibrium shape in about 500 μs.

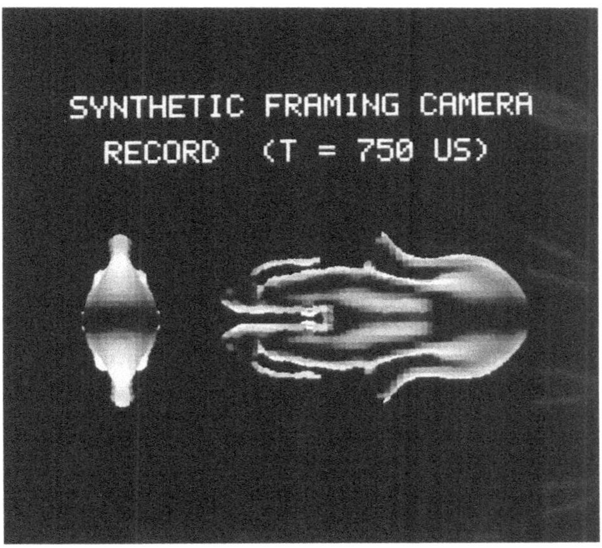

Figure 10.15. The synthetic framing camera record from the code calculation for density at time 750 μs.

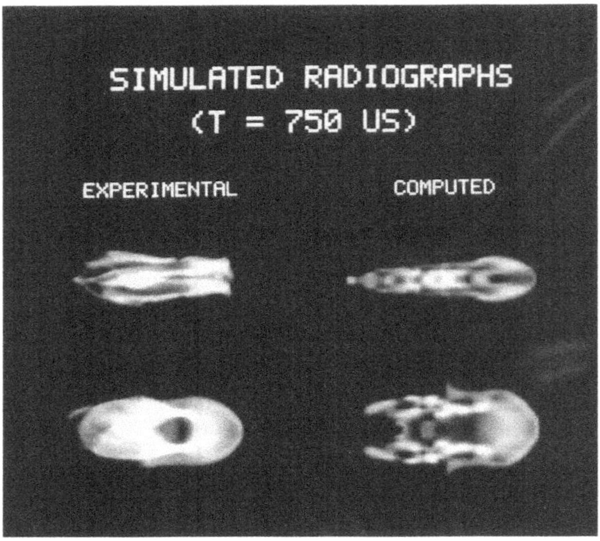

Figure 10.16. The results of radiographic experiments on two identical penetrators are shown on the left side. The two images correspond to radiographs taken in orthogonal views of the somewhat flattened penetrator (traveling from left to right). The top view was taken through the thick lateral dimension and the bottom view shows the thin dimension. In both cases, the EFPs were about 9 cm long and had a velocity of 2.3 km/s. The Eulerian 3D code calculations are shown on the right. The synthetic radiographs agree very well with those in the experiment in the forward portion of the body. The flattened nose of the EFP, shown at top left, resulted from an inadvertent experimental problem.

The results of radiographic experiments on two identical penetrators are shown on the left side of Fig. 10.16. The two images correspond to radiographs taken in orthogonal views of the somewhat flattened penetrator that travels from left to right. The top view was taken through the thick lateral dimension and the bottom view through the thin dimension. In both cases, the EFPs were about 9.0 cm long and had a velocity of 2.3 km/s. The Eulerian 3D code calculations are shown on the right. The synthetic radiographs agree very well with those in the forward portions of the body in the experiment. The flattened nose of the EFP, shown at top left, resulted from an inadvertent experimental problem. Figure 10.17 shows the X-ray radiographs at different time frames. The front views are shown at the left and the side view on the right.

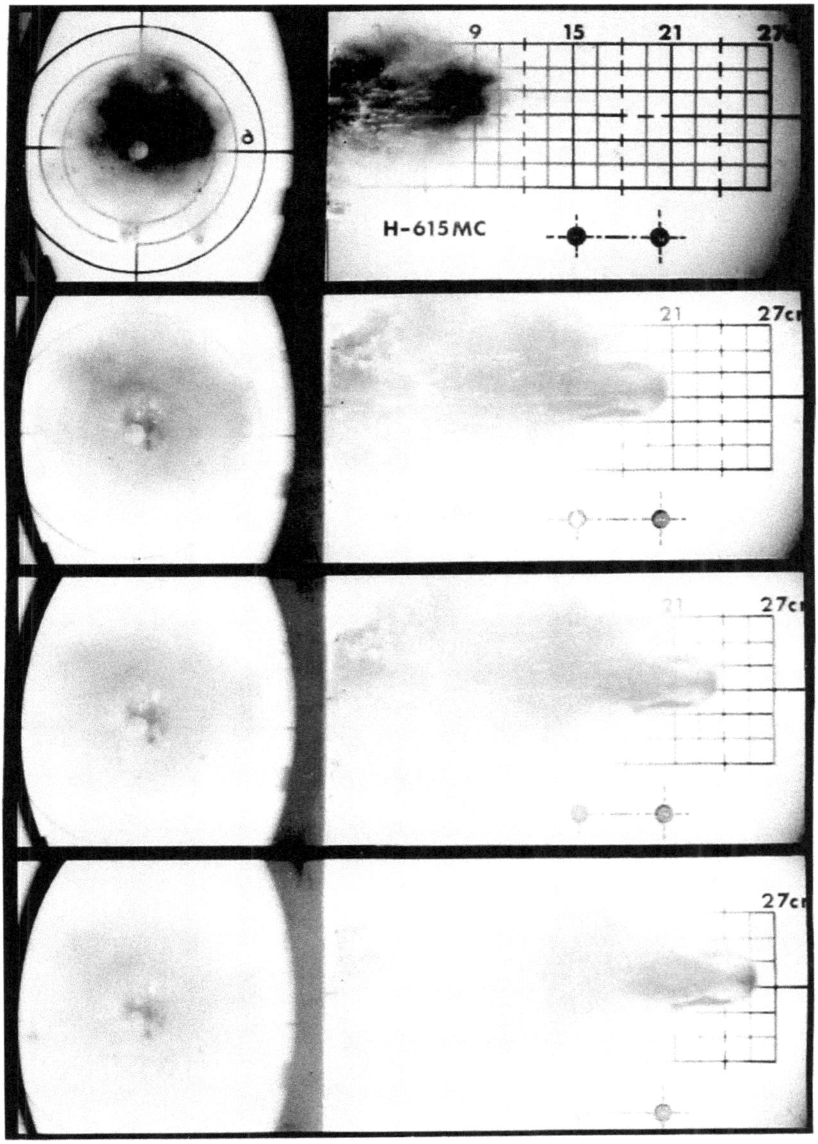

Figure 10.17. The X-ray radiographs at different time frames. The front views are shown at left and side views on the right.

CHAPTER 11

THE PENETRATION OF SHAPED CHARGE JET

11.1 Introduction

The penetration of the target from a shaped charge jet is primarily due to the penetration stress exceeding the material strength of the target. The heat generated during the penetration process also helps to soften the target and causes the erosion of the target and penetrator. Penetrators can be classified as conical, trumpet, bell, and hemispherical shaped charges and explosive formed projectiles (EFP). A conical shaped charge can penetrate the target very deeply but the hole diameter is generally small, while an EFP can create a large hole with a shallow penetration depth. A hemispherical shaped charge, somewhere between EFP and conical shaped charge, will form a thick jet with a tip velocity around 5–6 km/s. Therefore, a hemispherical shaped charge is suitable for long standoff targets. Some of the copper hemispherical charges have been observed with low density along the center core of the thick jet.

In the following sections, the calculations of the penetration due to high speed rods and the shaped charge jets will be discussed.

11.2 Calculations of Tungsten Rod Penetrating Aluminum Target

The first test problem is the calculation of a tungsten rod (diameter = 0.28 cm, length = 2.8 cm, and velocity = 3.4 km/s) penetrating a semi-infinite aluminum block. The experimental data are given by Hohler and Stilp [11.1], and the computer simulations have also been studied by Cullis and Nash [11.2] using HULL code.

The equation of state for both tungsten and aluminum is the quadratic form of Eq. (8.10). For tungsten, the coefficients are: $A_1 = 21.67419$; $A_2 = 14.93338$; $B_0 = 10.195827$; $B_1 = 12.263234$; $B_2 = 9.3051515$; $C_0 = 0.33388437$; $C_1 = 0.48248861$; $D_0 = 7.0$; and $\rho_0 = 19.17$. For aluminum, $A_1 = 1.1867466$; $A_2 = 0.762995$; $B_0 = 3.4447654$; $B_1 = 1.5450573$; $B_2 = 0.96429632$; $C_0 = 0.43381656$; $C_1 = 0.54873462$; $D_0 = 1.5$; and $\rho_0 = 2.806$.

The constitutive relations for the shear modulus G, and the flow stress Y, are given by Eqs. (8.20) and (8.21). For tungsten, the coefficients are: $G_0 = 1.6$; $Y_0 = 0.022$; $\beta = 7.7$; $n = 0.13$; $Y_{\max} = 0.04$; $b = 1.375$; $h = -0.0001375$; $q = 1.0$; $f = 0.001$; $g = 0.001$; $R' = 0.000008671$; $Tm_0 = 4520$; $\gamma - a = 0.27$; and $a = 1.4$. For aluminum, $G_0 = 0.276$; $Y_0 = 0.0029$; $\beta = 125.0$; $n = 0.1$; $Y_{\max} = 0.0068$; $b = 7.971$; $h = -0.0067159$; $q = 1.0$; $f = 0.001$; $g = 0.001$; $R' = 0.000008326$; $Tm_0 = 1220.0$; $\gamma - a = 0.49$; and $a = 1.7$.

Figure 11.1 shows the initial problem geometry with the tungsten rod of diameter 0.28 cm and length 2.8 cm located between $z = 1.7$ cm and $z = 4.5$ cm. The aluminum block has a dimension of 14.28 cm × 18.0 cm ($r \times z$) indicated by ABCD. Because this is an axisymmetrical problem, we only simulate the half plane between $r = 0.0$ cm and $r = 8$ cm with 160 zones or grid size $\Delta r = 0.5$ mm. The computational window in the

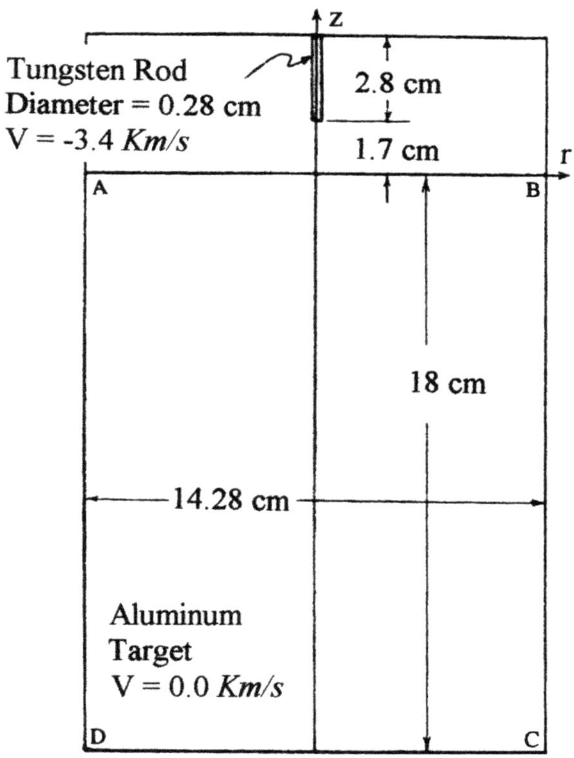

Figure 11.1. Initial conditions of the tungsten rod and aluminum target.

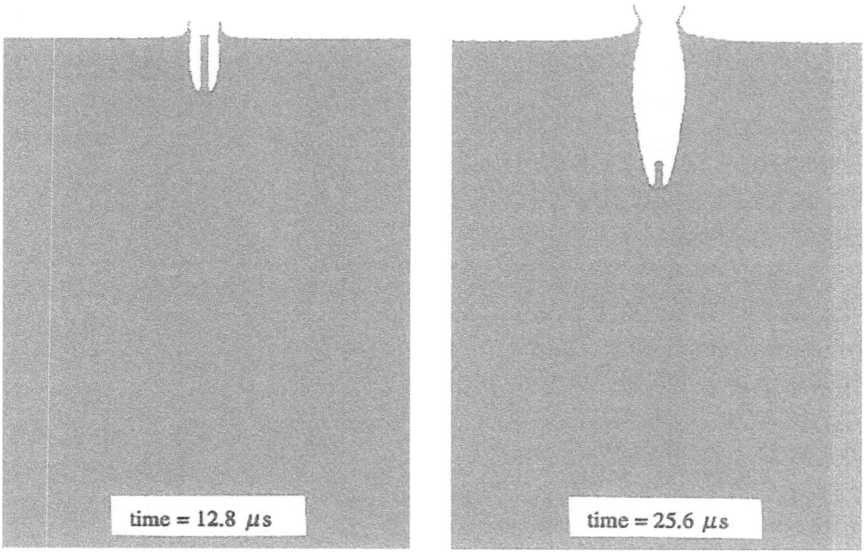

Figure 11.2. The crater sizes and the penetration depths at time 12.8 and 25.6 μs.

z-direction is between $z = -20$ cm and $z = 6.0$ cm with 260 zones, or grid size $\Delta z = 1$ mm.

The crater sizes and penetration depths at time $t = 12.8$, 25.6, 38.4, and 44.7 μs are shown in Figs. 11.2 and 11.3. Some instability of the crater surface near $z = -0.5$ cm starts to grow at $t = 44.7$ μs.

The maximum penetration depth reaches 10 cm at time 45 μs as compared with the experimental data of 10.1 cm and HULL calculation of 10.6 cm. The crater diameter also has a steady-state value of 1.4 cm that is in good agreement with the HULL calculation (also 1.4 cm) as well as experimental result of 1.27 cm. The total kinetic energy is approaching zero at a time greater than 80 μs, while the total internal energy reaches a maximum of 0.025 $\text{Mbar} - \text{cm}^3/\text{cm}^3$ (see Fig. 11.4).

11.3 Calculations of Copper Rod Penetrating Steel Plate

The second problem is a copper rod (diameter = 2 cm, length = 27 cm, and velocity = 5 km/s) penetrating a motionless steel disk that is 5 cm thick and 30 cm in diameter. The initial condition and setup are shown in Fig. 11.5. The equation of state for the copper is given by Eq. (8.10)

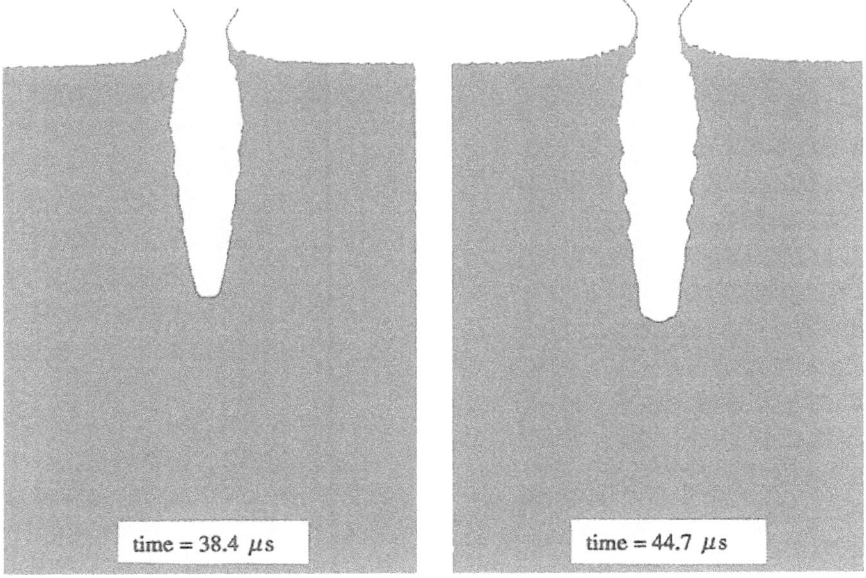

Figure 11.3. The crater sizes and the penetration depths at time 38.4 and 44.7 μs.

with the parameters: $A_1 = 4.9578323$; $A_2 = 3.6883726$; $B_0 = 7.4727361$; $B_1 = 11.519148$; $B_2 = 5.5251138$; $C_0 = 0.39492613$; $C_1 = 0.52883412$; $D_0 = 0.6000001$; and $\rho_0 = 8.899$. The formulations of the constitutive relations for shear modulus and yield strength are given by Eqs. (8.20) and (8.21) and the necessary parameters for copper are: $G_0 = 0.477$; $Y_0 = 0.0012$; $\beta = 36$; $n = 0.45$; $Y_{\max} = 0.0064$; $b = 3.14465$; $h = -0.000377$; $q = 1.0$; $f = 0.001$; $g = 0.001$; $R' = 0.00001164$; $Tm_0 = 1790.0$; $\gamma - a = 0.52$; $a = 1.5$; $\alpha_0 = 0.68462$; $\alpha_1 = -0.00868$; $\alpha_2 = 0.26429$; $\alpha_3 = 0.0068119$; $\beta_0 = 1.0$; $\beta_1 = 0.00903$; $\beta_2 = 0.170902$; and $\beta_3 = 0.0417657$.

For steel, the coefficients for the *Mei–Grüneisen* equation of state, i.e. Eq. (8.1), are: $\rho_0 = 7.9$; $C_0 = 4.57$; $s = 1.49$; and $\gamma_0 = 2.17$. The constitutive parameters used for Eqs. (8.40)–(8.41) for steel are: $G_0 = 0.77$; $Y_0 = 0.0034$; $\beta = 43.$; $n = 0.35$; $C = 2.26$; $m = 0.000455$; and $T_{melt} = 0.0$.

The initial problem setup for the PIC code has a window of 15 cm × 40 cm ($r \times z$) with a computational mesh of 75 × 200 zones (uniform grid size of 2 mm). Figure 11.6 shows the shapes of the projectile and the target at time 16 and 32 μs. At time 16 μs, the rugged surfaces on both the upper

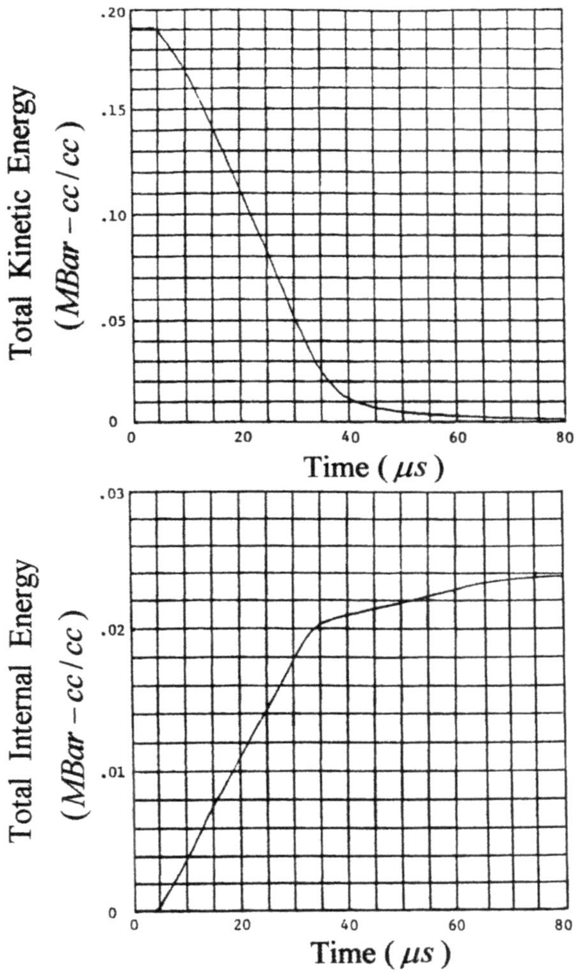

Figure 11.4. The total kinetic and internal energy.

and lower sides of the steel plate indicate that the spallation occurred near the surface of the plate. This phenomena is more pronounced at time 32 μs. The copper rod has penetrated through the steel plate at time 40 μs (see Fig. 11.7) with some perforated residue of steel. The code calculations were stopped at time 56 μs after the rod had punched through the plate. The length of the copper rod is only 15.26 cm with 11.74 cm eroded. There are

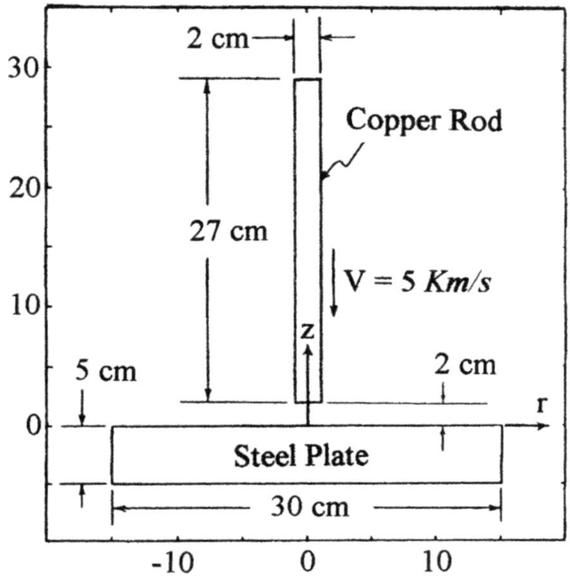

Figure 11.5. The initial conditions for the copper rod and the steel plate.

some branches at the upper part of the rod indicating that the rarefaction waves have produced spallation.

11.4 Tantalum Shaped Charge Penetrates into Steel Block

As shown in the upper plot of Fig. 11.8, the shaped charge uses RDX as explosive and uses the conical tantalum as liner, which is thicker near the apex section. The steel case is as thick as the high explosive that detonates at $r = 0.0\,\text{cm}$ and $z = -4.0\,\text{cm}$. The distance between the shaped charge and the steel block is 2.5 cm (stand-off distance). The constant pressure regions at time 30 μs are shown in the lower plot of Fig. 11.8 with the maximum pressure located inside the steel block and in the front of the jet tip as expected. The low pressure regions are located at the crater area inside the steel block and at the slug section of the tantalum jet. The penetration of the jet into the steel at time $t = 50.0\,\mu\text{s}$ is shown in the upper plot of Fig. 11.9 and the constant pressure regions are shown in the lower plot for time $t = 70.0\,\mu\text{s}$. The lowest pressure region is located near the free surface of the steel block between $z = 5.0\,\text{cm}$ and $z = 10.0\,\text{cm}$. This suggests that a spallation may take place inside this area.

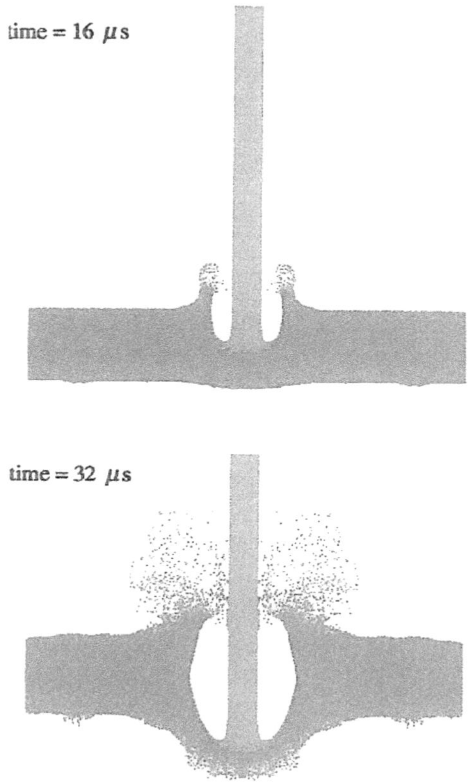

Figure 11.6. The shapes of the penetrator and the target at times 16 and 32 μs.

11.5 Tantalum Shaped Charge Penetrates into Steel Plate, Water, and Another Steel Plate

This tantalum shaped charge is identical to the one described in the previous section as shown in the upper plot of Fig. 11.10. The distance between the shaped charge and the first steel plate is also 2.5 cm. The thicknesses are 0.37 cm for the left steel plate, 2.69 cm for the water column, 0.65 cm for the right steel plate, and 4.91 cm for the air column located between $z = 5.09$ cm and $z = 10.0$ cm. The constant density regions at time $t = 18.0$ μs are shown in the lower plot of Fig. 11.10 along with the tantalum jet penetrating the left steel plate.

Figure 11.11 shows the constant pressure plot (upper one) at time $t = 28.0$ μs and the material type plot at time $t = 40.0$ μs (lower one). The size

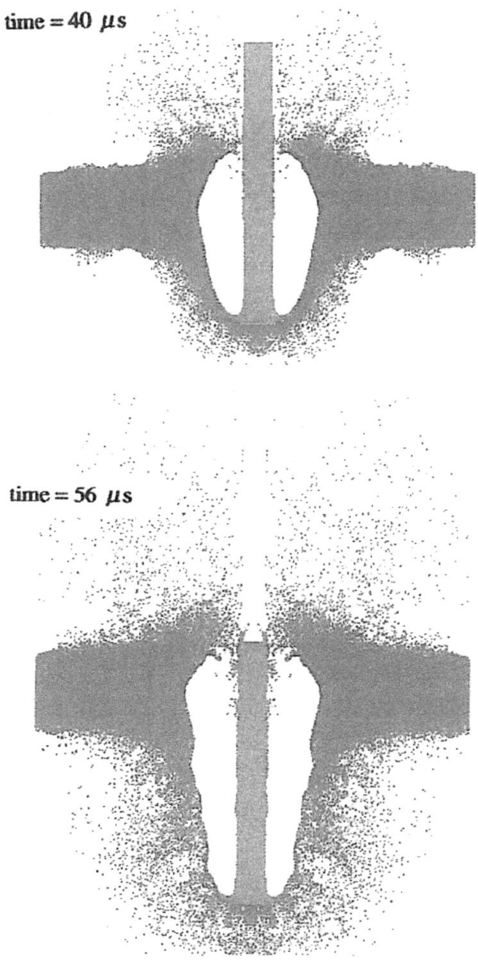

Figure 11.7. The shapes of the penetrator and the target at time 40 and 56 µs.

of the hole in the left steel plate is much smaller compared with that of the right steel plate.

11.6 Copper-lead Shaped Charge Penetrates into Steel Block

The upper plot of Fig. 11.12 shows the initial setup of the copper-lead liner shaped charge which has a hole near the apex region. The detailed

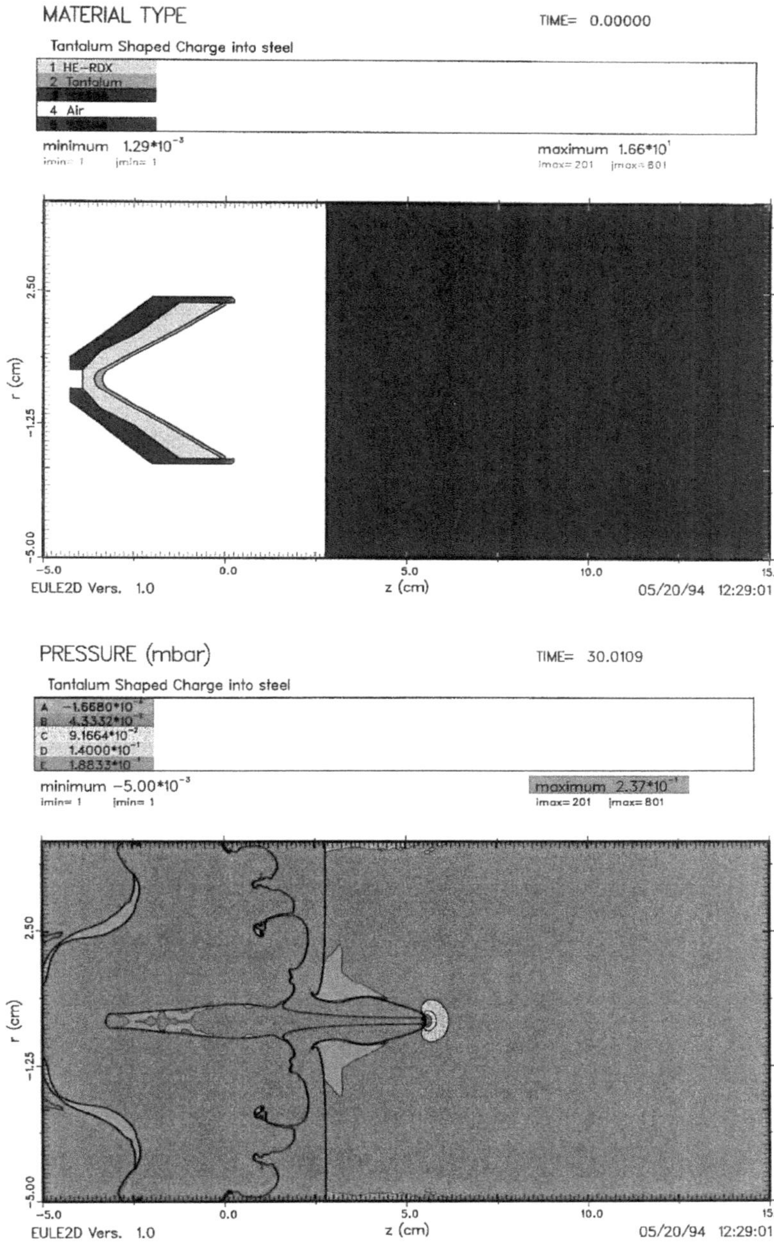

Figure 11.8. Tantalum shaped charge penetrates into the steel block with initial setup (top) and pressure contour at time 30 μs (bottom).

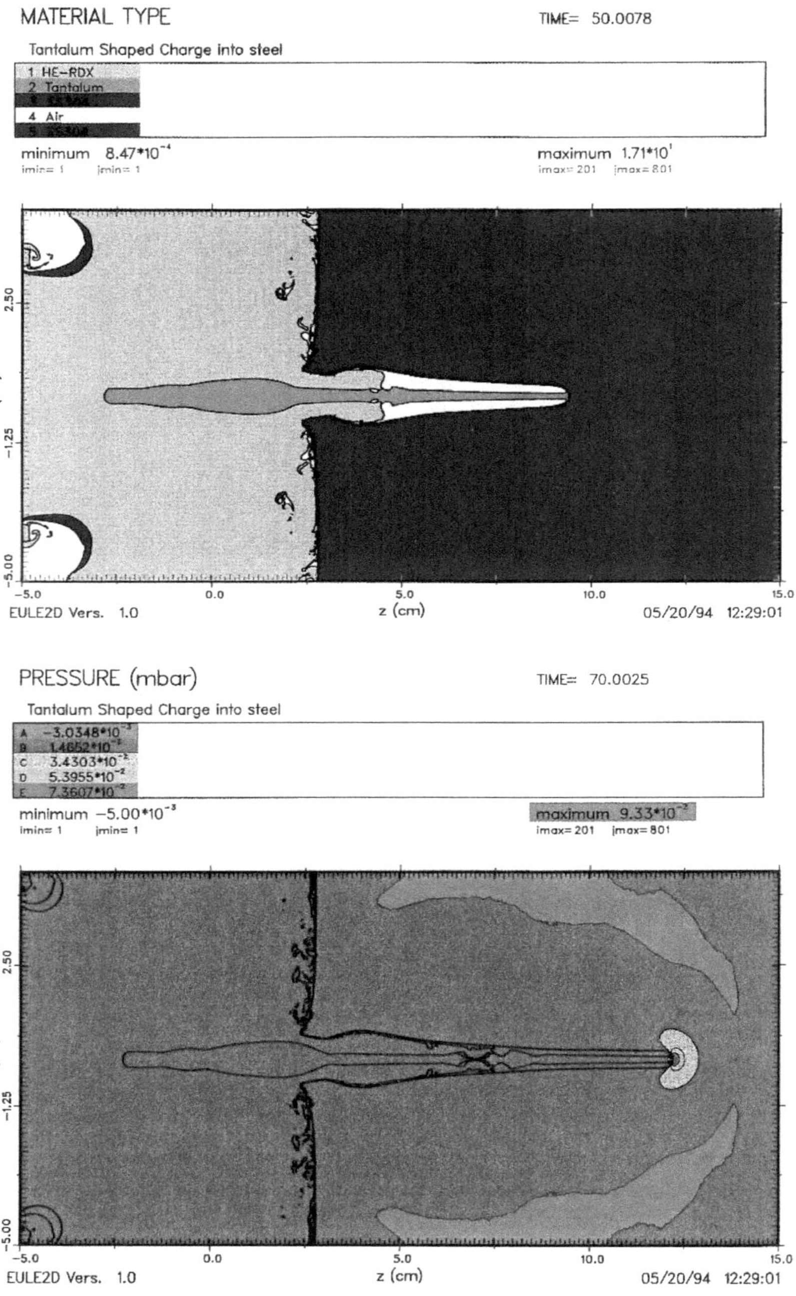

Figure 11.9. Tantalum shaped charge penetrates into the steel block at time 50 μs and pressure contour at time 70 μs.

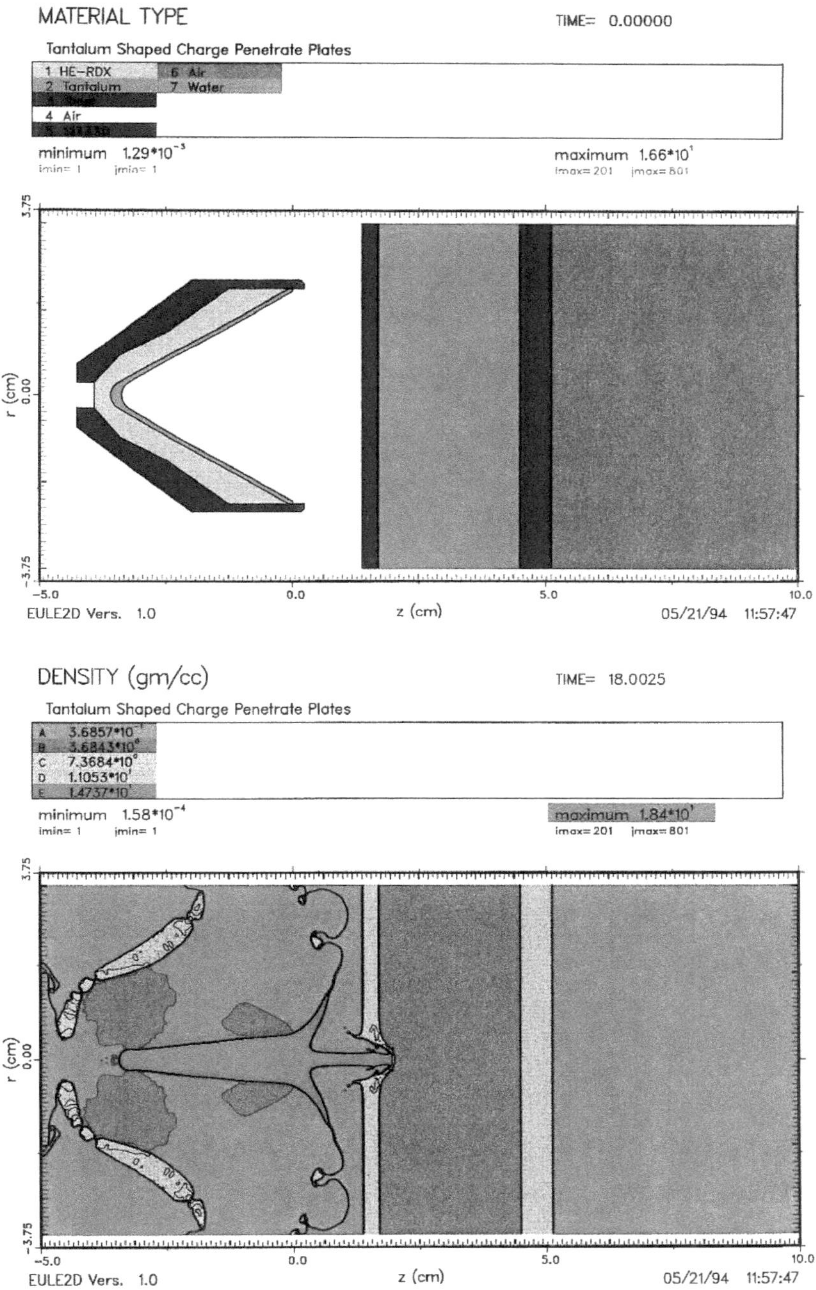

Figure 11.10. Tantalum shaped charge penetrates thin steel plate, water and thick steel plate with initial setup (top) and density contour at time 18 µs (bottom).

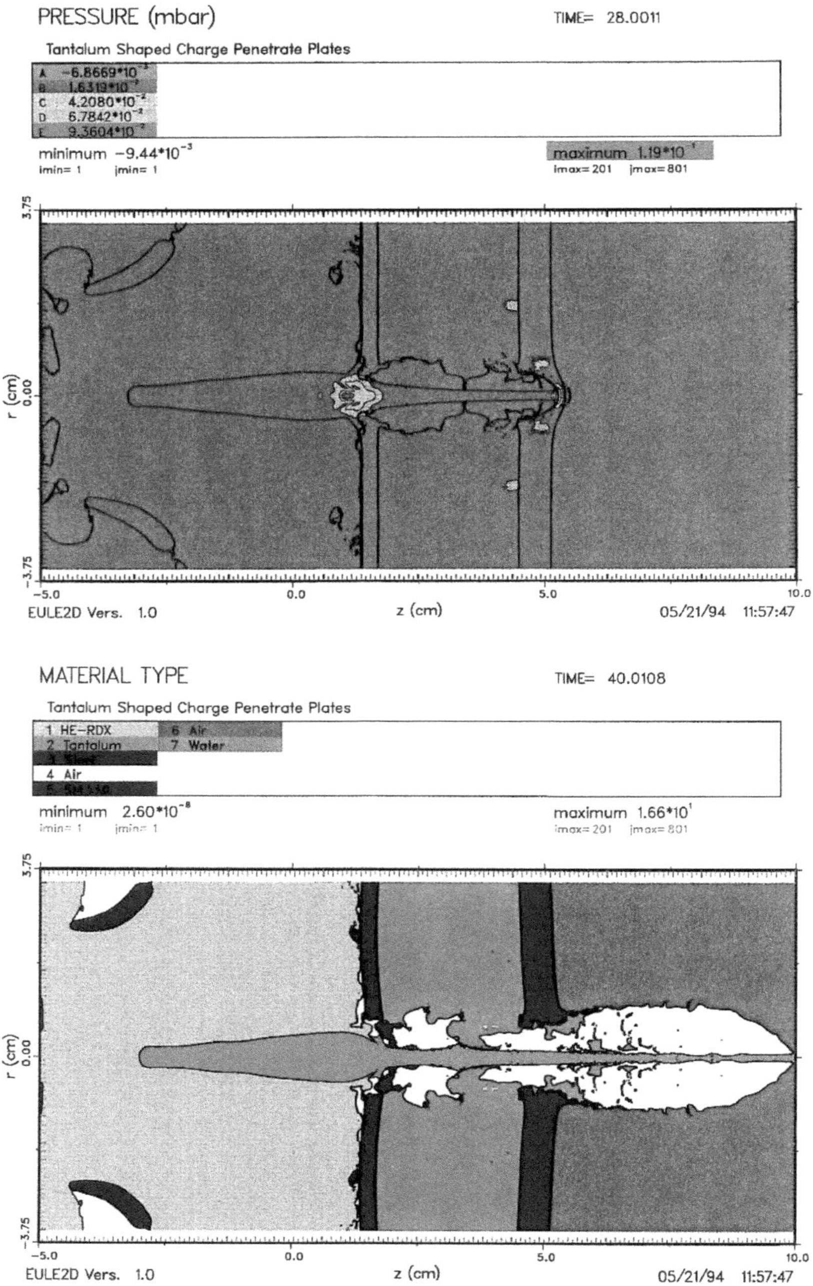

Figure 11.11. Tantalum shaped charge penetrates steel plates and water. Pressure contour at time 28 μs (top) and material type at time 40 μs (bottom).

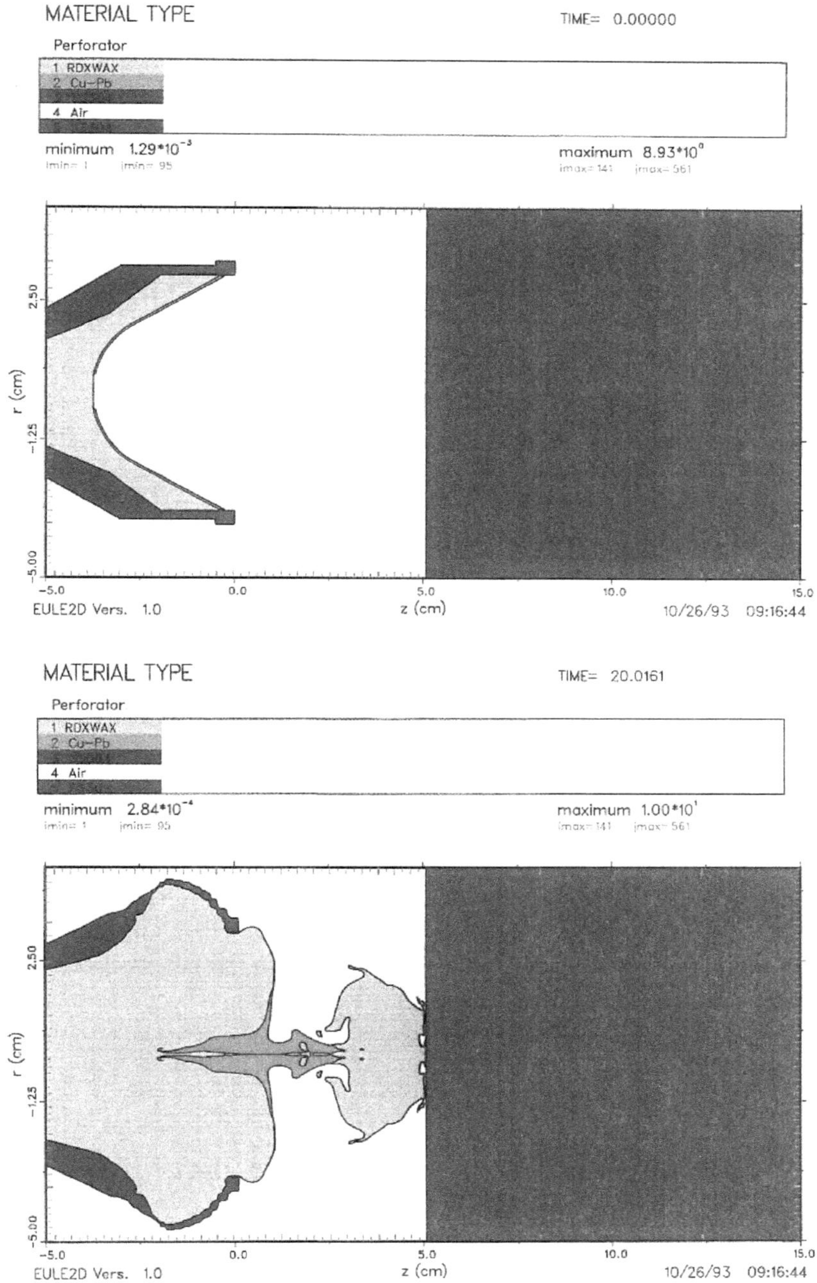

Figure 11.12. The stand-off is 5 cm with the initial setup showing in the upper plot and the jet formation at time 20 μs in the lower plot.

dimensions of this shaped charge are given in Chapter 12, Sec. 12.5 with the distance of 5 cm between the shaped charge and the steel block.

The jet formation at time $t = 20.0\,\mu s$ with a separation gap along the z-axis is given at the lower plot of Fig. 11.12. The penetrations of the steel block by the jet are given in Fig. 11.13 for time $t = 40.0\,\mu s$ (upper plot) and $t = 70.0\,\mu s$ (lower plot).

11.7 Bi-conical Copper-lead Powder Shaped Charge Penetrates into Steel Block

The liner of this bi-conical shaped charge is made of copper-lead powder with initial setup shown in the upper plot of Fig. 11.14. The RDX explosive is detonated at one point located at $r = 0.0$ cm and $z = -4.0$ cm with the stand-off of 2.5 cm from the steel block. The material type plot and the crater created by the jet penetration at time $t = 30.0\,\mu s$ are shown in the lower plot of Fig. 11.14. Figure 11.15 shows the jet penetrations at time $t = 60.0\,\mu s$ (upper plot) and $t = 100.0\,\mu s$ (lower plot).

11.8 Bi-conical Copper Shaped Charge Penetrates into Thin Steel Plate, Water, Thick Steel Plate and Rock

The upper plot of Fig. 11.16 shows the initial dimensions of the copper shaped charge and the steel plates that are identical to Fig. 11.10 except that, in this problem, the column between $z = 5.09$ cm and $z = 10.0$ cm is Berea rock and the liner is made of copper powder. The RDX explosive also detonated at the point located at $r = 0.0$ cm and $z = -4.0$ cm. The lower plot of Fig. 11.16 shows the material type at time $t = 15.0\,\mu s$. The double inverse velocity gradients produce a small hole in the left steel plate and a large hole in the right steel plate. Figure 11.17 shows the penetration of the copper jet into the steel plate and the Berea rock at time $t = 30.0\,\mu s$ (upper plot) and $t = 40.0\,\mu s$ lower plot).

11.9 Viper Shaped Charge Penetrates into Steel Block

The initial dimensions of the Viper shaped charge are described in Chapter 9, Sec. 9.4.1. In this penetration calculation, the Viper charge is set at 12.50 cm from the target, i.e. the steel block. At time $t = 30.0\,\mu s$ after the LX-14 detonated, the copper jet just touches the target as shown in the

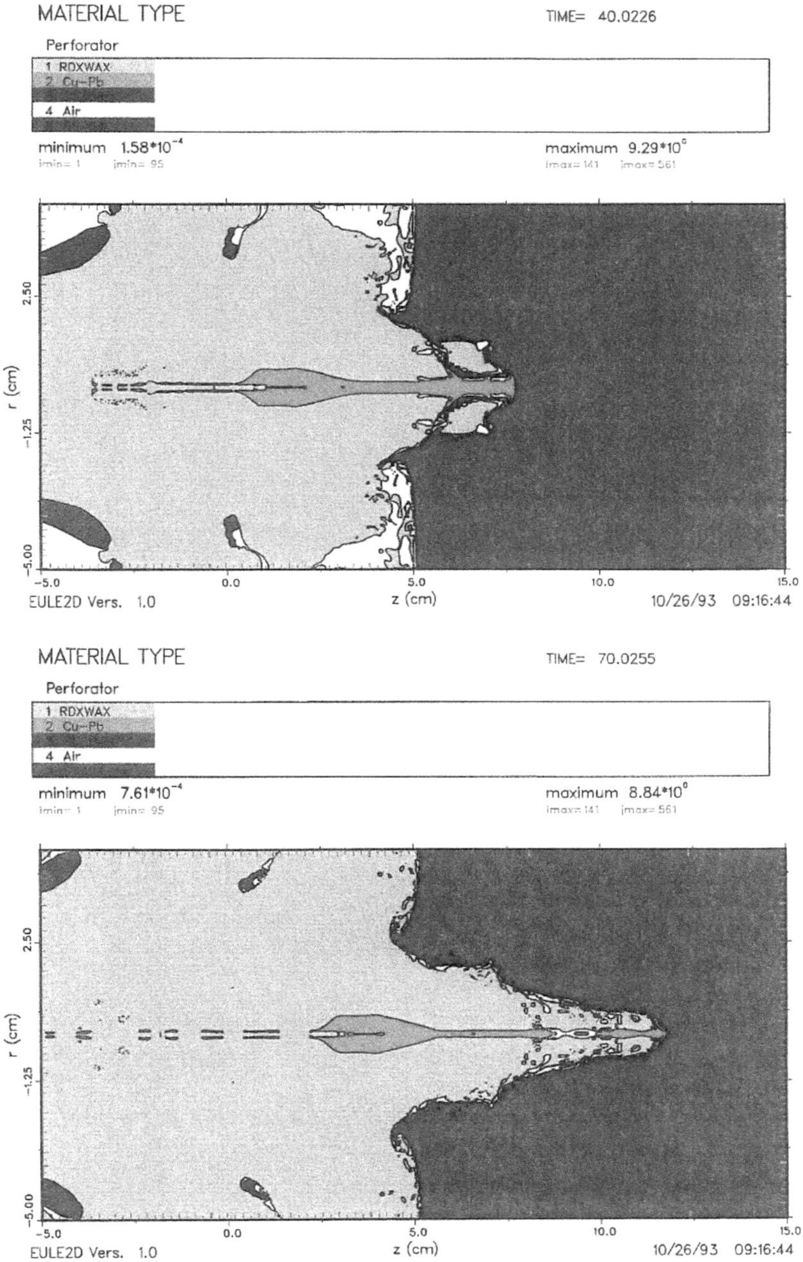

Figure 11.13. The same penetration problem as described in Fig. 11.12 at time 40 (top) and 70 μs (bottom).

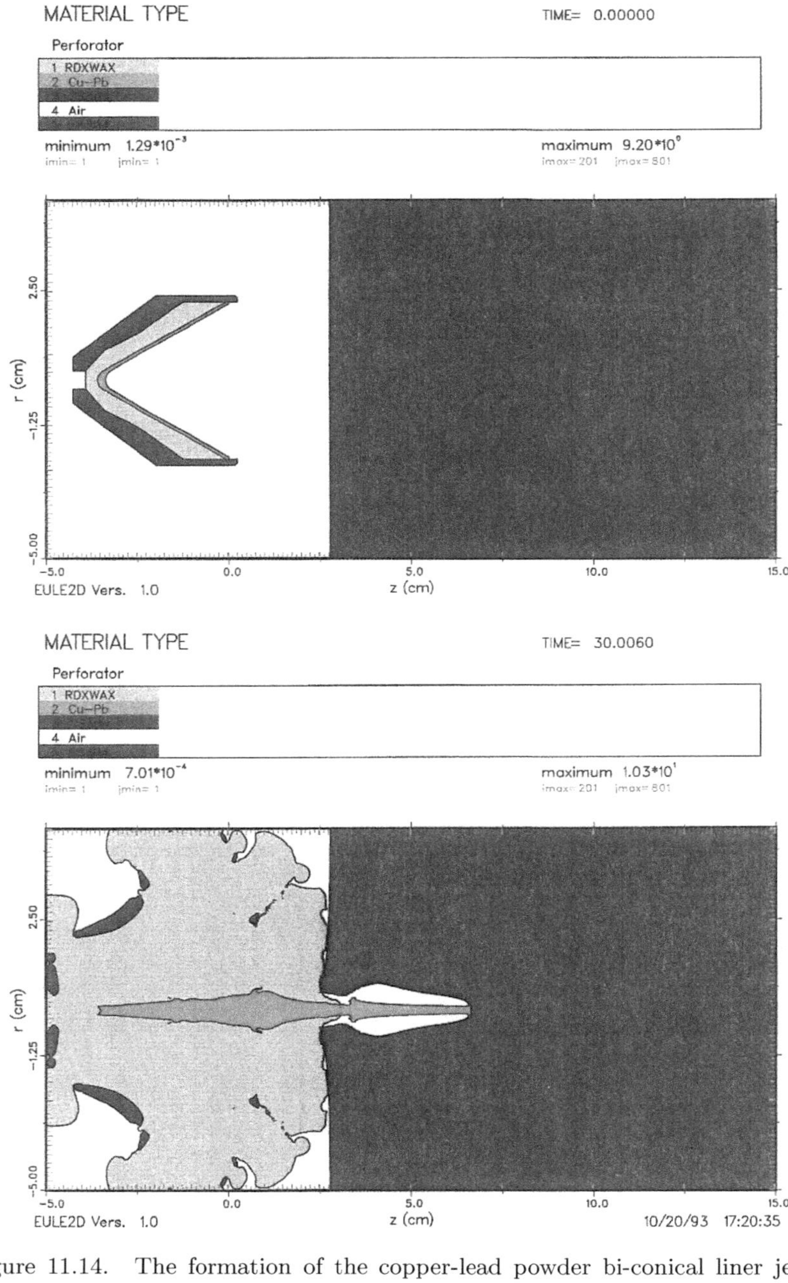

Figure 11.14. The formation of the copper-lead powder bi-conical liner jet is shown for the initial setup (top) and its penetration into the steel block at time 30 μs (bottom).

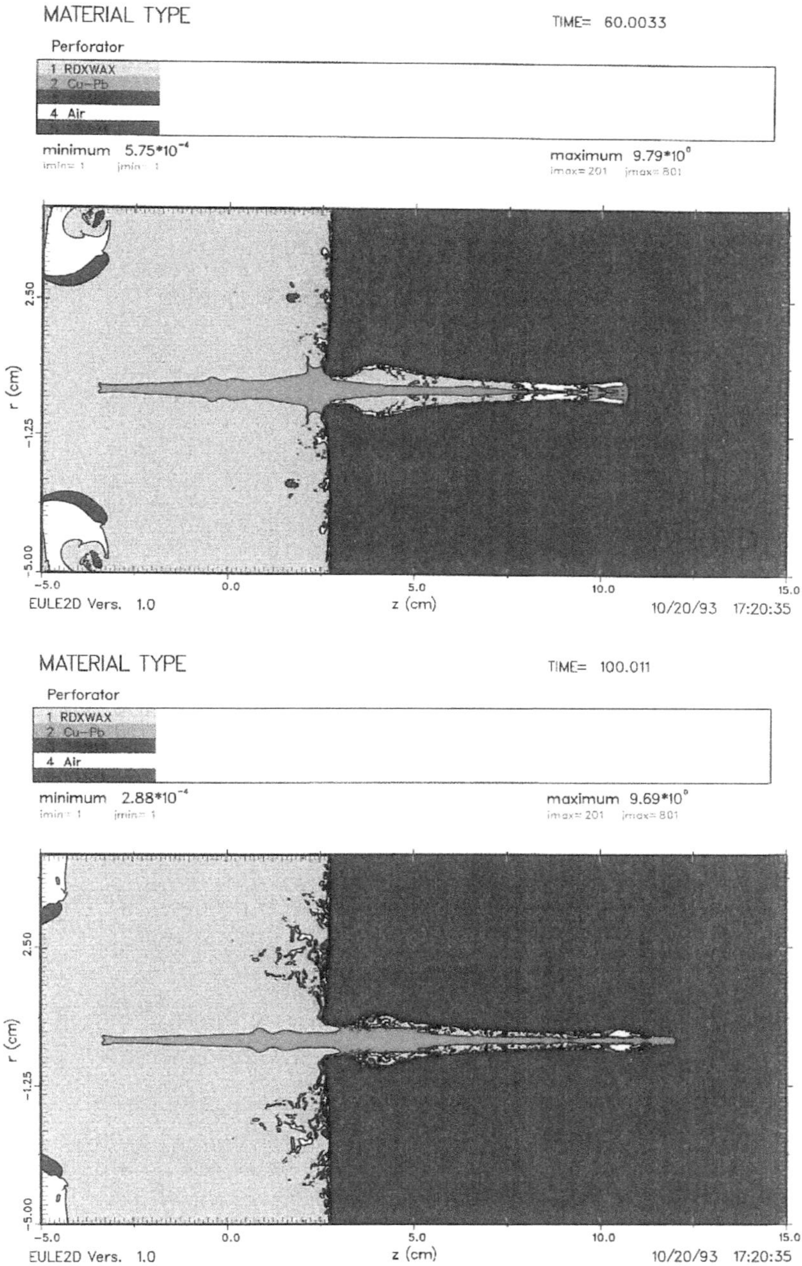

Figure 11.15. The same problem as described in Fig. 11.14 at times 60 (top) and 100 μs (bottom).

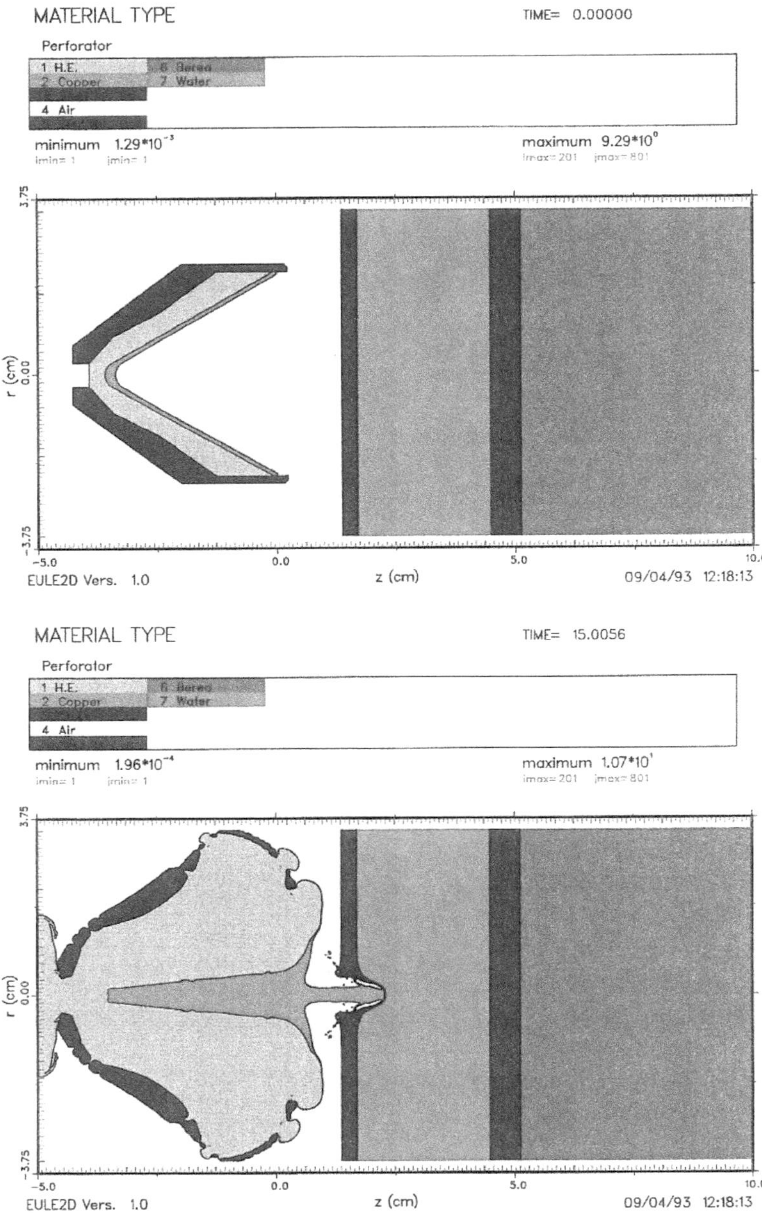

Figure 11.16. Bi-conical shaped charge with double inverse velocity gradient which produces a small hole in the gun wall and large hole in the casing wall with the initial setup (top) and the penetration at time 15 μs (bottom).

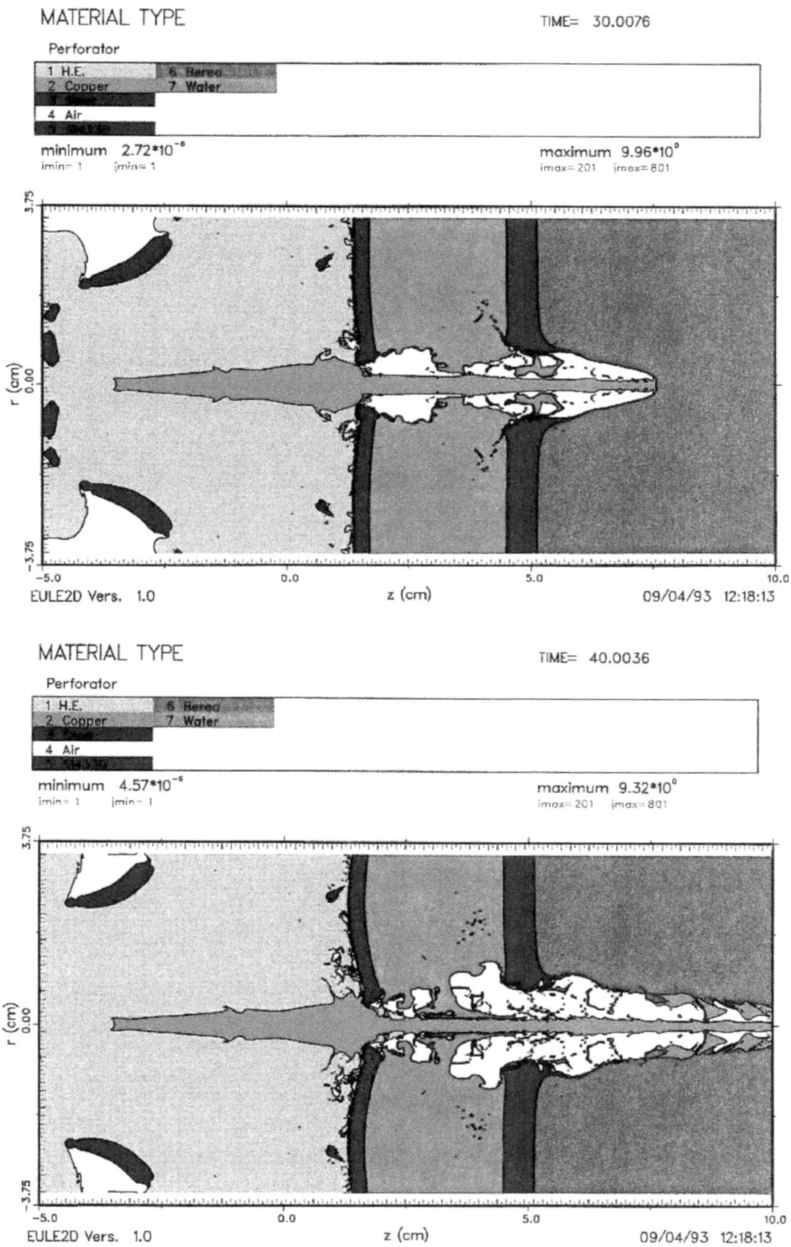

Figure 11.17. The penetrations of the jet into other materials for the same problem as described in Fig. 11.16 are shown for times 30 μs (top) and 40 μs (bottom).

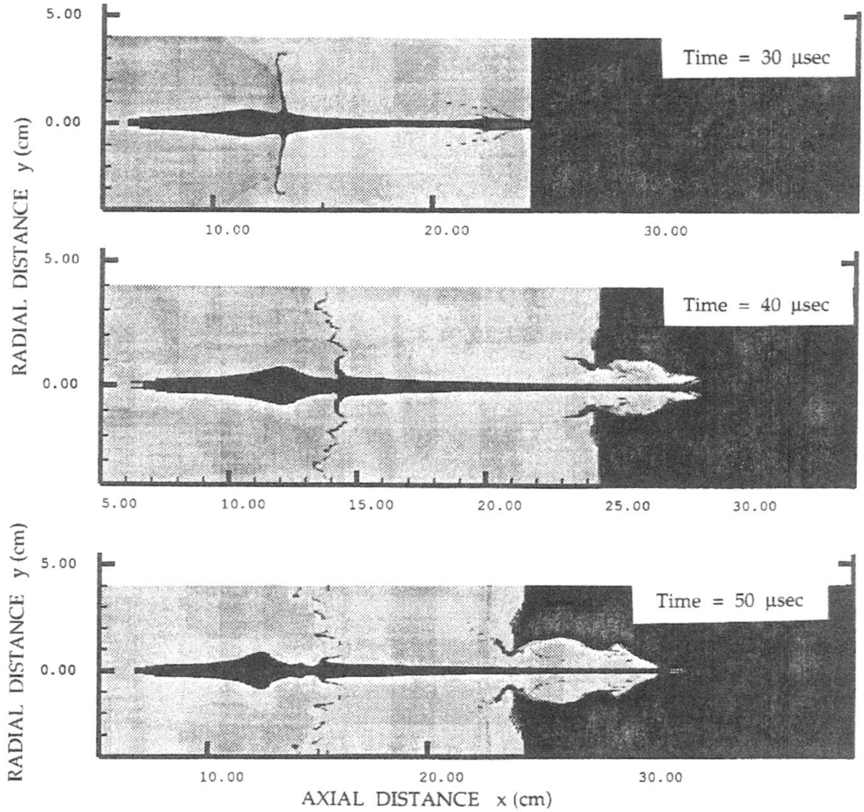

Figure 11.18. The penetrations of Viper shaped charge jet into the steel block at times 30 μs (top), 40 μs (center) and 50 μs (bottom).

top plot of Fig. 11.18. The penetrations of the jet into the steel block are shown for time $t = 40.0\,\mu s$ (central plot) and $t = 50.0\,\mu s$ (bottom plot).

11.10 Computational Assessment of LEAP Performance with Different Lethality Enhancements

The hydro code used, EULER3D, is a 3D Eulerian code developed for use on massively-parallel computers. The code was written using a data parallel strategy in Connection Machine FORTRAN (FORTRAN 90 with extension). It was developed for the CM-2 Connection Machine, and is now available on the CM-200's and CM-5.

EULER3D has been applied to simulations involving conventional munitions and armor, theater missile defense, high explosive safety of weapon systems, and civilian applications, such as oil well perforations. EULER3D has a variety of models to describe the behavior of explosives, gases, and solids subject to elastic/plastic deformation, damage, and fracture. Material interfaces are followed using an accurate Youngs-type interface reconstruction method [11.3]. The required small mesh sizes to resolve phenomena of interest, coupled with the large spatial extent of full-system simulations, force the use of EULER3D onto the largest and fastest massively-parallel computers.

The Army LEAP Kinetic Kill Vehicle (KKV) was developed by Hughes Missile Systems company under the Lightweight Exo-Atmospheric Projectile (LEAP) program, sponsored by the Ballistic Missile Defense Organization and executed by the U.S. Army Space and Strategic Defense Command. The current configuration Army LEAP KKV incorporates a long wave infrared (IR) seeker with a 128×128 HgCdTe staring focal plane array, high-density dual-processor electronics, and a fiber optic inertial measurement unit (IMU) into a lightweight guidance unit. While Hughes has successfully integrated this guidance unit with both liquid bi-propellant and solid propulsion system, this section deals with an analysis of only the solid propulsion driven unit.

Although a series of tests demonstrated that Hughes had met the challenge of integrating a lightweight autonomous kill vehicle, they were faced with the task of demonstrating that a kill vehicle can both be lightweight and lethal. Hughes has been conducting a detailed study of lethality in kinetic hit-to-kill intercepts since 1993. The goals of this study are to evaluate lethality of current LEAP design and to investigate methods of improving KKV lethality against theatre ballistic missile targets including stressing multiple submunitions. Based on the results of this study, Hughes plans to implement lethality enhancement in future LEAP KKV designs to ensure maximum kill capability.

Figure 11.19 shows the kinetic kill projectile with the rod attached to the main body of the projectile. The initial positions of the target and the kinetic kill vehicle are shown in Fig. 11.20 with an angle of attack of 21°. The chemical submunition bottles are assumed to be packed in the white section of the target, while the red section of the target primarily contains the guidance system. Since all rods are attached to the projectile, the kinetic kill vehicle behaves like a heavy single rod. Therefore, the crater of the target created by the impact of the projectile is small and

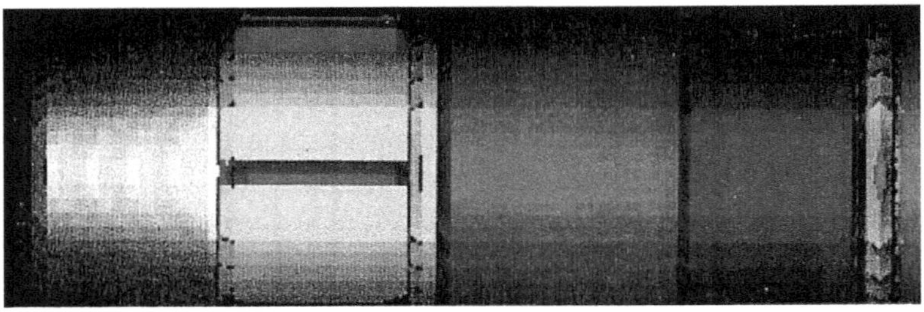

Figure 11.19. The kinetic kill projectile with all of the rods attached to the main body.

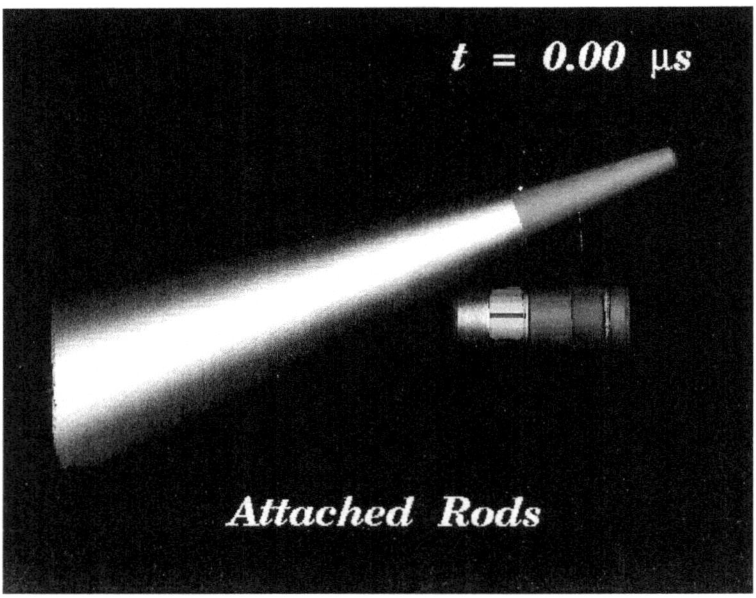

Figure 11.20. The initial positions of the target and the projectile are shown with the angle of 21° between the target and the projectile.

deep at time 300 μs as seen in Fig. 11.21. The red and yellow fragments shown in the lower section of Fig. 11.21 belong to the projectile. At time 600 μs, the crater of the target is larger with some perforation as shown in Fig. 11.22. Figure 11.23 shows that the target is almost broken into two pieces at time 1,000 μs. The blue fragments represent the severe damage

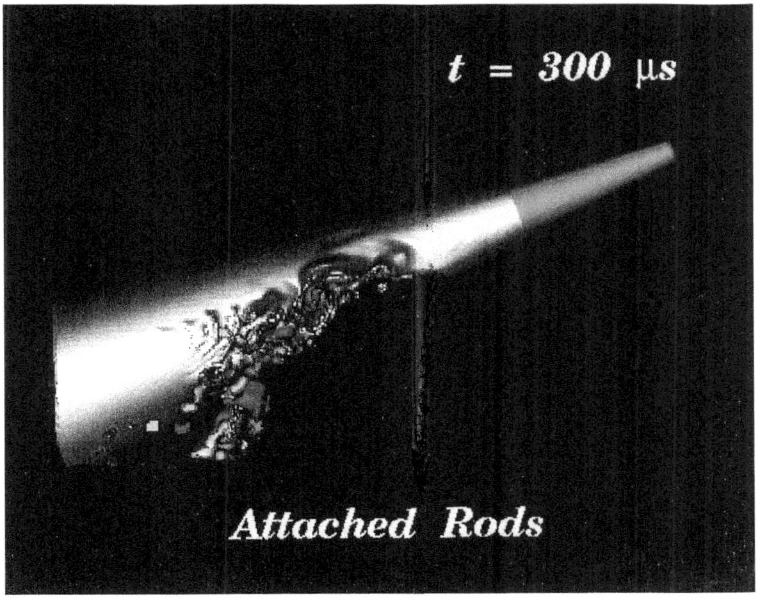

Figure 11.21. The damage of the target at time 300 μs.

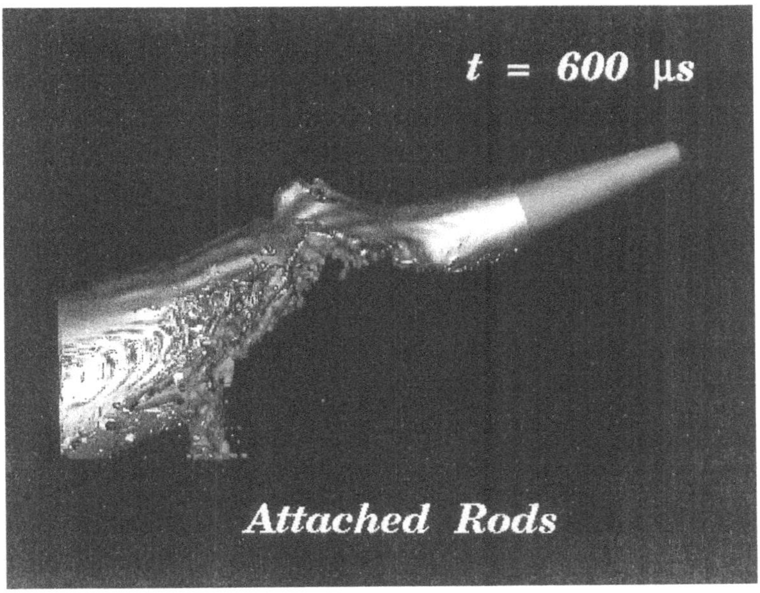

Figure 11.22. The damage of the target at time 600 μs.

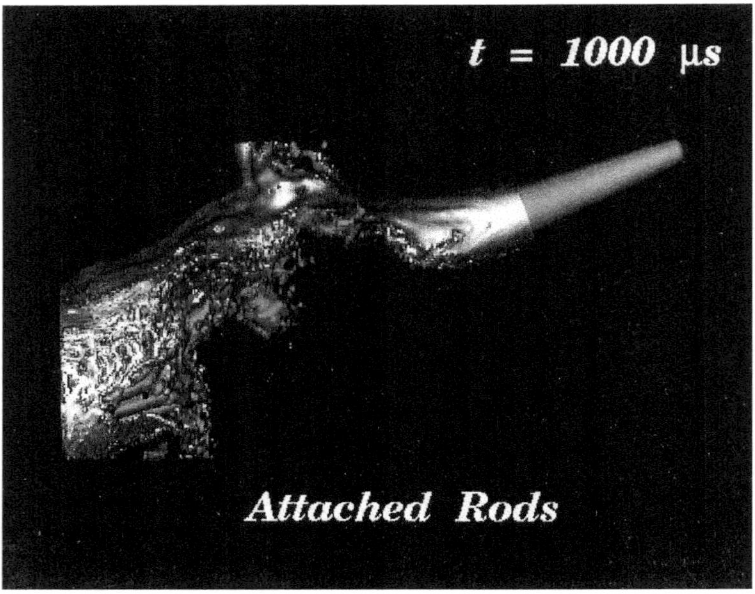

Figure 11.23. The damage of the target at time 1,000 μs.

to the chemical submunition from the target. Figure 11.24 shows the initial position of the target and the projectile with dispersed rods that are detached from the main body. Again, the angle between the target and the projectile is 21°. At time 300 μs, two small craters are formed in the center section of the target due to rod and projectile penetrations as shown in Fig. 11.25.

In Fig. 11.26, two larger craters are formed at time 600 μs, while a third crater near the lower section of the target is created from the damage by a separate rod. Figure 11.27 shows the damage of the target at time 1,000 μs with the broken submunition in blue color. The calculations end at time 1,040 μs with severe damage to the target as shown in Fig. 11.28.

11.11 Lethality Assessments for Ascent-phase Interceptor Impacts on a Generic Chemical Submunition Target

Two idealized hypervelocity (8 km/s) missile intercepts are simulated using a 3D Eulerian hydrodynamics code that runs on the massively parallel

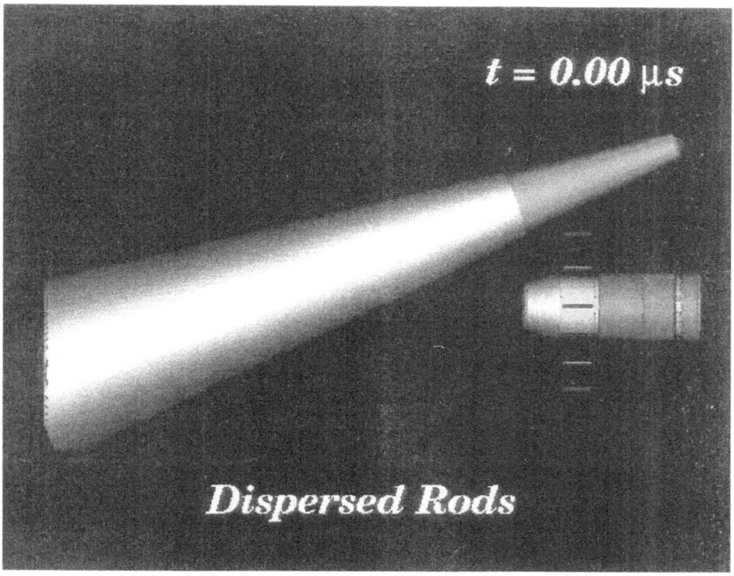

Figure 11.24. The initial position of the target and the kinetic kill projectile are shown with all of the rods detached from the main body.

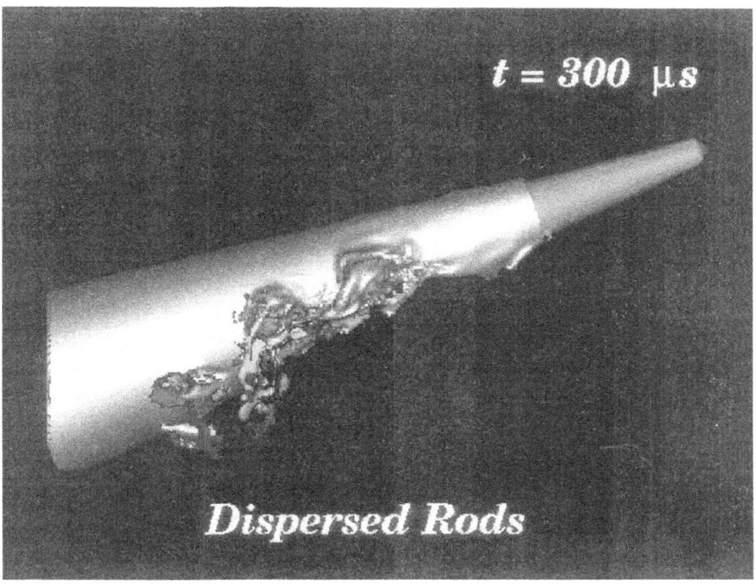

Figure 11.25. The damage of the target at time 300 μs.

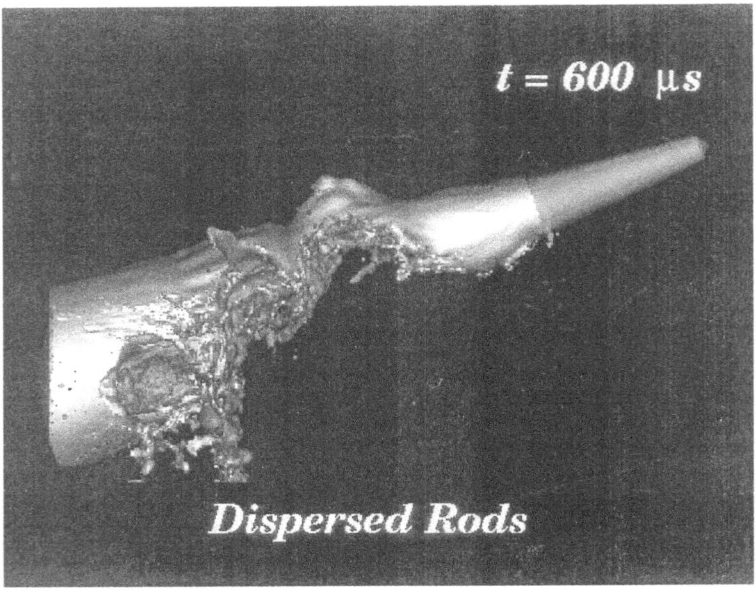

Figure 11.26. The damage of the target at time 600 μs.

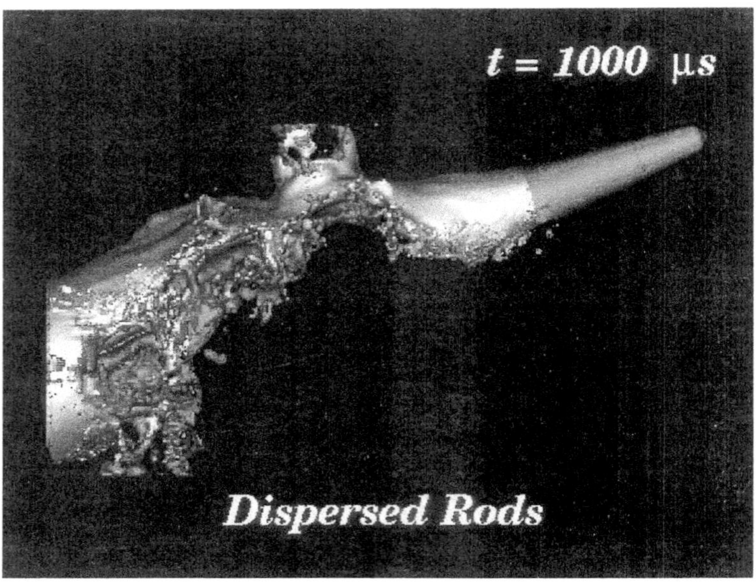

Figure 11.27. The damage of the target at time 1,000 μs.

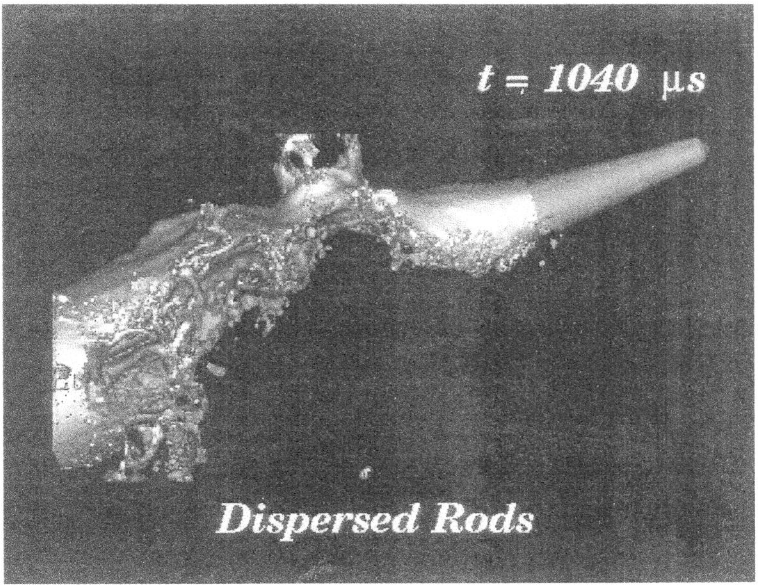

Figure 11.28. The damage of the target at time 1,040 μs.

CM-200 computer. Calculations for both intact and fragmented missiles were run from 0 to 500 μs against a hypothetical chemical-submunition target. Images of the 3D output for both calculations illustrate the evolution, extent, and distribution of the damage produced in the target. The calculations lead to the conclusion that roughly half of the incident transverse momentum is transferred to the target by the fragmented missile and that this value is approximately four times larger than the momentum transfer by the intact missile.

Figure 11.29 shows the damage of the target from an intact missile projectile. At time 0 μs, the missile has just touched the target, therefore, no penetration has occurred yet. The white circular area is the real end of the missile. At time 60 μs, about 1/3 of the missile has penetrated into the target. At time 120 μs, a bell-shaped crater is formed outlined by blue fragments representing the submunition canisters. At time 180 μs, the crater size is increasing and a slit-like perforation hole is visible. At time 240 μs, more canisters are destroyed by the projectile. At time 300 μs, both the perforation and the crater sizes are increasing.

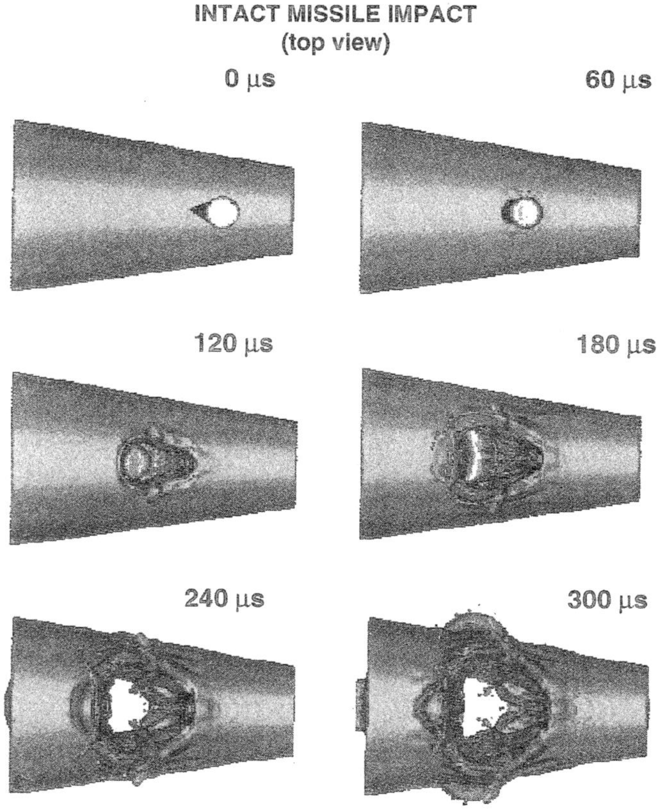

Figure 11.29. The top view of a single intact missile impacting the target at times 0, 60, 120, 180, 240 and 300 μs.

The side view of the same intact missile impacting the target as described in Fig. 11.29 is shown in Fig. 11.30. At time 120 μs, a crater is formed on the side of impact. At time 180 μs, target fragments are departing from the main body. At both times 240 μs and 300 μs, complete perforations are formed.

The cross-sectional view of a single intact missile impacting the target is shown in Fig. 11.31. At time 0 μs, the chemical submunition canisters are properly aligned. At time 60 μs, the missile has penetrated the target and damaged some of the canisters. At time 120 μs, a crater is formed and the missile has completely penetrated the target. At time 180 μs,

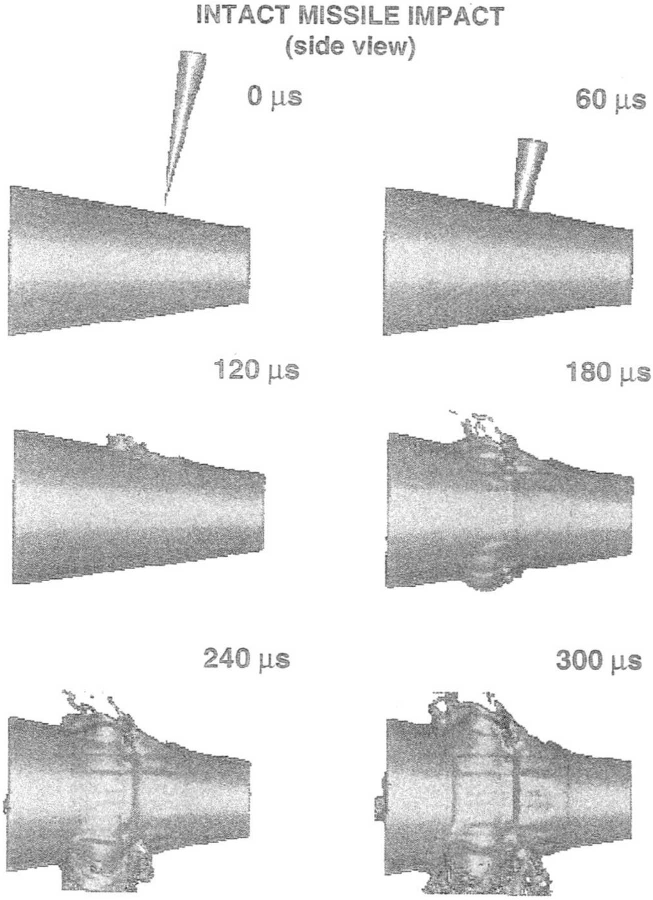

Figure 11.30. The side view of a single intact missile impacting the target at times 0, 60, 120, 180, 240 and 300 μs.

fragments are separated from the main body of the target. At time 240 μs, the missile has completely penetrated the target further damaging the canisters. At time 300 μs, the crater created by missile perforation becomes larger.

The side view of the fragmented missile impacting the target is shown in Fig. 11.32. At time 60 μs, there are four pieces of fragmented missiles

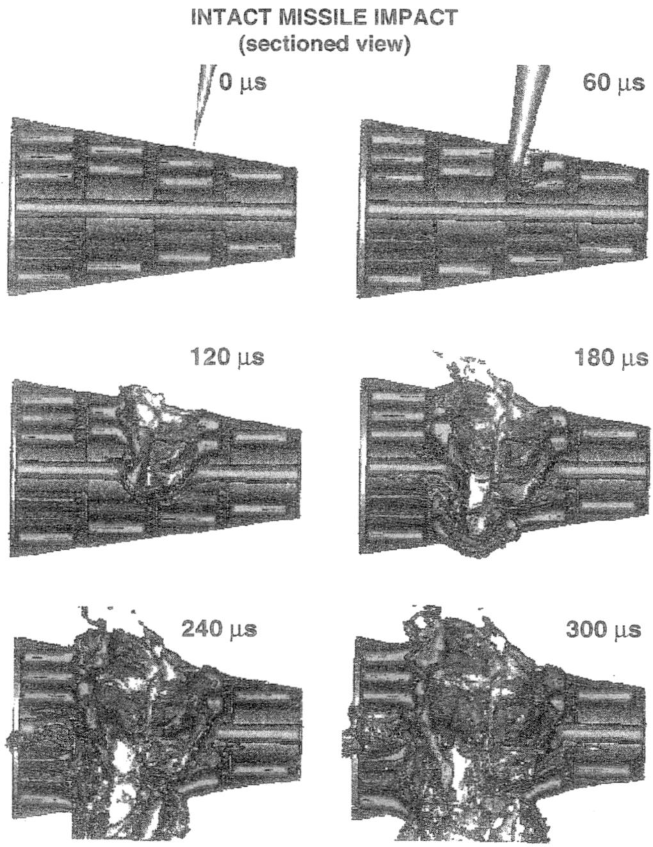

Figure 11.31. The cross-sectioned view of a single intact missile impacting the target at times 0, 60, 120, 180, 240 and 300 μs.

approaching the target. At time 120 μs, all fragments have penetrated the target. At time 180 μs, some of the chemical submunitions are visible through the opening of the damaged target. At both times 240 μs and 300 μs, the impact region becomes larger resulting in a perforation through the target.

The top view of the fragmented missile impacting the target is shown in Fig. 11.33. At time 0 μs, the fragmented missile impacts the surface of

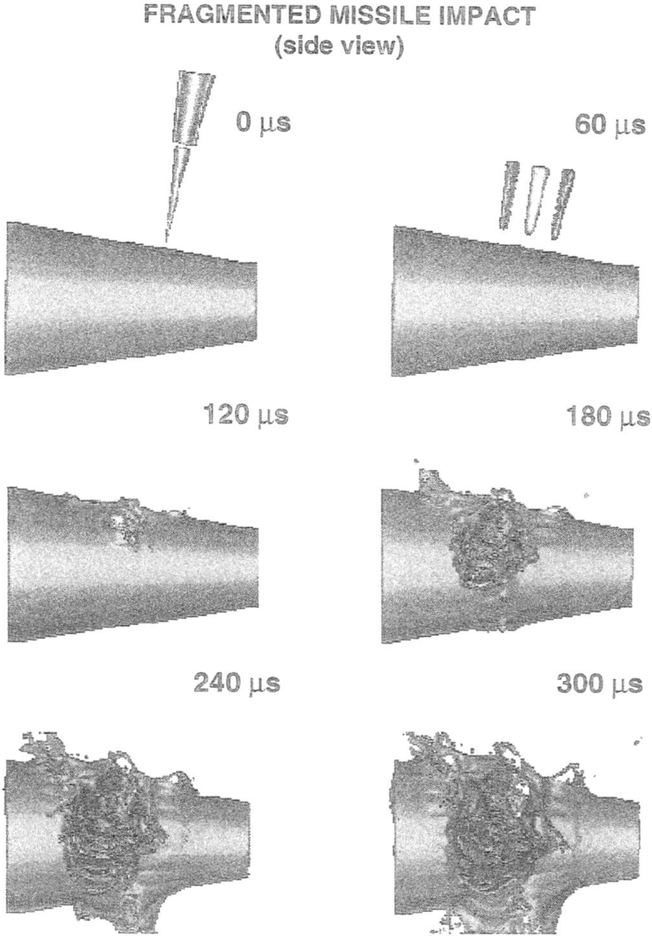

Figure 11.32. The side view of the fragmented missile impacting the target at times 0, 60, 120, 180, 240 and 300 μs.

the target. At time 120 μs, four craters are formed upon missile impaction. At time 180 μs, the chemical canisters can be seen along the horizontal opening. At time 240 μs and 300 μs, the craters are much larger with a perforation created by the right horizontal missile.

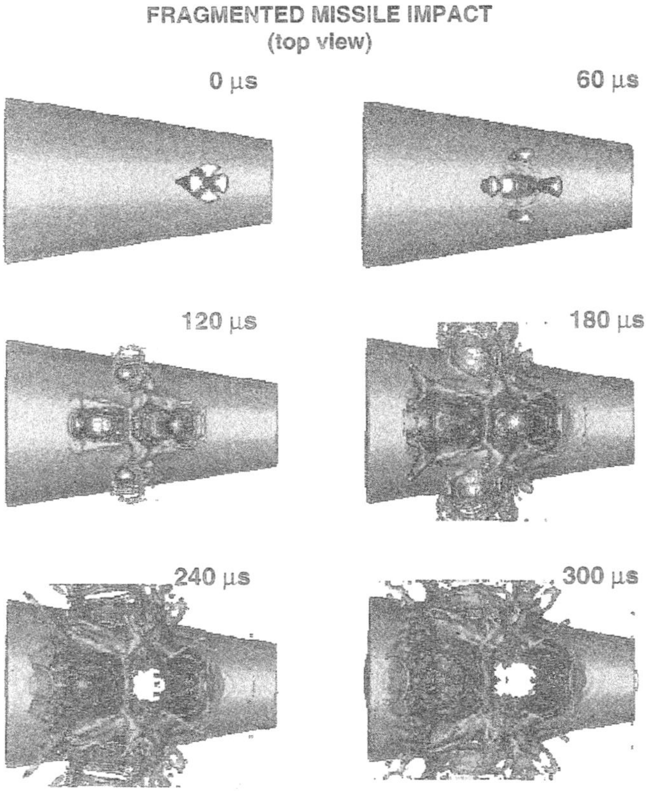

Figure 11.33. The top view of the fragmented missile impacting the target at times 0, 60, 120, 180, 240 and 300 μs.

The cross-sectional view of the fragmented missile impacting the target is shown in Fig. 11.34. At time 0 μs, the first fragmented missile has just touched the target. At time 60 μs, the first missile has completely penetrated the target just as the other three fragmented missiles are arriving. At time 120 μs, several chemical canisters are damaged by the four fragmented missiles. At time 180 μs, the first fragmented missile has reached the bottom of the target. At time 240 μs, perforation of the target has occurred resulting in many damaged canisters.

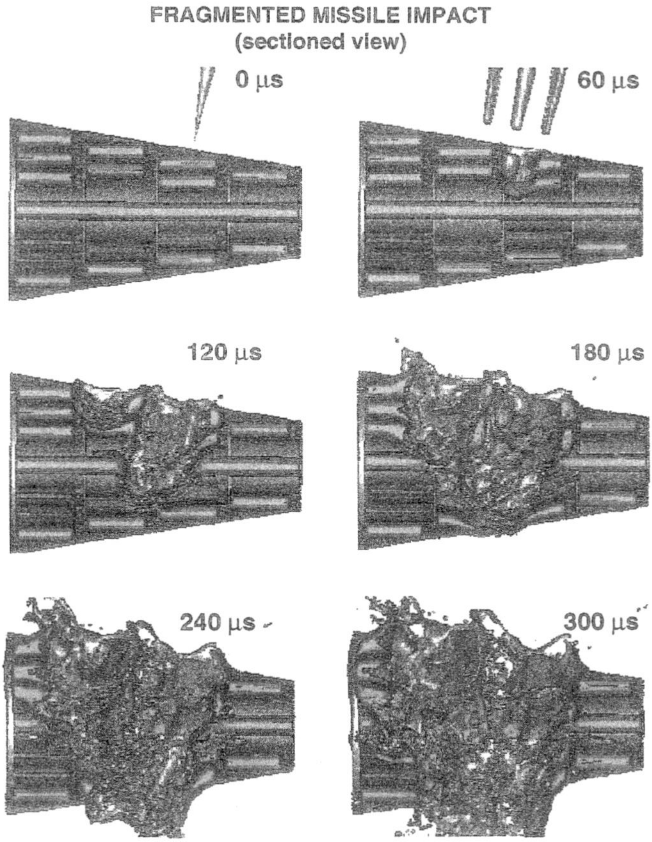

Figure 11.34. The cross-sectioned view of the fragmented missile impacting the target at times 0, 60, 120, 180, 240 and 300 μs.

At time 300 μs, more extensive damage to the chemical submunitions is noticeable.

The 3D interface plots of fluid payload assemblies at time 300 μs with intact missile are shown in the upper plot of Fig. 11.35. Most of the damaged canisters are located at the lower section of the payload due to the perforation of the target. The lower plot shows that the fragmented missiles have damaged the chemical submunitions more severely as compared to

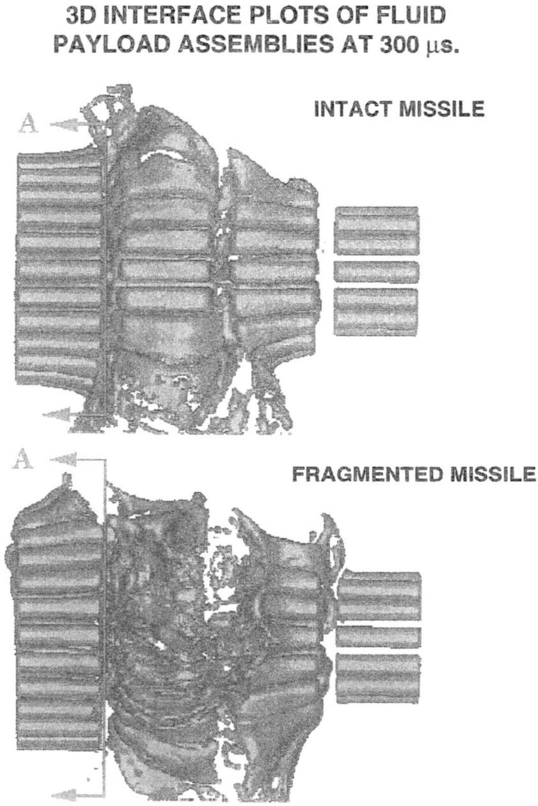

Figure 11.35. The 3D interface plots of fluid payload assemblies at time 300 μs with intact missile (upper) and fragmented missile (lower). The $\overline{AA}$ cross-sectional view will be shown in the next plot.

the intact missile. However, the perforation size at the lower end of the target is smaller for the fragmented missile.

The 3D interface plot of payload assemblies at time 300 μs with intact missile is shown on the upper plot of Fig. 11.36. The sectioned-view from the rear tier shows a large perforation at the lower end of the target. The lower plot of Fig. 11.36 shows that the fragmented missiles have damaged the interior section of the target more severely as compared with the intact missile.

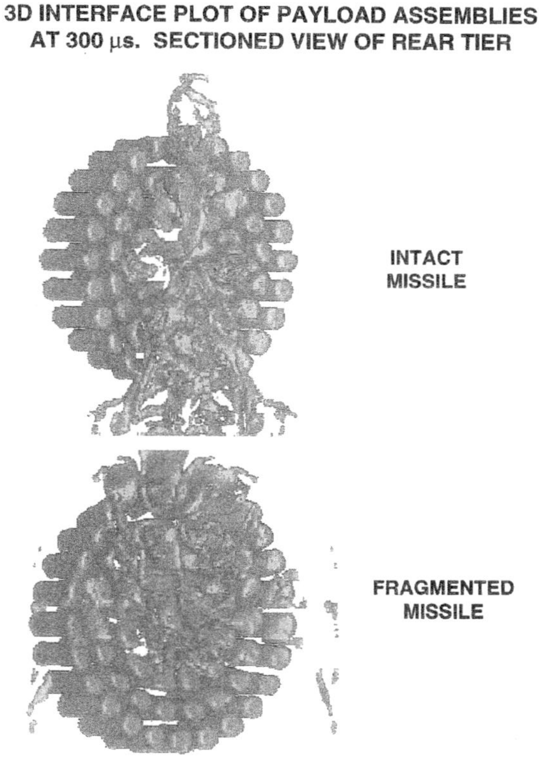

Figure 11.36. The 3D interface plot of payload assemblies at time 300 μs with intact missile (upper) and fragmented missile (lower).

References

11.1 Hohler, V and Stipl, A, Influence of the length to diameter ratio in the range from 1 to 32 on the penetration performances of rod projectiles, 8th Int. Symp. on Ballistics (1984).
11.2 Cullis, IG and Nash, MA, The importance of after flow in high impedance mismatch penetration, 9th Int. Symp. on Ballistics (1986).
11.3 Youngs, D, An interface tracking method for a 3D Eulerian hydrodynamics code, UK Atomic Weapons Establishment, AWRE/44/92/35 (1987).

CHAPTER 12

OIL WELL PERFORATOR DESIGN USING 2D EULERIAN CODE

Notations

a	Tillotson parameter
A	JWL parameter (Mbar)
A_1	Tillotson parameter
b	Tillotson parameter
B	JWL parameter (Mbar)
B_1	Tillotson parameter
C_0	parameter used in Mie-Gruneisen equation of state
BI	burn interval (μs)
BT	burn time (μs)
D	detonation velocity (cm/μs)
E	detonation energy per unit volume (Mbar-cm^3/cm^3)
G	shear modulus of elasticity (Mbar)
I	specific internal energy per unit mass (Mbar-cm^3/g)
I_s	Tillotson parameter
k, ℓ, m, n	Johnson-Cook parameters
k_1, ℓ_1, m_1, n_1	Steinberg-Guinan parameters
P	pressure (Mbar)
P_c, P_E	pressures used in Tillotson equation of state (Mbar)
r	radial coordinate (cm)
R_1, R_2	nonlinear coefficients used in the JWL equation of state
s	parameter used in the Mie-Gruneisen equation of state
S^{ij}	stress deviator tensor (Mbar)
S^{rr}, S^{zz}, S^{rz}, $S^{\theta\theta}$	stress deviator components (Mbar)
t, T	time (μs)
T^*	reduced temperature used in the Johnson-Cook model
u	velocity in r direction (cm/μs)
U_p	particle velocity used in the Mei-Gruneisen equation of state (cm/μs)
U_s	shock velocity used in the Mei-Gruneisen equation of state (cm/μs)

v	velocity in z direction (cm/μs)
V	specific volume (cm^3/g)
W	rate of energy source due to work hardening
Y^0	yield stress in simple tension (Mbar)
z	axial coordinate (cm)

Greek Letters

α, β	Tillotson parameters
γ	parameter used in the Mie-Gruneisen equation of state
Γ	energy source (Mbar-cm^3/g-μs)
ε_p	equivalent plastic strain
ε^*	nondimensional plastic strain rate
η	normalized density ($= \rho/\rho_0$)
μ	normalized density minus one ($= \eta - 1 = \rho/\rho_0 - 1$)
ρ	density (g/cm^3)
σ^{ij}	stress tensor (Mbar)
σ^{rr}, σ^{rz}, $\sigma^{\theta\theta}$, σ^{zz}	stress tensor components (Mbar)
ω	JWL parameter

Subscripts

0	initial value
e	elastic regime
p	plastic regime
r	derivative with respect to r coordinate
t	derivative with respect to time
z	derivative with respect to z coordinate

Superscripts

n	time at n time-step, i.e. $t^n = t_0 + n \cdot \Delta t$

12.1 Introduction

In the oil industry, a method of completing a well is to first run the casing pipe through the oil sand formation, set it below the producing horizon, and then cement it in place as shown in Fig. 12.1. Then the casing is perforated to allow the oil or gas to flow into the wellbore. Perforating is

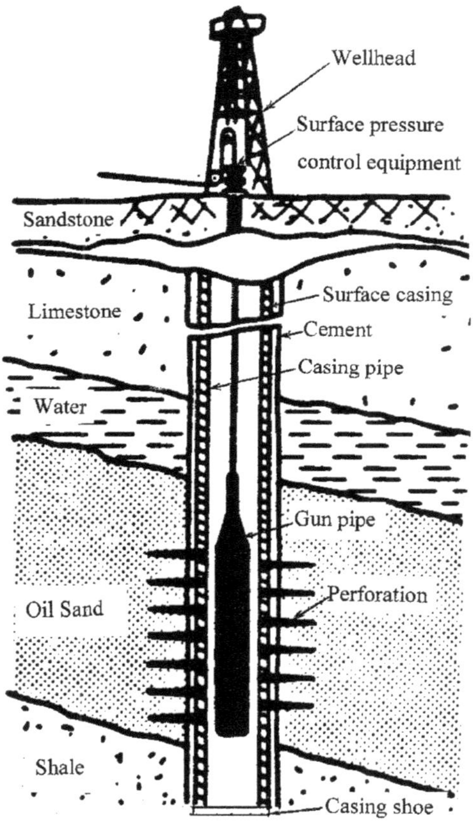

Figure 12.1. The setup of the well head, pressure control, gun, casing, cement, and shaped charges arrangement.

accomplished with a gun pipe equipped with shaped charges. Figure 12.1 shows the application of a typically one shaped charge per 2.54 cm in the pipe length direction. Any two-neighboring shaped charges are separated with an angle of 120° in the azimuthal direction. The gun, a circular container that fits inside the casing, has electrically fired cartridges that discharge small high-velocity jets. These jets penetrate the casing wall, the outside cement, and the oil sand. By doing this, channels are opened up through which the oil or gas can flow into the production pipe.

The main concept of modern perforators is to create a small hole in the gun pipe and a large hole in the casing wall. As shown in Fig. 12.2,

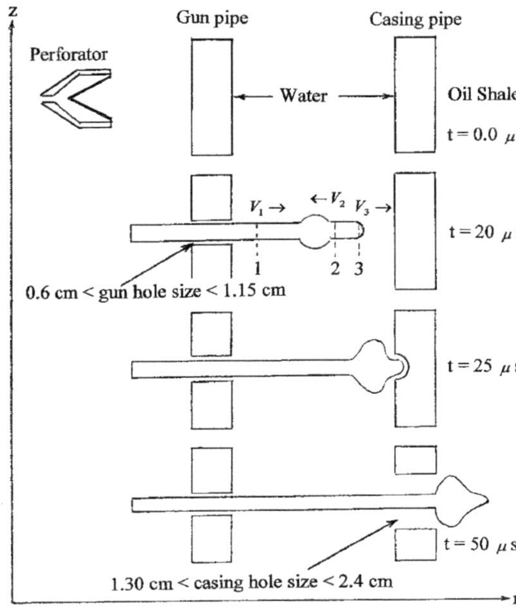

Figure 12.2. The formation of the double velocity inverse gradients and the hole sizes in the gun and casing walls.

at time $t = 0.0\,\mu$s, the perforator is located inside the gun with water between gun and casing walls. The outside of the casing is surrounded by oil shale or granite. At $t = 20\,\mu$s, the shaped charge jet has already punched through the gun wall with a hole size between 0.6 cm and 1.15 cm. The jet inside the water region has three different velocity vectors, i.e. $v_1 > 0$, $v_2 < 0$ and $v_3 > 0$. The resulting bulge forms between points 1 and 2. The bulge reaches the casing wall without impact at $25\,\mu$s which is also the maximum diameter of the bulge. At $t = 50\,\mu$s, the jet will penetrate through the casing wall and create a hole with a size between 1.30 cm and 2.40 cm.

The concept of using shaped charge for oil well casing perforation was introduced by McLemore [12.1] in 1946. For more than half a century, most of the shaped charge designs for the oil industry have been designed by trial and error through laboratory experiments. Recently, the two-dimensional Eulerian code similar to the particle code [12.2] has been used to simulate the shaped charge jet formations as well as the jet penetration into the casing wall and the surrounding rocks. The computed results are within 8%

error in comparison with the experimental data. Such successful simulations should be attributed to the development in numerical methods for solving a system of partial differential equations with a well-defined equation of state and constitutive law for real material in the last decade.

For an engineering problem which involves high explosive and a metal shell, e.g. shaped charge or an explosive formed projectile, one can use a good computer code to design a new device, provided this new device is of small deviation (e.g. less than 5%) from an existing one which has been well tuned with pertinent parameters against the experimental data. All of the shaped charges described in this chapter are based on this concept. These oil well perforators include the charge diameters of 3.6 cm, 4.4 cm, 6.5 cm, and 8.2 cm using copper as the liner and RDX as the main charge with smaller charge diameter (e.g. 3.6 cm) used for deep well-bore perforation. Most of the liners are made from metallic powder and pressed into conical, hemispherical, trumpet, or bell shapes. For the small conical liner angle, penetrations into the rock outside the casing are deeper. However, the hole diameters are smaller. Hemispherical and bell-shaped liners tend to produce a larger hole that is shallow in penetration depth. A trumpet liner is somewhat in-between conical and hemispherical ones. In general, a larger hole in the casing wall and the outside rock is more desirable since this may allow more oil or gas to flow into the well-bore. The thickness and the shape of the outside steel case for the shaped charge also have some effect on the performance of the copper jet.

Since the clearance between the gun wall and the casing wall is small, the stand-off for the shaped charge is limited. As a result, the copper liner has limited time to form a jet which must posses a double inverse velocity gradient. The high explosive also has to sustain a harsh environment; that is, the temperature in the deep down-hole is usually above 260°C. These restrictions make the design of the oil well shaped charge much more difficult compared with the one used in a conventional weapon.

In this chapter, eleven perforators are presented with their important characteristics pertinent to the optimum design for oil well perforation applications. Section 2 describes the 2D Eulerian code; section 3 gives the formula of the equation of state and constitutive relation; section 4 discusses the general requirements for a good perforator; section 5 presents comparisons of jet formation and penetration of some perforators between calculations and experiments; sections 6 and 7 present the penetration calculation of eight perforators; section 8 discusses the results; and the last section gives the conclusions.

12.2 Description of the 2D Eulerian Code

The present study uses Lagrangian phase, or the predictor step, and remap phase, or the corrector step, to solve the dependent variables such as density, velocity, internal energy, and deviatoric stress. For a cylindrical coordinate, the governing equations are split into three parts. The first part is the Lagrangian phase which simultaneously solves the dependent variables in both radial (r) and axial (z) directions. The second part solves the remap phase in radial direction and the third part solves the remap phase in axial direction. However, in order to maintain better accuracy, the calculations in parts 2 and 3 are alternated in directions for each computational cycle, e.g. $r-z-z-r$ or $z-r-r-z$. For example, the equations solved in the Lagrangian phase are:

Mass

$$\rho_t = -\frac{\rho}{r}(ru)_r - \rho v_z \,. \tag{12.1}$$

Momentum

$$u_t = -\frac{1}{\rho}P_r + \frac{1}{\rho}\left[S_r^{rr} + \frac{1}{r}(2S^{rr} + S^{zz}) + S_z^{rz}\right], \tag{12.2}$$

$$v_t = -\frac{1}{\rho}P_z + \frac{1}{\rho}\left(S_r^{rz} + \frac{1}{r}S^{rz} + S_z^{rz}\right). \tag{12.3}$$

Energy

$$I_t = -\frac{P}{\rho}\left[\frac{1}{r}(ru)_r + v_z\right] + \frac{1}{\rho}\Big[(S^{rr}u_r + S^{rz}v_r) - \frac{u}{r}(S^{rr} + S^{zz}) + S^{zz}v_z + S^{rz}u_z\Big] + \Gamma \,. \tag{12.4}$$

The equations of state are:
Pressure

$$P = g(\rho, I)\,. \tag{12.5}$$

Elastic regime

$$S_t^{rr} = \frac{2}{3}G\left(2u_r - \frac{u}{r} - v_z\right), \tag{12.6}$$

$$S_t^{zz} = \frac{2}{3}G\left(-u_r - \frac{u}{r} + 2v_z\right), \tag{12.7}$$

$$S_t^{rz} = G(v_r + u_z), \tag{12.8}$$

and

$$S^{\theta\theta} = -(S^{rr} + S^{zz})\,. \tag{12.9}$$

Plastic regime

$$\left(S_t^{ij}\right)_p = (S^{ij})_e - \left[\frac{3G}{(Y^0)^2}\right] \dot{W} S^{ij}\,. \tag{12.10}$$

Solving Eqs. (12.2)–(12.10) produces the time level $(n+1)$ quantities of $\tilde{u}$, $\tilde{v}$, $\tilde{I}$, and $\tilde{S}^{ij}$ for each cell, based on previous known quantities at the (n)th time step. The time-averaged velocities are

$$\bar{u} = \frac{1}{2}(u^n + \tilde{u})\,, \tag{12.11}$$

and

$$\bar{v} = \frac{1}{2}(v^n + \tilde{v})\,. \tag{12.12}$$

In part 2, the dependent variables are solved in the r-direction using the averaged velocity $\bar{u}$. They are

$$\rho_t = -\bar{u}\rho_r\,, \tag{12.13}$$

$$u_t = -\bar{u}u_r\,, \tag{12.14}$$

$$v_t = -\bar{u}v_r\,, \tag{12.15}$$

$$I_t = -\bar{u}I_r\,, \tag{12.16}$$

$$S_t^{rr} = -\bar{u}S_r^{rr}\,, \tag{12.17}$$

$$S_t^{zz} = -\bar{u}S_r^{zz}\,, \tag{12.18}$$

and

$$S_t^{rz} = -\bar{u}S_r^{rz}\,. \tag{12.19}$$

Similar equations are obtained for the z-direction in part 3 if the average velocity $\bar{u}$ is replaced by $\bar{v}$ and the derivative r is replaced by z in the right hand side of Eqs. (12.13)–(12.19). Some of the detailed procedures used in the remap phase are described in Ref. [12.2]. A modified SLIC [12.3] method is used for tracking the material interfaces.

The energy released from the high explosive is treated as a programmed burn with a JWL equation of state [12.4] which is

$$P = A\left[1 - \frac{\omega}{R_1 V}\right] \cdot e^{-R_1 V} + B\left[1 - \frac{\omega}{R_2 V}\right] \cdot e^{-R_2 V} + \frac{\omega(E - E_0)}{V}. \tag{12.20}$$

For the present study, we use RDX (with 2% weighted wax) as the main charge with the parameters, $A = 6.45614$ (Mbar); $\omega = 0.3$; $R_1 = 0.04833$; $B = 7.2$ (Mbar); $R_2 = 2.24$; $E_0 = 0.0562$ (Mbar-cm^3/cm^3); $\rho_0 = 1.6$ (g/cm^3); $D = 0.81$ (cm/μs). In a high-explosive programmed burn, the basic assumption is that the detonation wave front travels in all directions at the Chapman-Jouguet detonation velocity. Information concerning the energy released from the high explosive such as burn time (BT) and burn interval (BI), is pre-calculated and stored in the code at time $t = 0\,\mu$s. During the run, when the problem time T^n, at cycle n becomes greater than the BT value of a high-explosive cell but less than $(BT + BI)$, a fraction of the specific energy for the particular HE is deposited in the cell. This fraction is given by $(T^n - BT)/BI$. On the next cycle, at T^{n+1}, if T^{n+1} is still less than $(BT + BI)$, another fraction $(T^{n+1} - T^n)/BI$ of specific energy is deposited in the cell. This process continues until the cell is completely burned. The burn calculations depend on the Huygen's construction for the burn distance. In Fig. 12.3, if A is the detonation point, then the burn distances to point 1 and 2 are the line-of-sight distances from the detonation

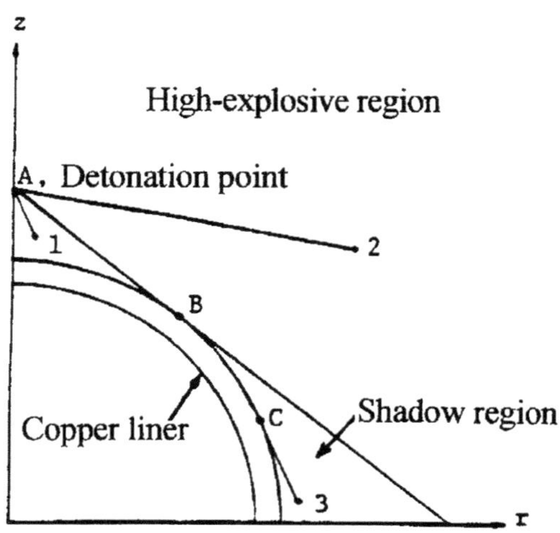

Figure 12.3. Huygen's construction for burn distances.

point, i.e. lines $\overline{A1}$ and $\overline{A2}$. However, the calculation of the burn distance to point 3 is more complex because point 3 is located in the shadow region from point A. In this case, the burn distance is obtained from the shortest distance of the high-explosive wave paths including a new spherical wave centered at point B. One possible solution of the burn distance for point 3 is the total distances of line $\overline{AB}$, chord $\overline{BC}$, and line $\overline{C3}$.

12.3 Equation of State and Constitutive Relation

In the present studies, we use the *Mie-Gruneisen* EOS (Equation of State) [12.5], for all materials except oil shale. The *Mie-Gruneisen* EOS is

$$U_s = C_0 + sU_p\,, \tag{12.21}$$

$$\gamma = \frac{1}{\rho}\left(\frac{\partial P}{\partial I}\right)_\rho\,, \tag{12.22}$$

and

$$\gamma_0 = \frac{1}{\rho_0}\left(\frac{\partial P}{\partial I}\right)_{\rho_0}\,. \tag{12.23}$$

The *Mie-Gruneisen* parameters are given in Table 12.1.
For the oil shale, we use Tillotson EOS [12.6] which is

$$P = P_c = \left[a + \frac{b}{\frac{I}{I_0\eta^2} + 1}\right] I\rho + A_1\mu + B_1\mu^2, \quad \text{for } I < I_s\,, \tag{12.24}$$

$$P = P_E = aI\rho + \left[\frac{bI\rho}{\frac{I}{I_0\eta^2} + 1} + A_1\mu e^{-\beta\left(\frac{\rho_0}{\rho}-1\right)}\right] e^{-\alpha\left(\frac{\rho_0}{\rho}-1\right)^2}, \quad \text{for } I > I_s'\,, \tag{12.25}$$

Table 12.1. Parameters for the *Mie-Gruneisen* EOS.

Material	ρ_0(g/cm^3)	C_0(km/s)	s	γ_0
Stainless steel 304	7.9	4.57	1.49	2.17
Copper OFHC 1/2 hard	8.93	3.94	1.49	1.96
Water	0.9979	2.393	1.333	0.50
Granite	2.672	3.712	1.086	0.90
Bronze	8.733	3.814	1.452	2.12

and

$$P = [(I - I_s)P_E + (I'_s - I)P_c]/(I'_s - I_s), \quad \text{for } I_s < I < I'_s. \qquad (12.26)$$

The parameters used for oil shale are $a = 0.5$; $b = 1.0$; $A_1 = 0.28$; $B_1 = 0.11$; $\alpha = 5.0$; $\beta = 5.0$; $I_0 = 0.11$; $I_s = 0.032$; $I'_s = 0.16$; and $\rho_0 = 2.3$ (g/cm^3). The constitutive relation for the granite and the oil shale is the Johnson-Cook [12.7] model which is:

$$Y = Y_0(1 + k\varepsilon^\ell)(1 + n \cdot \ln \varepsilon^*)(1 - T^{*m}). \qquad (12.27)$$

The parameters are $Y_0 = 0.00048$ (Mbar); $k = 1.0$; $\ell = 1.0$; $n = 0.0$; $\varepsilon^* = \dot{\varepsilon}_p/\dot{\varepsilon}_0$; $\dot{\varepsilon}_0 = 1.0\,\mathrm{s}^{-1}$; $m = 0.5$; and $T^* = 300\,\mathrm{K}$.

The constitutive relation for other material is the Steinberg-Guinan [12.8] model which is

$$Y = Y_0(1 + k_1\varepsilon_p)^{\ell_1}[1 + m_1 P\eta^{1/3} - n_1(T - 300)], \qquad (12.28)$$

and

$$G = G_0[1 + m_1 P\eta^{1/3} - n_1(T - 300)]. \qquad (12.29)$$

The parameters used in the present studies are given in Table 12.2.

12.4 Shaped Charge Jet Characteristics

A good oil well perforator should produce a bulged or high-density section along the liner jet between 20 and 25 μs after the main charge ignites. It is assumed that the average water gap between the gun wall and the casing wall is about 2.0 to 2.5 cm, and the stand-off between the shaped charge and gun wall is approximately 1.0 cm. The conical perforator *P-A*, shown in Fig. 12.4, uses copper as liner, RDX as main charge, and steel as case. The double-inversed velocity gradients or the bulged sections are indicated by *A* and *B* as shown in Fig. 12.5 for time 20 and 25 μs. The liner density

Table 12.2. Parameters for Steinberg-Guinan model.

Material	Y_0(Mbar)	G_0(Mbar)	k_1	ℓ_1	m_1	n_1
Stainless steel 304	0.0034	0.77	43.	0.35	2.26	0.000455
Copper OFHC 1/2 hard	0.0012	0.477	36.	0.45	2.83	0.000377
Bronze	0.00123	0.409	39.	0.27	3.03	0.000411

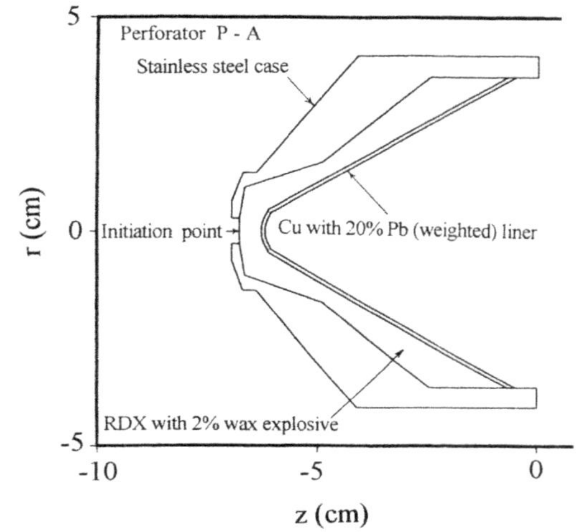

Figure 12.4. The initial setup for the conical perforator *P-A* with copper liner, RDX main charge, and steel case.

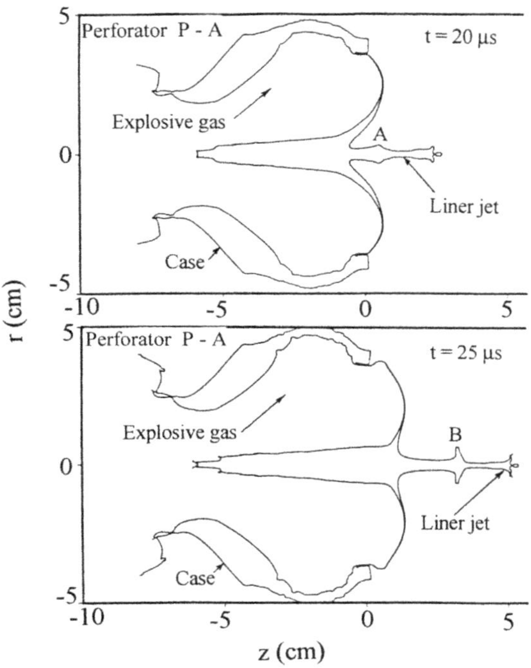

Figure 12.5. The liner jet formation at time 20 and 25 μs with the bulged sections indicated by *A* and *B*.

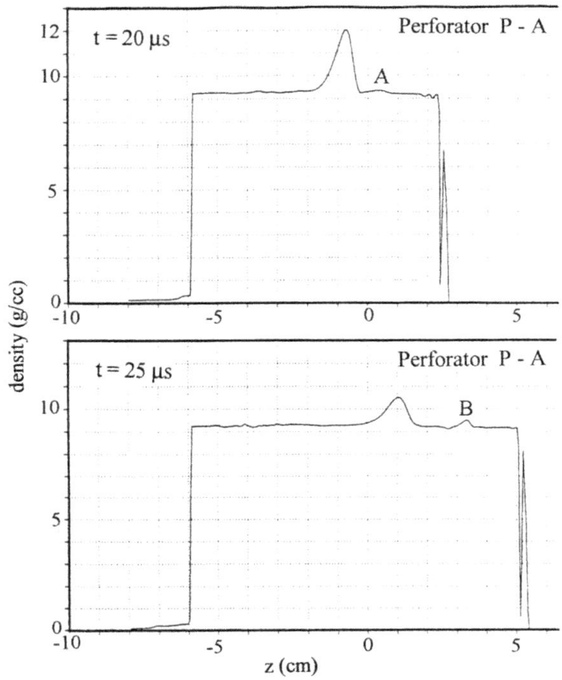

Figure 12.6. The liner density distributions along the z-axis (i.e. $r = 0$) for time 20 and 25 μs with the bulged sections indicated by A and B.

distributions along the z-axis (where $r = 0.0$ cm) for time 20 and 25 μs are shown in Fig. 12.6 with the bulged sections marked by A and B.

Figure 12.7 shows another popular oil perforator P-B, using the same material as described in Fig. 12.4 for the case, the explosive, and the liner. This bell-shaped liner has a non-uniform thickness with thicker liner near sections A and C and a thinner one at section B. The liner jet formations are shown in Fig. 12.8 for time 20 and 25 μs with the bulged sections indicated by A and B which are also shown in Fig. 12.9 for the density distributions along the z-axis.

12.5 Jet Formation and Penetration as Compared with Experimental Data

Figure 12.10 shows a 6.5-cm-charge diameter perforator with copper-lead as liner, RDX as high explosive, and steel as case. The apex section of the

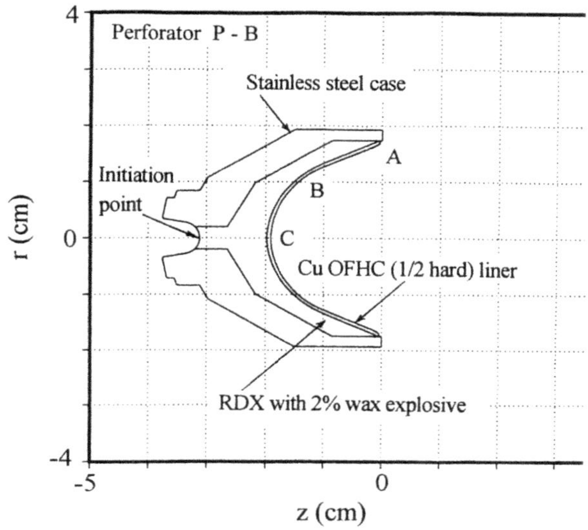

Figure 12.7. The initial setup for the bell-shaped perforator with copper liner, RDX main charge, and steel case.

liner has a hole with diameter of 0.927 cm. The thickness of the liner at the end point B is 0.05202 cm, where at A, the thickness is 0.14136 cm.

High-explosive burn is initiated at point D ($r = 0.0\,\text{cm}$, $z = -6.4\,\text{cm}$). The formations of the liner jet at time 24, 34, and 44 μs are shown in Fig. 12.11. In general, the lengths of the jet from the code calculations match the data quite well. However, the location of the puff-shaped jet tip from the data is slightly different from the calculation. At time 44 μs, the experimental result shows a broken section of jet formed near $z = 4.5\,\text{cm}$ which is not seen from the calculation. The jet velocities from the computation are in good agreement with the data. The penetration of the liner jet into the steel block is shown in Fig. 12.12 for code calculation (solid line) at time 200 μs and the experimental data (dashed line). The initial configuration and the initial location of the perforator are also shown. The stand-off, i.e. the distance between the perforator and the steel block, is 5.0 cm. The penetration depth is 8.3 cm as obtained by both experiment and calculation. The crater diameter at $z = 5.0\,\text{cm}$ is approximately 4.0 cm for the experiment and 5.0 cm for the calculation. Although the crater size at $z = 5.0\,\text{cm}$ has a 25% discrepancy compared with the data, the penetration depth is in excellent agreement. The copper liner (with 20% weighted

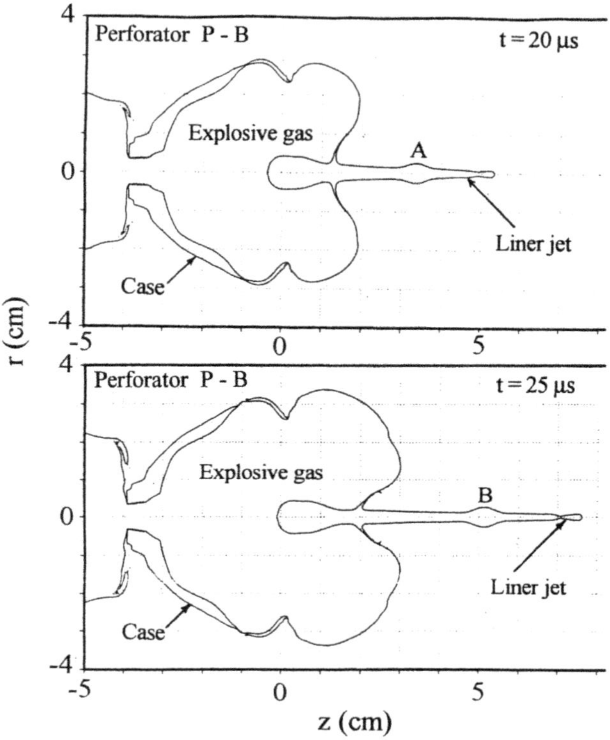

Figure 12.8. The liner jet formation at time 20 and 25 μs with the bulged sections indicated by A and B.

lead) of the perforator as shown in Fig. 12.13 comprises three linear segments, i.e. $\overline{AB}$, $\overline{BC}$, and $\overline{CD}$. The angles between these segments and the z-axis are slightly different with $\angle ABE > \angle CDG > \angle BCF$. Each liner segment has uniform thickness except the apex section which is composed by two hemispherical segments of thicker liner. The case of the perforator is made of stainless steel where the explosive is of RDX with 2% weighted wax and the HE is initiated at point H. The stand-off between the perforator and the steel block is 2.76 cm. Both the code calculation and the experimental data give the same penetration depth and crater size. However, the bulged section of the penetration hole is located at K ($z = 4.0$ cm) by the calculation and at M ($z = 5.5$ cm) by the experiment. Also, the code calculation shows another bulged section at N ($z = 11.3$ cm) which is not seen in the experiment.

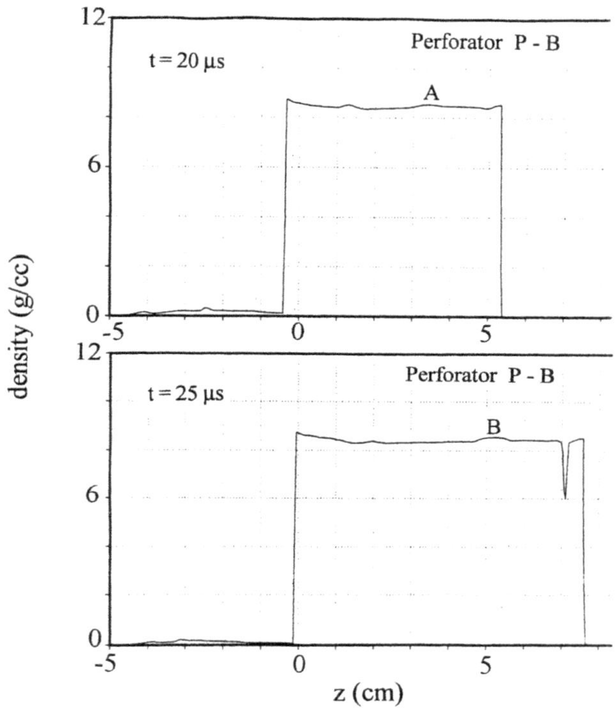

Figure 12.9. The liner density distributions along the z-axis for time 20 and 25 μs with the bulged sections indicated by A and B.

12.6 Perforators of 4.4 cm Charge Diameter

In this section, four different perforators of 4.4-cm-charge diameter are presented with their performance of penetration into the casing wall. The liner section AXB of the perforator P-E, shown in Fig. 12.14, is made of bronze (0.9 weighted Cu + 0.1 weighted Sn). Near the apex section, the spherical surface BC (0.5 cm in radius and centered at $r = 0.0$ cm, $z = 2.47$ cm) is the bare surface of the explosive RDX (with 2% weighted wax) which is initiated at $r = 0.0$ cm and $z = -3.77$ cm. In the upper plot of Fig. 12.14, the red color portion AXB is the liner, the yellow color is the high explosive, the green color DEFWVU is the iron case, the dark blue color HKGLM is the gun wall, dark blue color KNPLG is the casing wall, and the green color NRSP is the Westerly granite. A blowup plot of the perforator is given in the lower portion of Fig. 12.14 which shows a tapered liner, thicker at point A and thinner near point B. There are seven blue markers inside the liner

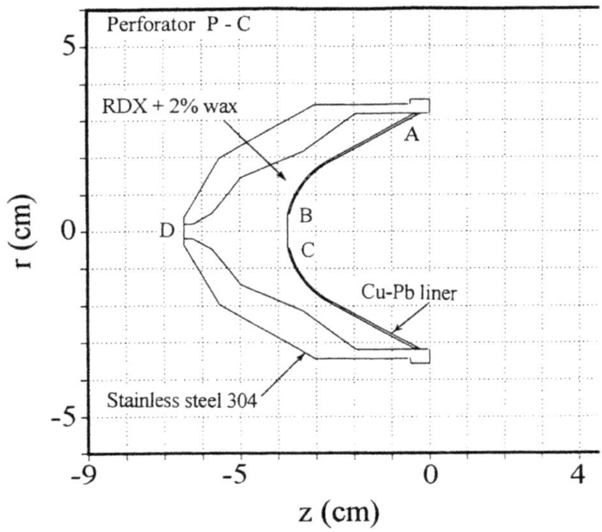

Figure 12.10. The initial setup for a copper-lead liner consisting of a spherical apex section with a hole along the axis of symmetry and a tangent truncated conical section.

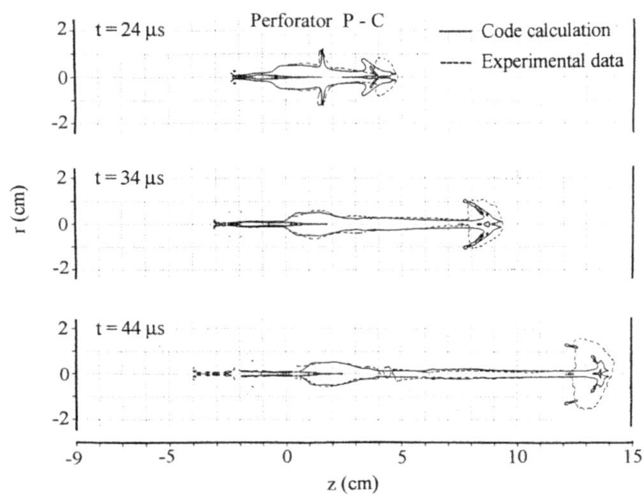

Figure 12.11. The jet formation of the perforator *P-C*, shown in Fig. 12.10, at times 24, 34, and 44 μs with code calculation in solid line and experimental data in dashed line.

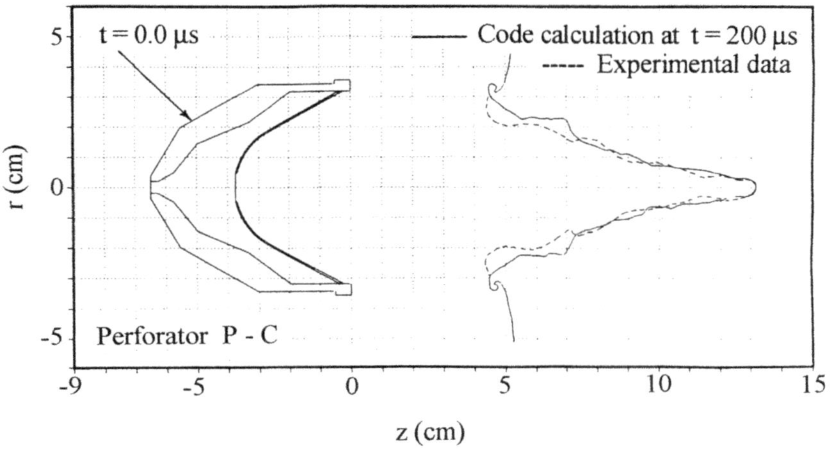

Figure 12.12. The initial location of the perforator *P-C* with 5 cm in stand-off (left). The formation of the crater as calculated by code (solid line) at $t = 200\,\mu$s, and experimental data (dashed line).

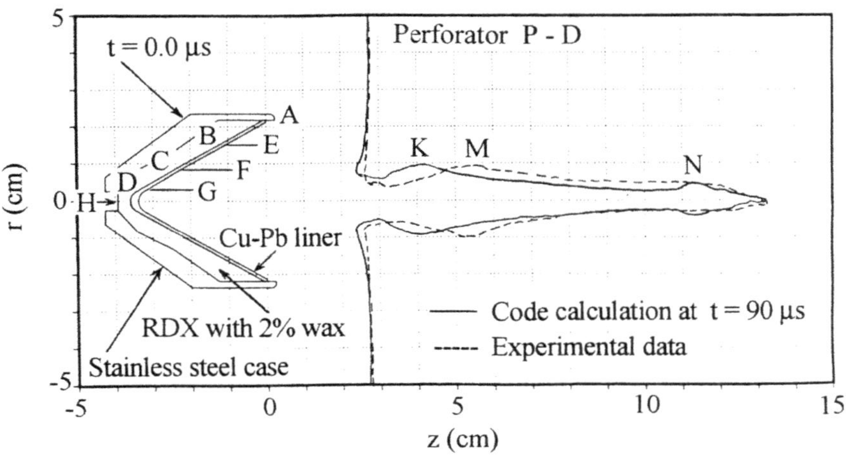

Figure 12.13. The initial location of the perforator *P-D* with 2.76 cm in stand-off (left). The formation of the crater as calculated by code (solid line) at $t = 90\,\mu$s and experimental data (dashed line).

for tracing the motion of the liner during the jet formation. Figure 12.15 shows the formation of the jet and the jet penetration into the gun wall, the casing wall, and the granite outside the casing wall. There is no clearance between the gun and the casing walls except the small concave gap on the

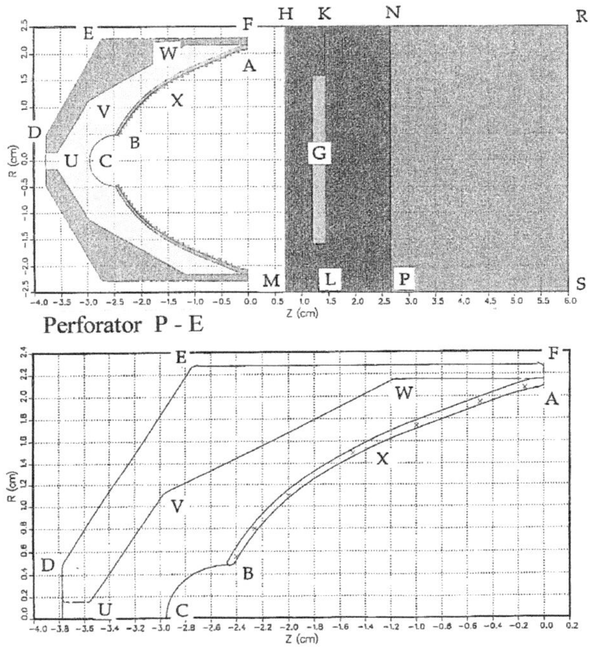

Figure 12.14. This is a 4.4-cm charge diameter bronze liner with a 1.0-cm diameter hole located at the apex region, and the stand-off distance is 0.685 cm. The high explosive is RDX with 2% wax. The perforator case is made of iron, whereas the gun and outside casing tube are steel. The material outside the well casing is Westerly granite. Problem setup is shown on the upper plot with the perforator configuration on the lower plot.

gun wall to reduce the clogging with the casing wall. The top plot shows the jet formation at time 10 μs. The middle plot gives the penetration of both walls at time 30 μs. The lower plot shows the jet penetration into the granite and the broken jet near $z = 2.0$ cm at time 50 μs. The liner of the perforator *P-F*, shown on the top portion of Fig. 12.16, is made of two different materials with section AXY of copper (OFHC 1/2 hard) and section BXY of bronze (0.9 weighted Cu + 0.1 weighted Sn). Spherical surface BC (0.5 cm in radius and centered at $r = 0.0$ cm, $z = 2.47$ cm) is the bare surface of the explosive RDX (with 2% weighted wax) that is initiated at point I ($r = 0.0$ cm, $z = -3.77$ cm). The iron case is indicated by DEFSWVUT. There are eight markers inside the liner which trace the liner location during the jet formation: four for copper and four for bronze. The liner of the

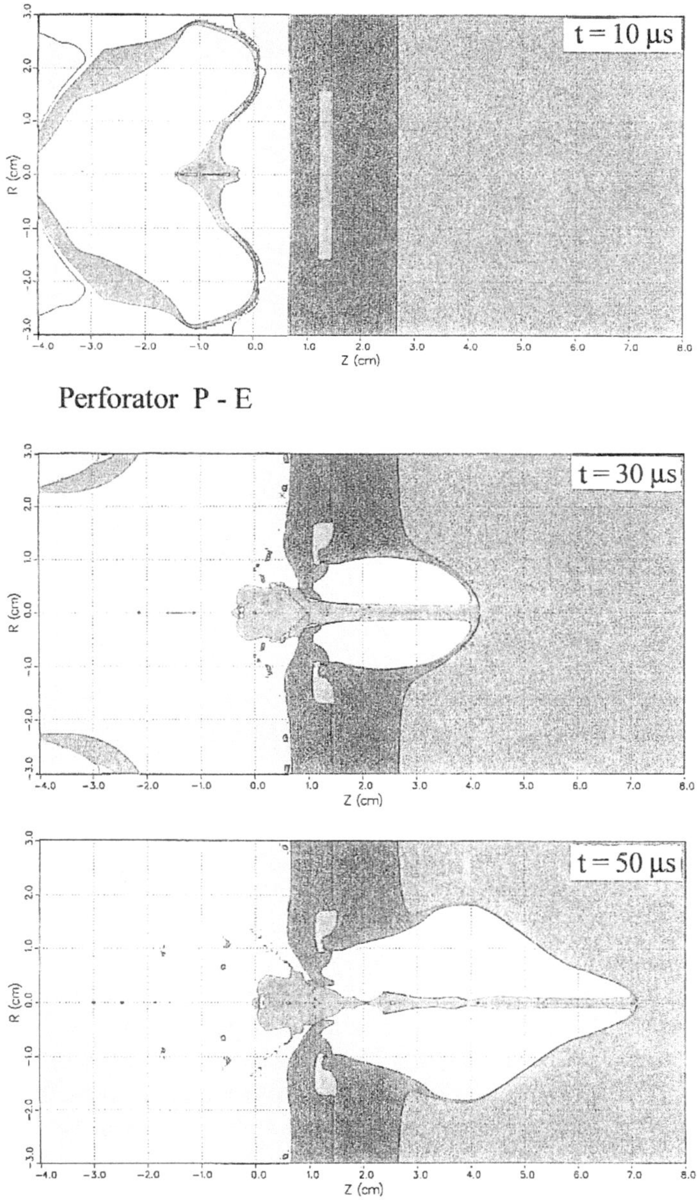

Figure 12.15. The formation of the jet and the jet penetration into the gun, casing walls, and the granite is shown at time $10\,\mu s$ (top), $30\,\mu s$ (middle), and $50\,\mu s$ (bottom).

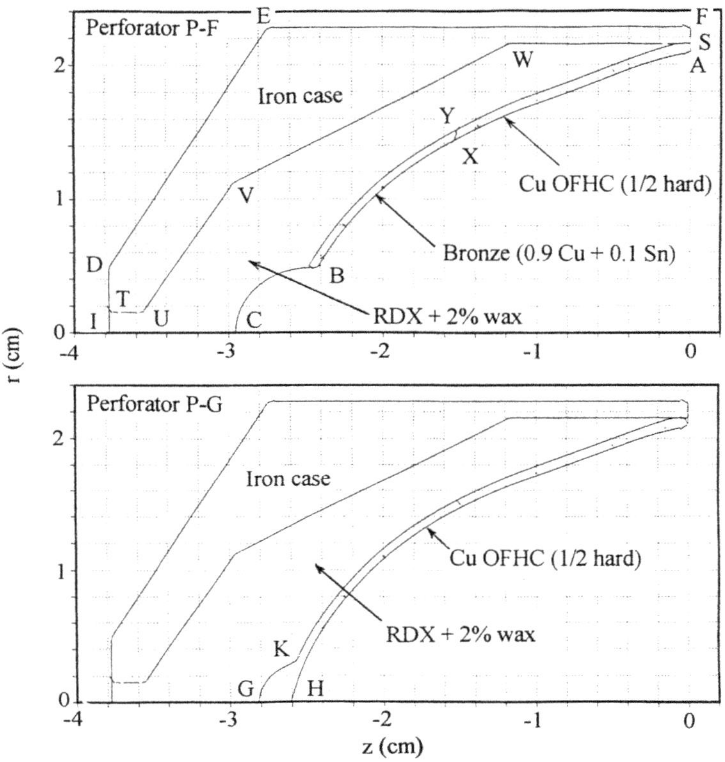

Figure 12.16. The initial configurations of perforators *P-F* (upper plot) and *P-G* (lower plot).

perforator *P-G*, shown in the lower plot of Fig. 12.16, has only one material, i.e. copper (OFHC 1/2 hard). The dimension and the thickness of this liner is the same as that of perforator *P-F* except that the convex portion as indicated by KGH near the apex section is different. Figure 12.17 shows another perforator *P-H* which has the same configuration for the case, the explosive, and the liner as perforator *P-F* shown in the upper plot of Fig. 12.16, except that the copper liner AXY is shorter and the bronze liner BXY is longer. The total masses of the case, the explosive, and the liner for perforator *P-E*, *P-F*, *P-G*, and *P-H* are provided in Table 12.3.

If we replace the perforator *P-E* in the upper plot of Fig. 12.14 by perforator *P-F* and use the same stand-off distance, i.e. 0.685 cm, then we will get a set of calculations similar to Fig. 12.15. The same calculations are

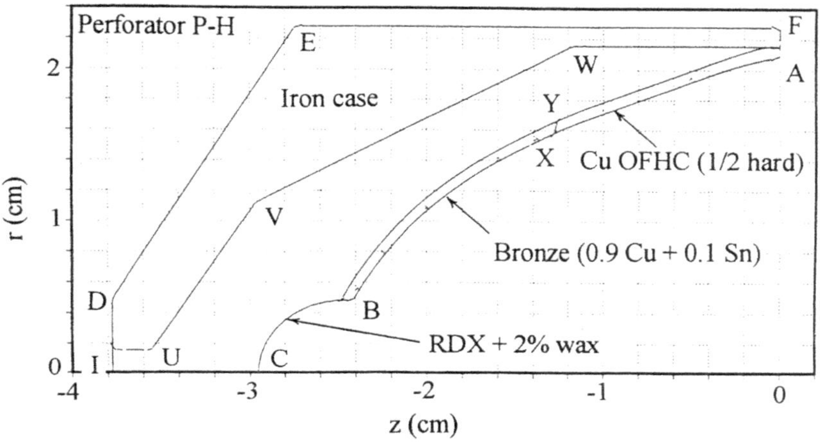

Figure 12.17. The initial configuration of perforator *P-H*.

Table 12.3. Masses (g) of the case, the explosive, and the liner.

Perforator	*P-E*	*P-F*	*P-G*	*P-H*
Case (Iron)	143.198	143.198	142.881	142.881
Explosive (RDX)	19.902	19.902	20.0	19.97
Liner (Bronze)	15.919	4.068	16.89	6.169
Liner (Copper)		11.8		10.169

performed for perforator *P-G* and *P-H*. The casing hole sizes versus the water gaps between the gun wall and casing wall are plotted in Fig. 12.18 for the perforator *P-E*, *P-F*, *P-G*, and *P-H*. From Fig. 12.18, it is obvious that the perforator *P-F* should be used if only one type of perforator is all permitted. However, if a spacer (not much deviates from 1.27 cm) is designed to set the water gap, then one can use the perforator *P-G* for that particular direction (i.e. water gap of 1.27 cm) and use the perforator *P-F* for the rest directions. This will produce the best perforation for the casing wall.

12.7 Perforators of 3.6 cm Charge Diameter

Four perforators, i.e. *P-I*, *P-J*, *P-B*, and *P-K*, of 3.6-cm charge diameter are described in this section with their penetrations into the casing

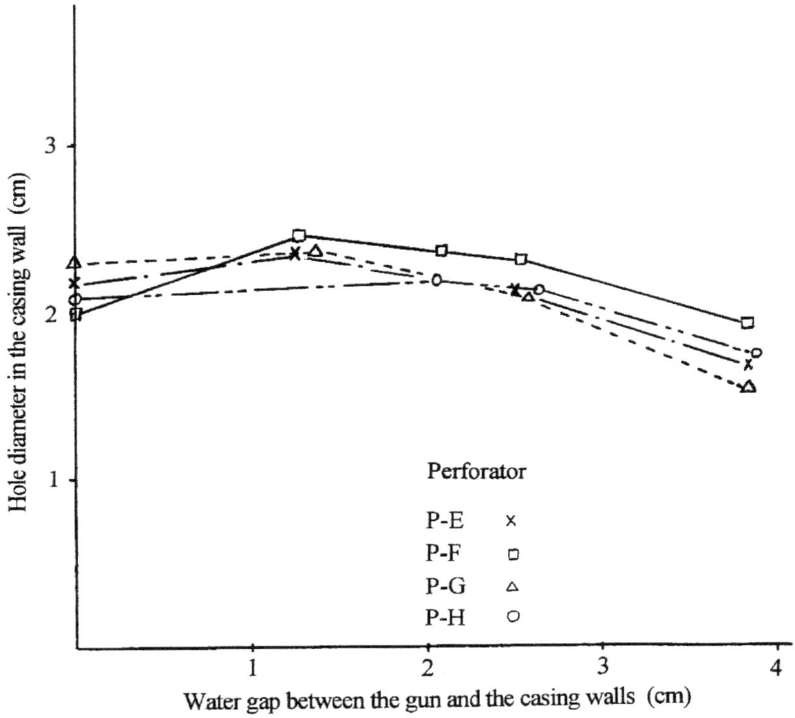

Figure 12.18. The hole diameter in the casing wall as a function of water gap for perforators *P-E*, *P-F*, *P-G*, and *P-H*.

wall. These four perforators use stainless steel for the case, RDX (with 2% weighted wax) for the explosive, and copper (OFHC, 1/2 hard) for the liner. The upper plot of Fig. 12.19 shows the initial setup of the perforator *P-I* with the steel gun wall QYZLM, the steel casing wall YNPLZ, and the oil shale NRSP. The stand-off, i.e. the distance between the perforator and the inside wall of the gun, is 1 cm. The gap between the gun wall and casing wall is zero except that a slot, indicated by Z, on the outside surface of the gun wall is made to reduce the clogging with the casing wall. A blowup plot of the perforator *P-I* is given in the lower portion of Fig. 12.19 where AX is the thicker section of the liner, XB is the thinner one, and BC is a much thicker spherical section. Section DEFGHKWVUT is the stainless steel case. Between the case and the liner is the RDX explosive which is initiated at point *I*. The best design of the 3.6 cm charge diameter perforator is the one shown on the top portion of Fig. 12.20. This perforator, *P-J*, has

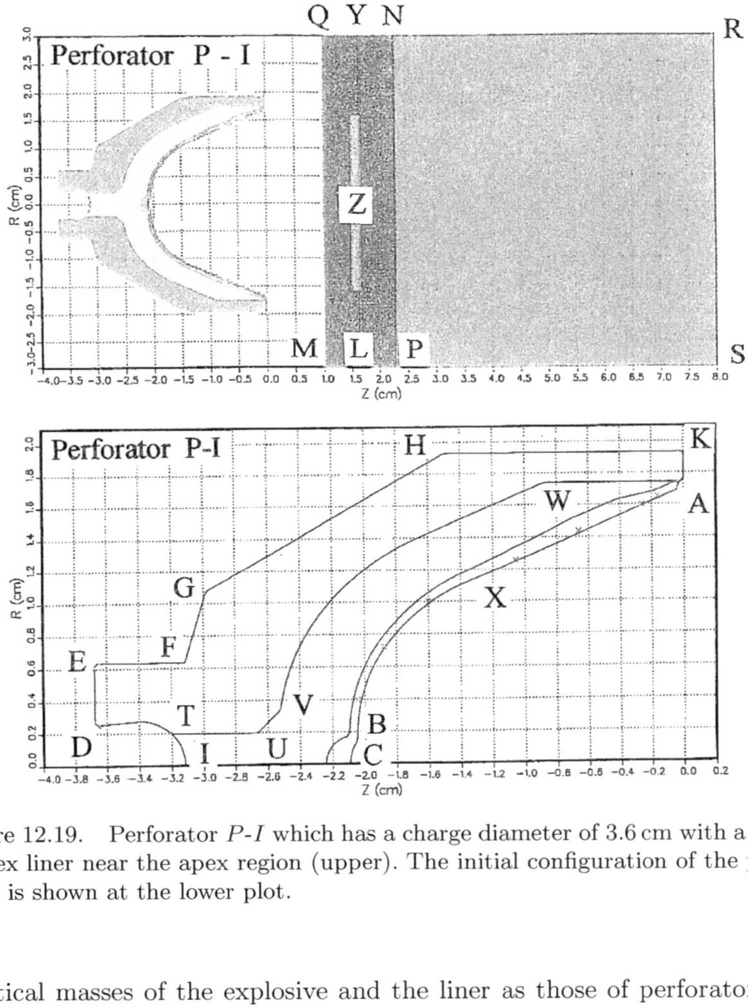

Figure 12.19. Perforator *P-I* which has a charge diameter of 3.6 cm with a small convex liner near the apex region (upper). The initial configuration of the perforator is shown at the lower plot.

identical masses of the explosive and the liner as those of perforator *P-I* except that the stainless steel case of the perforator *P-J* is slightly thicker. Perforator *P-B*, shown in the lower portion of Fig. 12.20, is a widely used device. Figure 12.21 shows another perforator *P-K* with less mass of liner but more mass of the steel case as compared with the perforator *P-J*. The masses of the case, the explosive, and the liner are given in Table 12.4 for perforators *P-I*, *P-J*, *P-B*, and *P-K*.

If we replace perforator *P-I* by *P-J*, *P-B*, or *P-K* and repeat the same calculations for different water gaps, then we will get the results plotted in Fig. 12.22. From Fig. 12.22, one will choose perforator *P-J* if only one

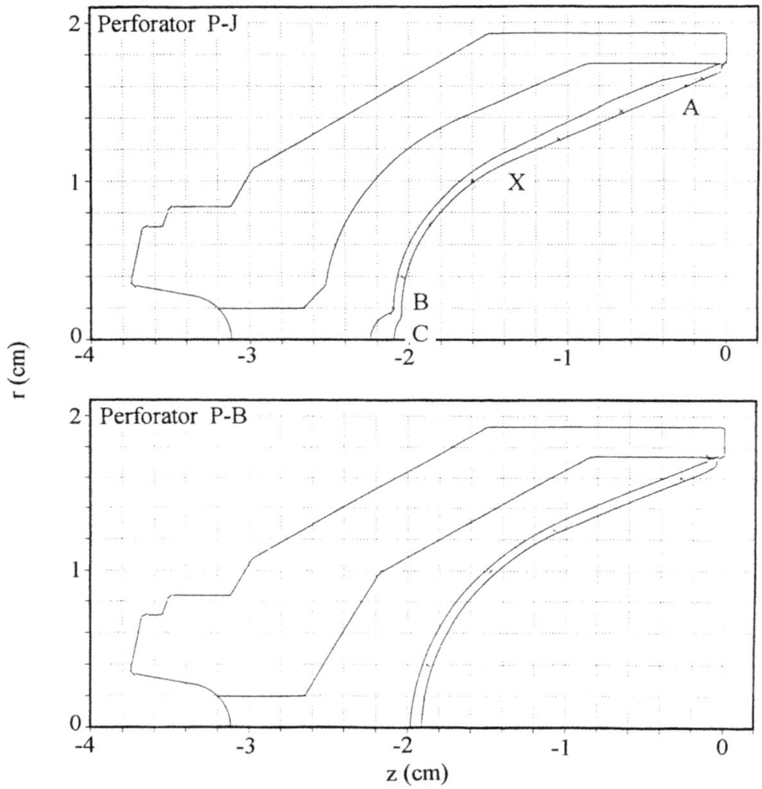

Figure 12.20. The initial configurations of perforators $P\text{-}J$ and $P\text{-}B$.

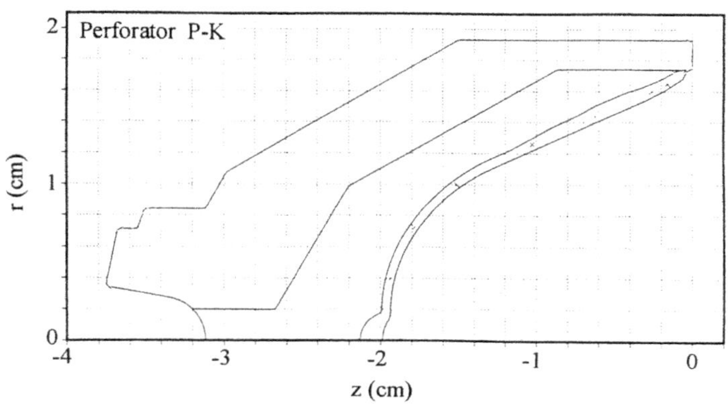

Figure 12.21. The initial configuration of perforator $P\text{-}K$.

Table 12.4. Masses (g) of the case, the explosive, and the liner.

Perforator	*P-I*	*P-J*	*P-B*	*P-K*
Case (Stainless steel)	143.198	143.198	142.881	142.881
Explosive (RDX, 2%wax)	19.902	19.902	20.0	19.97
Liner (Copper)	12.4656	12.4656	9.37998	11.8067

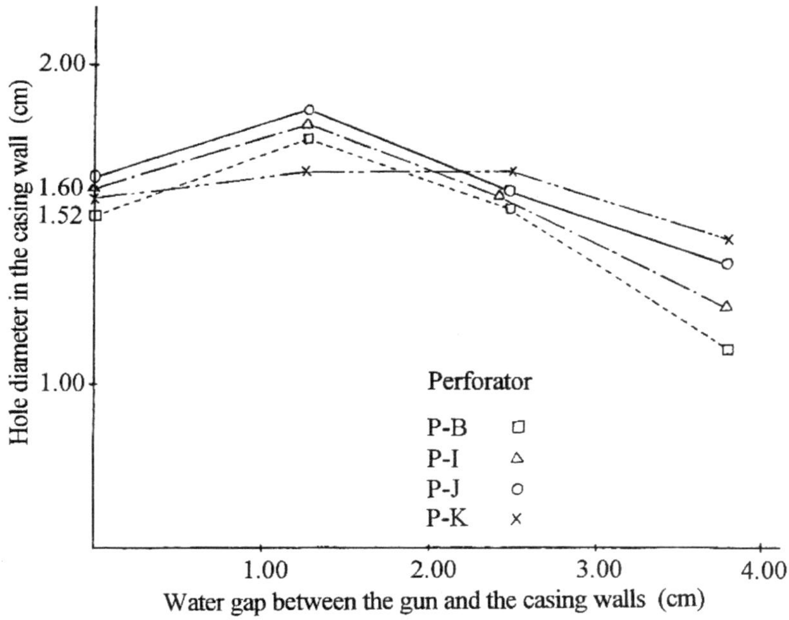

Figure 12.22. The hole diameter in the casing wall as a function of water gap for perforators *P-B*, *P-I*, *P-J*, and *P-K*.

type of device is allowed. However, as described in the previous section, one can use perforator *P-K* for the water gap (space between gun and casing walls) greater than 2.54 cm and use perforator *P-J* for the gap smaller than 2.54 cm. Since the liner masses of perforator *P-I*, *P-J*, and *P-K* are approximately equal to 12.0 g and their explosive masses are identical, one can thus plot the casing hole sizes versus the masses of the stainless steel case. This is shown in Fig. 12.23 with the hole size in the vertical coordinate and case mass in the horizontal one. From this plot, it is obvious, on the average, that the perforator *P-J* is better than *P-I* and *P-K*.

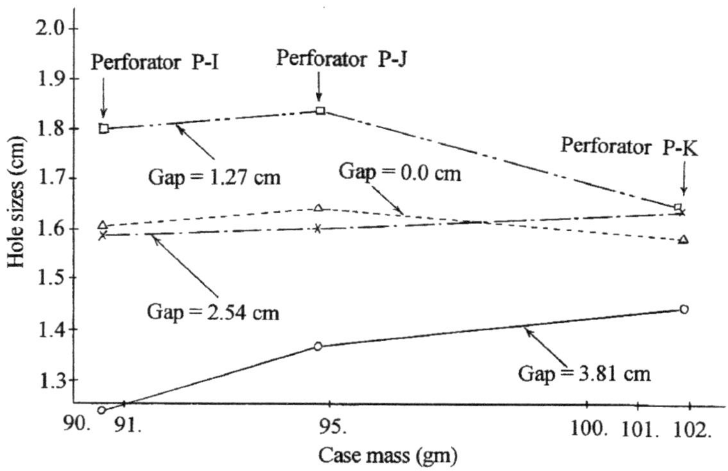

Figure 12.23. The hole diameter in the casing wall as a function of case mass where the masses of the explosive and the liner are fixed for perforators *P-I*, *P-J*, and *P-K*.

12.8 Discussions

The design philosophy between oil well perforator and conventional shaped charge is quite different. For designing an oil well perforator, one has to consider the following stringent factors:

(1) The stand-off is small since the space is limited inside the gun pipe.
(2) The high explosive should be able to survive in a harsh environment.
(3) The cost of the perforator should be low.
(4) A good perforator requires varied thickness along the length of the liner. It is not easy to achieve the desired tolerance since most of the liners are made by pressing the liner material powder into the designed shape.
(5) A good perforator should produce a double velocity gradient between 20 and 25 μs after the main charge ignites. This can be done by using a liner with various thickness, while the conventional shaped charges do not usually have this requirement.

The important items for designing a good oil well perforator are

(1) The perforated hole in the oil casing wall should be large.
(2) The perforated hole in the gun wall should be small.

(3) The high explosive should be able to sustain a harsh environment, e.g. with temperature higher or equal to 260°C.
(4) The penetration into the surrounding rock or granite should be as deep as possible.
(5) The slug portion of the liner should not plug the oil casing hole.

If one only considers items 1 and 4, then the problem of designing an oil well perforator is the same as that for a conventional weapon, i.e. shaped charge. However, besides items 1 and 4, the oil well perforator also has to take items 2, 3, and 5 into account, which makes the design very complicated. For meeting the requirements of items 1 and 2, one has to optimize the shape, the thickness, and the dimension of the liner. This has been done by trial and error through the laboratory experiments for the last fifty years. Fortunately, due to the advanced development of computer software, most of the tedious design work can be handled by 2D or 3D Eulerian codes today.

As shown in Fig. 12.18, for the perforators of 4.4 cm charge diameter, perforator *P-F* will be the best choice for item 1. Also, perforator *P-J* meets item 1 the best as shown in Fig. 12.22 for the perforators of charge diameter 3.6 cm. For item 2, the present studies show that the perforated hole size in

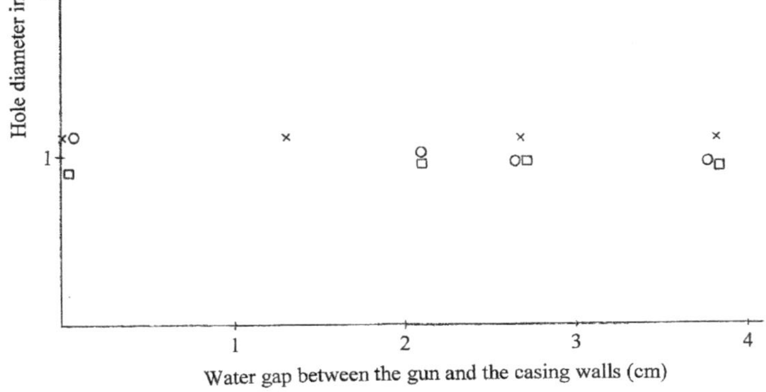

Figure 12.24. The hole diameter in the gun wall as a function of water gap for perforators *P-E*, *P-F*, and *P-H*.

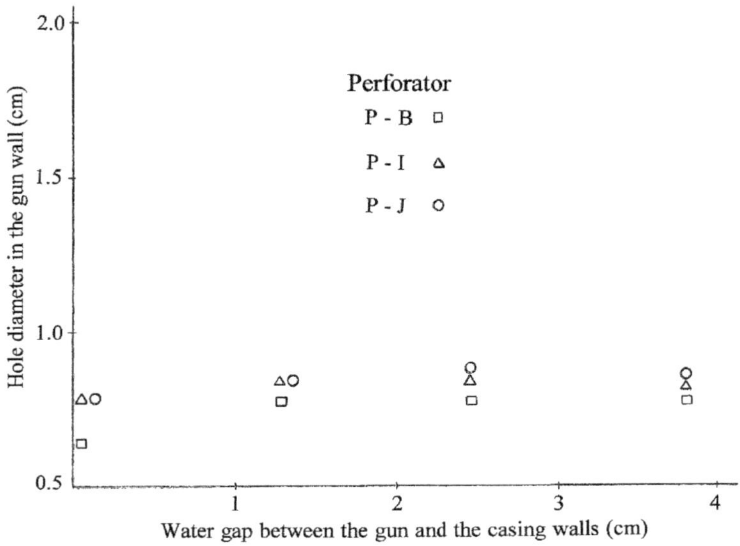

Figure 12.25. The hole diameter in the gun wall as a function of water gap for perforators *P-B*, *P-I*, and *P-J*.

the gun wall is not sensitive to either water gap between the gun and the casing walls or the type of perforators as shown in Fig. 12.24 for perforators *P-E*, *P-F*, and *P-H* and in Fig. 12.25 for perforators *P-B*, *P-I*, and *P-J*.

Item 3 deals with the harsh environment near the wellbore location. In principle, energetic explosive is preferred due to its high chemical energy contents. But, for safety reasons, the explosive should be able to sustain high temperature. One of the industrial design parameters for producing the double velocity gradients is varying the explosive density along the axial axis. However, in the current study, we use uniform density and no material strength for the explosive. In modeling the high explosive burning process, a reactive burn model is better than the programmed burn (present studies) due to the 90° turning angle near point U in Fig. 12.14. At this turning point, the detonation velocity of the explosive burning front will be slower to accommodate the divergent cross-section, i.e. $\overline{UV}$. Since the reactive burn model requires fine mesh for calculations, it may become impractical for industrial application. In this case, one may consider the modification of the detonation velocity due to the boundary curvature using a simple method [12.9].

Item 4 requests a deep penetration into the granite surrounding the casing wall. From the present studies, if a perforator will produce a large hole in the casing wall, it will also penetrate deeper into the granite. Perforator *P-E* is a good device for creating a large hole in the casing wall (see Fig. 12.18) and for penetrating deeply into the granite (see Fig. 12.26). All of the liners in this study is made by pressing the powder liner material with a small amount of wax. Therefore, the slug portion of the liner jet will not plug the casing hole as it has been verified by experiments. This will satisfy the requirement of item 5.

It is interesting to note that the perforator *P-E*, shown in Fig. 12.14, with a 1-cm-diameter hole (without liner) at the apex region, will produce good size hole in the casing wall. Due to the inertia effect, the 1-cm-diameter cavity section results in higher velocity for the material near the apex area. This is why perforator *P-E* is still considered a good device. If the surface of the explosive cavity, i.e. the 1-cm-diameter hole, is coated with a thin layer of liner material, the perforation in the casing wall will be much better. The calculation of this design has been done, but the result is not discussed here.

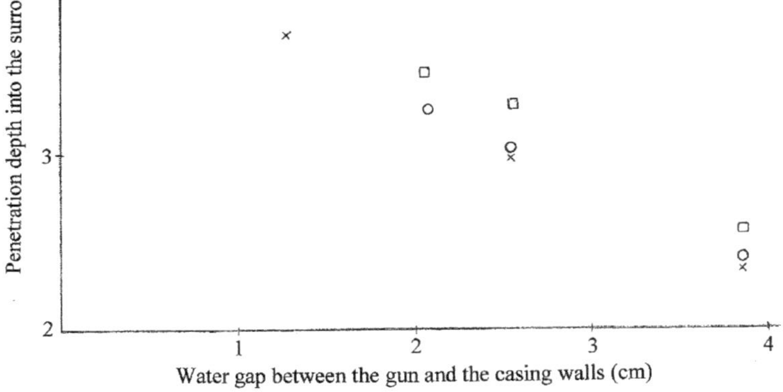

Figure 12.26. The penetrating depth into the surrounding granite as a function of water gap for perforators *P-E*, *P-F*, and *P-H*.

The thickness of the perforator case as well as the liner will affect the performance of the liner jet formation and the penetration into the oil well casing. Since the detonation waves (compression waves) from the high explosive, the rarefection waves from the free surfaces of the liner, and the perforator case react with each other, the situation becomes so complicated that it is impossible to calculate them by analytical method. The detailed physics can only be obtained by using multi-dimensional code simulations. In the lower plot of Fig. 12.14, the liner near point A is the thickest one while point B is the thinner one and point X is somewhat in-between. Liner surface AX comprises a straight line where XB is made up of four different circular arcs. These dimensions and thickness come from many experiments and calculations obtained by trial and error.

12.9 Conclusions

In this study, we use a 2D Eulerian code to simulate the performance of the oil perforator and present the character of the double velocity inverse gradient which is related to the perforated hole sizes in the gun and casing walls.

Experimental testing is still the best way to obtain a good oil perforator. However, due to the advances of computer hardware and software, it is possible today to design a perforator using a multi-dimensional code with well-defined material properties.

The objective of well-bore perforation is to increase the conductivity between the reservoir and well-bore so that the oil or gas may flow into the well. However, after production for some time, the permeability of rock formation around the well-bore will decrease due to the migration of mud particles and fines into rock pores (also known as formation damage). In order to restore the conductivity, the general practice in the industry is to introduce a hydraulic fracture from the well-bore penetrating into the rock formation. The perforated tunnels, in this operation, will serve as initiators for this massive hydraulic fracture. It is clearly demonstrated in the previous sections that the performance of a shaped charge gun can be greatly enhanced by a properly designed liner. A properly designed gun would produce a clean tunnel with great depth.

In general, the requirement for a good gun is its ability to produce a large and clean surface in the rock medium in the vicinity of well-bore for oil and gas to flow with least amount of resistance into the well-bore. This leads to the development of linear shaped charge (LSC) gun. A LSC gun would produce a narrow and long slot (approximately 0.64 cm by 20 cm) along the well casing and a narrow elliptical opening (yet to be verified) extending from well-bore into the surrounding rock medium. The elliptical opening has much larger exposed area to the rock medium than the cylindrical tunnel produced by the perforators described in the previous sections. Furthermore, a narrow elliptical opening would be a much more effective initiator for hydraulic fracture. It becomes clear that a properly designed linear shaped charge (LSC) gun would produce a better result than its conical or bell-shaped counterpart.

References

12.1 Mclemore, RL, World Oil, July 8 (1946).
12.2 Lee, WH, and Painter, J, Material void-opening computation using particle method, Int. J. Impact Eng., 22, pp. 1–22 (1999).
12.3 Noh, WF, and Woodward, P, SLIC (simple line interface calculation), Report UCRL-77651, Lawrence Livermore National Laboratory (1976).
12.4 Dobratz, BM, LLNL Explosive handbook — Properties of chemical explosive and explosive simulants, UCRL-52997 (1981).
12.5 McQueen, RG, Marsh, SP, Tayler, JW, Fritz, JN, and Carter, WJ, The equation of state of solids from shock wave measurements, High Velocity Impact, edited by R. Kinslow, Academic Press, New York (1970), Chapter 7.
12.6 Tillotson, JH, Metallic equations of state for hypervelocity impact, General Atomic Report GA-3216, July (1962).
12.7 Johnson, GR, and Cook, WH, A constitutive model and data for metals subjected to large strains, high strain rate, temperatures and pressures, Proc 7th Int. Symp. on Ballistics, the Netherlands, The Hague, pp. 541–547 (1983).
12.8 Steinberg, DJ, Equation of state and strength properties of selected materials, UCRL-MA-106439, Lawrence Livermore National Laboratory, February (1991).
12.9 Aslam, TD, Bdzil, JB, and Stewart, DS, Level set methods applied to modeling detonation shock dynamics, J. Compu. Phys., 126, pp. 390–409 (1996).

APPENDIX A

ONE DIMENSIONAL RADIATION HYDRODYNAMICS

Notations

B	Planck function
c	speed of light (3×10^{10} cm/s)
C_v	specific heat at constant volume
e	defined by Eq. (A.15.3)
E, E_R	radiation energy density
f	function defined as $f = P_{rr}/E$
$\bar{F}, \bar{F}_R$	radiation flux vector
$\bar{g}$	arbitrary vector
G_i	defined by Eq. (A.90)
H_i	defined by Eq. (A.91)
$I(\bar{r}, \nu, \bar{\Omega}, t)$	specific intensity of the radiation field defined as the rate of energy flow per unit frequency and solid angle across a unit area oriented normal to the direction of propagation at point $\bar{r}$, frequency ν, in the direction $\bar{\Omega}$, at time t.
m	mass(g), mass of electron in Eqs. (A.15.1) and (A.15.2)
N_e	free electron density
P_m	material pressure
P_{rr}	r-r component of the radiation pressure tensor
$\bar{\bar{P}}, \bar{\bar{P}}_R$	radiation pressure tensor
$\dot{q}$	external energy source density
Q	defined by Eq. (A.19)
$\bar{r}$	position vector
r_0	electron radius
$r^{\alpha-1}$	for $\alpha = 1, 2, 3$, the equation is for planar, cylindrical and spherical geometries respectively
t	time
T	defined by Eqs. (A.18)
W	used by Eq. (A.24), i.e. $W = h\nu/\theta$

Greek Letters

α defined by Eq. (A.15.2)
γ defined by Eq. (A.15.1)
ε defined by Eq. (A.140)
θ temperature ($^\circ K$)
κ Rosseland opacity (cm^2/g)
λ defined by Eqs. (A.111), (A.112), (A.115) (cm)
λ_R defined by $\lambda_R = \mu_R^{-1}$
μ_a absorption coefficient (1/cm)
μ'_a absorption coefficient corrected for stimulated emission (1/cm)
μ_s scattering coefficient (1/cm)
ν frequency (1/sec)
ζ number of mean free path, defined by Eq. (A.108) (no unit)
ξ_1 defined by Eq. (A.15.4)
ξ_2 defined by Eq. (A.15.5)
ρ density (g/cm^3)
τ specific volume (cm^3/g)
ϕ azimuthal angle
$\bar{\Omega}$ unit vector in the direction of the photon transport

Subscripts

0 initial value
i finite difference at grid location i
ij finite difference at grid location i and frequency group j
t derivative with respect to time
ν for a particular frequency ν

Superscripts

n time at n time-step, i.e. $t^n = t_0 + n \cdot \Delta t$

A.1 Introduction

This Appendix discusses the physical, mathematical, and computational aspects of one dimensional radiation hydrodynamic transport, especially under conditions of high temperature. Lagrangian coordinate is used for solving problems in both planes, and spheres and under certain assumptions, in cylinders.

The major aspect of this Appendix lies in the radiative transport approximation which combines unconditional stability with treatment of retardation and Compton scattering. In the Variable Eddington (VE) approximation, it is assumed that there exist a linear relationship between the radiation energy density and the r-r component of the pressure tensor. For one-dimensional geometries in planes and spheres, this assumption is sufficient to close the system of moment equations of the radiation transport operator and hence render them soluble. The VE approximation uses both the Rosseland and Planck mean opacities and is capable of describing the radiation transport in optically thick and thin systems for which retardation may be important. VE approximation is also used to describe grey or multi-group calculations. Energy exchange between the material and the radiation field due to Compton scattering and the attendant spectrum changes are also taken into account in an approximation valid for $h\nu \leq 50\,\text{keV}$.

The advantages of the VE approach can be summarized as follows:

(1) The number of operations performed and the number of quantities stored are significantly smaller for the VE code than for a full transport code, thereby contributing to code economy and efficiency.
(2) The VE equations can be formulated numerically in an unconditionally stable fashion so that large time intervals are feasible.
(3) The equations give the transport solutions in optically thick and thin regions and show promise of useful accuracy in intermediate regions.

The assumptions and derivation of the constitutive equations are documented in Section A.2. The difference approximations for these equations and the numerical analysis are included in Section A.3. Sections A.4 and A.5 describe the form of the 1D first moment equation and the stability and damping of the wave equation.

A.2 Mathematical Theory of Radiation Hydrodynamics

A.2.1 Introduction

The principal phenomena to be considered in this appendix are hydrodynamic motion and radiation transport. The emphasis will be on effects of interest at high temperatures in which the material is partially or wholly ionized and Compton scattering occurs. The solution of the radiative-hydrodynamic equations for a gas from its microscopic properties can be divided into two parts:

(1) The evaluation of the cross-sections of the pertinent interactions from atomic physics.
(2) The solution of various moment quantities of the Boltzmann equation.

This Appendix is concerned with the second aspect of the problem. The conservation laws of mass, momentum, and energy will be developed in a form convenient for numerical solution.

The solutions to the radiation transport equation that have been found in many cases requiring sophisticated mathematical analysis correspond to highly ideal physical situations. Consequently, new applications require greater faithfulness to the materials and configurations of practical problems. The following is a discussion and analysis of an approximation to the radiation transport equation which contains the physical effects of interest and yet is numerically efficient and stable.

The diffusion approximation to the equation of radiative transfer has proven to be a valuable and economical method for obtaining solutions to a wide range of nonlinear frequency and time-dependent problems. The numerical analysis of the diffusion approximation, as applied to astrophysical problems, has been generalized [A.1] to take account of the deviation from equilibrium between the radiation field and the material through which the radiation is diffusing. In the latter treatment, the Eddington diffusion approximation, $\nabla; \bar{\bar{P}} = \frac{1}{3}\nabla E$, relating the radiation pressure tensor ($\bar{\bar{P}}$) and the radiation energy (E) is employed. The conventional diffusion approximation differs from the latter scheme in two respects: (1) it is assumed in several terms that the radiation energy has the blackbody value associated with the material temperature, and (2) the term in the first moment equation arising from retardation is neglected. The Eddington diffusion approximation, in contrast, is capable of describing the propagation of waves with velocity $c/\sqrt{3}$, deviations of the spectrum from Planckian, and transient and spatial effects associated with the equilibration of radiation and material.

An extension of the above method to problems in which the Eddington diffusion approximation is not suitable, such as for streaming radiation, has been developed and is called the Variable Eddington (VE) approximation. This approximation assumes that the r-r component of the radiation pressure tensor P_{rr} is related to E through a time-, frequency-, spatially-varying function f; $P_{rr} = fE$. The evaluation of this function allows one to take account of the streaming properties of the radiation field. Moreover, in the limit $f = 1$, light waves assume the correct velocity c. If the function $f(\bullet)$ is accurately known then one is no longer solving a diffusion approximation to the transport equation, but is, in fact, solving the exact transport equations.

The remainder of this Appendix discusses the theoretical basis for the radiative hydrodynamic transport. In Section A.2.2, the defining equations of non-equilibrium diffusion are developed in the VE approximation. The Compton scattering terms to the non-equilibrium diffusion equations are developed in Section A.2.3. The effects of solving the radiation equations in an inertial frame are considered in Section A.2.4. The conservation laws of mass, momentum, and energy are shown in Section A.2.5.

A.2.2 Radiation Transport Equation

The interaction of the photons with the material must be taken into account in order to describe the way the retardation intensity changes in space and time along the photon ray. Contributions are made to the rate of change of I, $\left(\frac{1}{c}\frac{\partial}{\partial t} + \bar{\Omega}\cdot\nabla\right)I(\bar{\Omega}, \nu, \bar{r}, t)$, by absorption and scattering processes in the material.

To obtain these contributions, certain assumptions are made. Polarization of the photons is assumed to be negligible. All of the processes in which a photon is absorbed are described by the absorption coefficient $\mu_a(\theta, \rho, \nu)$. The absorption coefficient is a property of the material which can be calculated assuming local thermodynamic equilibrium. The arguments θ and ρ are the material temperature and density. The scattering processes are those in which a photon is deflected from its original direction $\bar{\Omega}$, into the direction $\bar{\Omega}'$. In the Thomson limit, these processes are characterized by a coefficient of scattering,

$$\mu_s = \frac{8}{3}\pi r_0^2 N_e \tag{A.1}$$

where N_e is the free electron density and r_0 is the classical electron radius (2.8×10^{-13} cm). Therefore, the deflected scattering term is

$$\mu(\Omega, \Omega') = \frac{3}{16\pi}\mu_s[1 + (\bar{\Omega}\cdot\bar{\Omega}')^2]\,. \tag{A.2}$$

In local thermodynamic equilibrium the radiation source term is given by the Planck function B where $B_\nu(\theta) = (2h/c^2)\nu^3/(e^{h\nu/\theta} - 1)$.

Thus, the transport equation in the Thomson limit of the Compton cross-section for scattering is found to be

$$\left(\frac{1}{c}\frac{\partial}{\partial t}+\bar{\Omega}\cdot\nabla\right)I(\bar{\Omega},\nu,\bar{r},t)=\mu_a'(B-I)-\mu_s I+\frac{3\mu_s}{16\pi}\int d\Omega'[1+(\bar{\Omega}\cdot\bar{\Omega}')^2 I(\bar{\Omega}')] \tag{A.3}$$

where μ_a' is the absorption corrected for stimulated emission, i.e.

$$\mu_a'=\mu_a(1-e^{-h\nu/\theta})\,. \tag{A.4}$$

To complete the specifications of the transport problem, the inward intensity at all points on the boundary of the system must be known.

The first three moments of I, $E_\nu, \bar{F}_\nu, \bar{\bar{P}}_\nu$, referred to as the radiative energy density, flux vector, and pressure tensor respectively, are defined as

$$E_\nu=\frac{1}{c}\int I d\Omega\,, \tag{A.5}$$

$$\bar{F}_\nu=\int I\bar{\Omega}d\Omega\,, \tag{A.6}$$

$$\bar{\bar{P}}_\nu=\frac{1}{c}\int I\overline{\Omega\Omega}d\Omega\,, \tag{A.7}$$

The equations for the rate of change of E_ν and $\bar{F}_\nu$ are obtained by integrating Eq. (A.3) over $d\Omega$ and $\bar{\Omega}d\Omega$ for the contributions to the energy and momentum equations per unit photon energy. Thus one has

$$\frac{\partial E_\nu}{\partial t}+\nabla\cdot\bar{F}_\nu=c\mu_a'\left(\frac{4\pi B_\nu}{c}-E_\nu\right), \tag{A.8}$$

and

$$\frac{1}{c}\frac{\partial \bar{F}_\nu}{\partial t}+c\nabla\cdot\bar{\bar{P}}_\nu=-(\mu_a'+\mu_s)\bar{F}_\nu\,. \tag{A.9}$$

Higher moment equations could be generated. But in each case, the number of dependent variables exceeds the number of equations. Consequently, the system of equations must be closed by some approximation. The Variable Eddington approximation assumes that the r-r component of the radiation pressure tensor is related to the energy density by

$$P_{rr}=fE\,, \tag{A.10}$$

where f is some function of space, frequency and time. The function f must be constrained to a limit of $1/3$ in a diffusion regime thereby recovering the Eddington approximation. It must approach the value 1 in a region of unidirectional photon flow.

The expansion of Eqs. (A.8) and (A.9) into component form in orthogonal curvilinear coordinates is carried out in Section A.4. Using Eq. (A.10) and writing the results in plane or spherical ($\alpha = 1$, or 3) one-dimensional geometry, one has

$$\frac{\partial E_\nu}{\partial t} + \frac{1}{r^{\alpha-1}}\frac{\partial\left(r^{\alpha-1}F_\nu\right)}{\partial r} = c\mu_a'\left(\frac{4\pi B_\nu}{c} - E_\nu\right), \tag{A.11}$$

and

$$\frac{1}{c}\frac{\partial F_\nu}{\partial t} + c\left[\frac{\partial(fE_\nu)}{\partial r} + \frac{\alpha-1}{2r}(3f-1)E_\nu\right] = -(\mu_a' + \mu_s)F_\nu \tag{A.12}$$

The situation for cylindrical geometry is more complicated since the components of the pressure tensor cannot be eliminated by a single function of E. However, as discussed in Section A.4, if the intensity I is independent of the azimuthal angle ϕ or if $\int_0^{2\pi} I\sin^2\phi d\phi = \frac{I}{2}$, then, Eq. (A.11) and (A.12) with $\alpha = 2$ for the cylindrical case is valid. Notice that in the case of radiation streaming in vacuo with $f = 1$, Eqs. (A.11) and (A.12) are reduced to the correct wave equations with velocity c:

$$\frac{\partial E_\nu}{\partial t} + \frac{1}{r^{\alpha-1}}\frac{\partial\left(r^{\alpha-1}F_\nu\right)}{\partial r} = 0, \tag{A.13}$$

$$\frac{1}{c}\frac{\partial F_\nu}{\partial t} + c\frac{1}{r^{\alpha-1}}\frac{\partial(r^{\alpha-1}E_\nu)}{\partial r} = 0. \tag{A.14}$$

To solve the problem, the boundary conditions for this coupled set of partial differential equations must be determined. Section A.3.2 discusses this point.

A.2.3 Non-equilibrium Diffusion Equation with Compton Scattering

The effect of energy exchange through Compton scattering can also be taken into account in the non-equilibrium diffusion equations in the Variable Eddington approximation. To determine the form of the scattering terms in the defining equations, it is necessary to form the zeroth and first moments of the radiative transfer equation in which Compton scattering is included.

Compton scattering of soft x-rays from free electrons has been treated in the Fokker-Planck approximation and has been specialized to plane or spherical geometry. The result is

$$\left(\frac{1}{c}\frac{\partial}{\partial t}+\bar{\Omega}\cdot\nabla\right)I_1=\mu_a'(B-I_1)-\mu_s\left\{(1-2\gamma)I_1-\frac{3}{16}\int d\mu' I_2\right.$$

$$\times\left[3-\mu^2+(3\mu^2-1)(\mu')^2\right]-\left(1+\frac{h^2c^2I_1}{e^3}\right)$$

$$\times\int d\mu'\left(3\gamma I_2-\frac{\partial\gamma^2 I_2}{\partial\gamma}\right)\xi_1+\alpha\left[\int d\mu' I_2\xi_2\right.$$

$$\left.\left.+\int d\mu'\left(\frac{\partial^2\gamma^2 I_2}{\partial\gamma^2}-6\frac{\partial\gamma I_2}{\partial\gamma}+4I_2\right)\xi_1\right]\right\} \quad (A.15)$$

where

$$\gamma=\frac{h\nu}{mc^2}\,, \quad (A.15.1)$$

$$\alpha=\frac{\theta}{mc^2}\,, \quad (A.15.2)$$

$$e=h\nu\,, \quad (A.15.3)$$

$$\xi_1=3-\mu^2+\mu(3\mu^2-5)\mu'+(3\mu^2-1)(\mu')^2+\mu(3-5\mu^2)(\mu')^3 \quad (A.15.4)$$

$$\xi_2=2(3\mu^2-1)[1-3(\mu')^2]+4\mu(1-3\mu^2)\mu'+4\mu(5\mu^2-3)(\mu')^3 \quad (A.15.5)$$

and $I_1=I(\gamma,\mu)$ and $I_2=I(\gamma,\mu')$ are specific intensities. The frequency subscripts are retained hereafter only where required for clarity.

The first two moments with respect to μ of Eq. (A.15) for one-dimensional geometry are now

$$\frac{\partial E}{\partial t}+\frac{1}{r^{\alpha-1}}\frac{\partial(r^{\alpha-1}F)}{\partial r}=c\mu_a'\left(\frac{4\pi B}{c}-E\right)+T \quad (A.16)$$

and

$$\frac{1}{c}\frac{\partial F}{\partial t}+c\left[\frac{\partial(fE)}{\partial r}+\frac{\alpha-1}{2r}(3f-1)E\right]=-(\mu_a'+\mu_s)F \quad (A.17)$$

The scattering contribution to the photon energy (zeroth moment) equation is

$$T = -\frac{c\mu_s}{mc^2}\left\{eE - \frac{\partial e^2 E}{\partial e} - \theta\left(4E - 6\frac{\partial eE}{\partial e} + \frac{\partial^2 e^2 E}{\partial e^2}\right)\right.$$
$$+ \frac{3c^3h^3}{32\pi e^3}\left[(3E - P)\left(3eE - \frac{\partial e^2 E}{\partial e}\right) + (3P - E)\left(3eP - \frac{\partial e^2 P}{\partial e}\right)\right.$$
$$\left.\left. + \frac{1}{c^2}(3Q - 5F)\left(3eF - \frac{\partial e^2 F}{\partial e}\right) + \frac{1}{c^2}(3F - 5Q)\left(3eQ - \frac{\partial e^2 Q}{\partial e}\right)\right]\right\}. \tag{A.18}$$

In Eqs. (A.16), (A.17), and (A.18), an additional moment of the radiation intensity has been introduced, namely

$$Q = 2\pi \int I\mu^3 d\mu \tag{A.19}$$

In the zeroth moment equation, considering T, there are no terms of zero order in γ or α; consequently, terms of first order are retained which account for the (small) exchange of energy by Compton and inverse Compton scattering events. In the first moment equation, a zero order term is present representing the large effect of scattering on the transport mean free path. This term was present in the derivation in Section A.2.2 of the Thomson limit. Higher order terms in the first moment equation are neglected compared to this term.

Only the quadratic terms representing the effects of induced scattering contain higher moments of the radiation field than the first. In order to close the calculation, it is necessary to adjoin an expression for Q as well as for P. In diffusion regions $3F = 5Q$, while in streaming regions (radiation traveling in a beam, distant from a localized source), $F = Q$.

These limiting cases are contained in the following formula containing f, constituting an interpolation expression for which f is simply a conveniently available quantity.

$$Q = \frac{1}{5}(3f + 2)F \tag{A.20}$$

This expression, together with the Variable Eddington approximation for P, substituted in Eqs. (A.16) and (A.17), might represent a logical extension of the Variable Eddington approximation to the Compton scattering terms. It is clear, however, that these terms are quite expensive to form. In view of the fact that under most circumstances these terms only contribute a few percent to T, it is very doubtful that the added computational investment is justified. Consequently, several coarser approximations to the induced terms are given below.

First, derivatives of f are dropped from Eqs. (A.16) and (A.17). Next, rearrangement of the Variable Eddington approximation shows

$$T = -\frac{c\mu_s}{mc^2}\left\{eE - \frac{\partial e^2 E}{\partial e} - \theta\left(4E - 6\frac{\partial eE}{\partial e} + \frac{\partial^2 e^2 E}{\partial e^2}\right)\right.$$
$$+ \frac{3c^3h^3}{32\pi}\left[\left(\frac{3}{2} - f + \frac{3}{2}f^2\right)\left(\frac{E^2}{e^2} - \frac{\partial E^2/e}{\partial e}\right)\right.$$
$$\left.\left. + \frac{1}{10c^2}(-17 + 6f - 9f^2)\left(\frac{F^2}{e^2} - \frac{\partial F/e}{\partial e}\right)\right]\right\}. \quad \text{(A.21)}$$

In Eq. (A.21) when $f = 1/3$ and $F = 0$ the induced terms are simplified to give

$$T = -\frac{c\mu_s}{mc^2}\left\{eE - \frac{\partial e^2 E}{\partial e} - \theta\left(4E - 6\frac{\partial eE}{\partial e} + \frac{\partial^2 e^2 E}{\partial e^2}\right)\right.$$
$$\left. + \frac{c^3h^3}{8\pi}\left(\frac{F^2}{e^2} - \frac{\partial F/e}{\partial e}\right)\right\}. \quad \text{(A.22)}$$

when $f = 1$ and $F = cE$, as occurs in streaming, the induced terms completely cancel out. It is desirable to try to attain these limiting expressions without evaluating the additional term of Eq. (A.21) containing F, since the latter would require additional time-consuming interpolations. A scheme for achieving the indicated behavior is

$$T = -\frac{c\mu_s}{mc^2}\left\{eE - \frac{\partial e^2 E}{\partial e} - \theta\left(4E - 6\frac{\partial eE}{\partial e} + \frac{\partial^2 e^2 E}{\partial e^2}\right)\right.$$
$$+ \frac{3c^3h^3}{32\pi}\left(\frac{E^2}{e^2} - \frac{\partial E^2/e}{\partial e}\right)\left[\left(\frac{3}{2} - f + \frac{3}{2}f^2\right)\right.$$
$$\left.\left. + \frac{F^2}{c^2E^2}(-1.7 + 0.6f - 0.9f^2)\right]\right\}. \quad \text{(A.23)}$$

Even simpler, the quantity $(-1.7 + 0.6f - 0.9f^2)$ could be given its value when $f = 1$, namely -2.0.

In summary, the moment equations to be simulated numerically are Eqs. (A.16) and (A.17) in which T of Eq. (A.23) is to be added to the right-hand side of the energy equation.

A.2.4 Non-equilibrium Diffusion Equations in an Inertial Frame

In deriving the previous moment equations the assumption was inherently made such that the local macroscopic velocity was zero. This assumption is required if the absorption coefficient is to be isotropic. The radiation transport equation is valid only in an inertial frame of reference. Consequently, the transformation between the co-moving frame, in which the absorption coefficient μ_0' is known, and the inertial frame moving with relative velocity u must be performed.

In the following discussion, terms of order u^2/c^2 will be neglected. For simplicity the effects of Compton scattering will not be considered. Only the pure absorption and emission processes are taken into account in the LTE approximation. The equation of radiative transfer in an inertial frame with the absorption coefficient evaluated in the co-moving frame, is given by

$$\begin{aligned}\left(\frac{1}{c}\frac{\partial}{\partial t}+\bar{\Omega}\cdot\nabla\right)&I(\bar{\Omega},\nu)\\ &=\left[B_\nu^0-I(\bar{\Omega},\nu)\right]\left[\mu_0'(\nu)-\frac{\bar{u}\cdot\bar{\Omega}}{c}\times\left(\mu_0'(\nu)+\nu\frac{\partial\mu_0'(\nu)}{\partial\nu}\right)\right]\\ &\quad+W\frac{\bar{u}\cdot\bar{\Omega}}{c}\frac{\mu_0'(\nu)B_\nu^0}{(1-e^{-W})}\end{aligned}\tag{A.24}$$

where

$I(\bar{\Omega},\nu)$ = specific intensity in inertial frame
B_ν^0 = Planck function evaluated in the co-moving frame
$\mu_0'(\nu)$ = absorption coefficient corrected for induced emission evaluated in the co-moving frame
W = $h\nu/\theta$ with the frequency evaluated in the inertial frame

The correction terms containing u/c arise from the Doppler shift of the radiation resulting from the change in frequency which occurs when the two frames moving relative to each other are interrelated.

Using this transport equation one can examine the magnitude of the errors involved in using Eqs. (A.16) and (A.17). For simplicity, scattering effects will be ignored and f will be assumed to be 1/3. Taking the zeroth and first moment of the above transport equation one finds

$$\frac{\partial E_\nu}{\partial t} + \nabla \cdot \bar{F}_\nu = c\mu_0'(\nu)\left(\frac{4\pi B_\nu^0}{c} - E_\nu\right) + \frac{\bar{u}}{c}\bar{F}_\nu\left[\mu_0'(\nu) + \nu\frac{\partial \mu_0'(\nu)}{\partial \nu}\right], \tag{A.25}$$

$$\begin{aligned}\frac{1}{c^2}\frac{\partial \bar{F}_\nu}{\partial t} + \nabla \cdot \bar{\bar{P}}_\nu = {} & -\frac{\bar{F}_\nu \mu_0'(\nu)}{c} - \frac{4\pi}{3}\frac{\bar{u}}{c}B_\nu^0\left[\mu_0'(\nu) + \nu\frac{\partial \mu_0'(\nu)}{\partial \nu}\right] + \frac{\bar{u}}{c}\bar{\bar{P}}_\nu \\ & \times\left[\mu_0'(\nu) + \nu\frac{\partial \mu_0'(\nu)}{\partial \nu}\right] + \frac{4\pi}{3}\frac{\bar{u}}{c}B_\nu^0\,\mu_0'(\nu)W(1 - e^{-W})^{-1},\end{aligned} \tag{A.26}$$

in which the pressure tensor is to be replaced by the Variable Eddington approximation. This replacement, however, is appropriate only in the co-moving frame of the material and not in the inertial frame of reference. Consequently, it is desirable to relate the quantity P measured in the inertial frame to the corresponding quantity P^0 measured in the co-moving or rest frame of the material. Two steps are required: transformation to the rest frame with a corresponding change in frequency and expansion to restore the frequency to its original value. These modifications are made in the context of Eqs. (A.25) and (A.26) in which terms up to the first order in u/c are retained.

The following first order transformations for I, ν, and $\bar{\Omega}$ are employed,

$$I(\bar{\Omega}, \nu) = I^0(\bar{\Omega}^0, \nu^0)\left(1 + \frac{\bar{u}}{c}\bar{\Omega}^0\right)^3, \tag{A.27}$$

$$d\Omega = d\Omega^0\left(1 - \frac{\bar{u}}{c}\bar{\Omega}^0\right)^2, \tag{A.28}$$

$$\bar{\Omega} = \left(\bar{\Omega}^0 + \frac{\bar{u}}{c}\right)\left(1 - \frac{\bar{u}}{c}\bar{\Omega}^0\right)^2, \tag{A.29}$$

$$\nu = \nu^0\left(1 + \frac{\bar{u}}{c}\bar{\Omega}^0\right). \tag{A.30}$$

Using the definitions of the various moments one can then show the first order in u/c

$$\bar{\bar{P}}(\nu) = \bar{\bar{P}}^0(\nu) + \frac{1}{c^2}[\bar{u}\bar{F}(\nu) + \bar{F}(\nu)\bar{u}] - \left(1 + \nu\frac{\partial}{\partial \nu}\right)\frac{\bar{u}}{c^2}\cdot\bar{\bar{Q}}(\nu) \tag{A.31}$$

and

$$E_\nu = E_\nu^0 + \left(1 - \nu\frac{\partial}{\partial \nu}\right)\frac{\bar{u}}{c^2}\cdot\bar{F}_\nu \tag{A.32}$$

The diffusion approximation constitutes an assumption about the angular distribution of the radiation field in the co-moving frame of reference in the form

$$I^0(\bar{\Omega}, \nu) = I_0 + I_1 \cos\theta \,, \tag{A.33}$$

where θ is the angle between the flux and $\bar{\Omega}$. In the one-dimensional geometries, the direction of the flux is the direction along the coordinate axis. We express the two constants I_0 and I_1 in terms of E^0 and F^0:

$$E^0 = \frac{4\pi I_0}{c} \,, \tag{A.34}$$

$$F_i^0 = \frac{4\pi I_1}{3}\delta_{i1} \,, \tag{A.35}$$

where the 1-direction is taken along the direction of the flux and the 2- and 3-direction are mutually perpendicular. All of the higher moments of the radiation distribution in the diffusion approximation can be expressed in terms of E^0 and F_1^0 as follows:

$$P_{ij}^0 = \frac{E^0}{3}\delta_{ij} \tag{A.36}$$

$$Q_{111}^0 = \frac{3}{5}F_1^0 \tag{A.37}$$

$$Q_{122}^0 = Q_{133}^0 = Q_{212}^0 = Q_{313}^0 = Q_{221}^0 = Q_{331}^0 = \frac{1}{5}F_1^0 \tag{A.38}$$

All of the elements of Q not specified are zero.

We are now prepared to make the diffusion approximation in Eqs. (A.25) and (A.26). These equations, with the substitutions derived above, constitute two relations for E and F. Equation (A.25) is exact. Equation (A.26), however, requires the replacement of P_ν in two places. The second term is of the first order in u/c and can be replaced directly by Eq. (A.36) since the correction terms are of higher order and consistently should be neglected. The first term is of zero order and requires the use of Eqs. (A.31) and (A.32) in addition to Eq. (A.36) to define the higher radiation moments in the diffusion approximation. This gives

$$P_{ij} = \frac{1}{3}E\delta_{ij} + \frac{u_i}{c^2}F_j + F_i\frac{u_j}{c^2} - \left(1 - \nu\frac{\partial}{\partial\nu}\right)\frac{u_i}{c^2}F_j - \left(1 + \nu\frac{\partial}{\partial\nu}\right)\frac{u_k}{c^2}Q_{kij} \tag{A.39}$$

The leading term is the familiar diffusion term. All the first order correction terms contain the flux; terms involving the frequency derivative of the flux account for the transformations required in relating the rest and laboratory frames.

The moment equations that are written out in detail containing terms of u/c become extremely difficult to handle numerically. Consequently, terms of order u/c will be ignored in the moment equations.

A.2.5 Conservation Laws

In this section, the effect of the radiation field on the conservation laws for mass, momentum, and energy are discussed. Using first order in u/c, one finds for

Conservation of mass:

$$\rho \frac{D\tau}{Dt} - \nabla \cdot \bar{u} = 0\,, \tag{A.40}$$

where

$$\tau = \frac{1}{\rho}\,. \tag{A.40.1}$$

Conservation of momentum:

$$\rho \frac{D\bar{u}}{Dt} + \nabla P_m + \frac{1}{c}\left(\frac{1}{c}\frac{\partial \bar{F}_R}{\partial t} + c\nabla \cdot \bar{\bar{P}}_R\right) = 0\,, \tag{A.41}$$

Rate of change of internal energy:

$$\rho \frac{DE}{Dt} + \rho P_m \frac{D\tau}{Dt} - \dot{q} + \left(\frac{\partial E_R}{\partial t} + \nabla \cdot \bar{F}_R\right) - \frac{\bar{u}}{c}\left(\frac{1}{c}\frac{\partial \bar{F}_R}{\partial t} + c\nabla \cdot \bar{\bar{P}}_R\right) = 0\,, \tag{A.42}$$

where $\frac{D}{Dt} = \frac{\partial}{\partial t} + \bar{u} \cdot \nabla$ is the substantive derivative, ρ is the material density, P_m is the material pressure, E is the material specific internal energy, u is the coordinate velocity, and $\dot{q}$ is an external energy source density. The quantities $E_R, \bar{F}_R, \bar{\bar{P}}_R$ are the total radiative energy density, flux velocity, and pressure tensor defined by

$$E_R = \int_0^\infty E_\nu d\nu \tag{A.43}$$

$$\bar{F}_R = \int_0^\infty \bar{F}_\nu d\nu \tag{A.44}$$

and

$$\bar{\bar{P}}_R = \int_0^\infty \bar{\bar{P}}_R d\nu \tag{A.45}$$

In order to solve the radiative-hydrodynamic problem numerically, it is desirable to place the equations in conservative form so that integrations can be performed accurately. The form $\frac{D}{Dt}$ and $\nabla \cdot \bar{g}$ can be integrated over volume without approximation since

$$\int dv \frac{Df}{Dt} = \frac{D}{Dt} \int f dv \,, \tag{A.46}$$

and

$$\int \nabla \cdot \bar{g} dv = \int \bar{g} \cdot d\bar{s} \,, \tag{A.47}$$

where f and $\bar{g}$ represent an arbitrary scalar and vector.

The radiative terms can be expressed in these forms to incorporate them into difference equations. To obtain the equation for conservation of energy the dot product of $\bar{u}$ and Eq. (A.41) is formed and added to Eq. (A.42), replacing $\frac{D\tau}{Dt}$ via Eq. (A.40).

Conservation of energy:

$$\rho \frac{D}{Dt}\left(E + \frac{1}{2}u^2\right) + \nabla \cdot (P_m \bar{u}) + \frac{\partial E_R}{\partial t} + \nabla \cdot \bar{F}_R - \dot{q} = 0 \,. \tag{A.48}$$

The Eulerian derivative can be eliminated as follows:

$$\begin{aligned} \frac{\partial E_R}{\partial t} &= \frac{DE_R}{Dt} - \bar{u} \cdot \nabla E_R \\ &= \rho \frac{D}{Dt}(E_R \tau) - E_R \rho \frac{D\tau}{Dt} - \bar{u} \cdot \nabla E_R \,. \end{aligned} \tag{A.49}$$

Via Eq. (A.40) one has

$$\frac{\partial E_R}{\partial t} = \rho \frac{D}{Dt}(E_R \tau) - \nabla \cdot (E_R \bar{u}) \,. \tag{A.50}$$

Thus one obtains the energy equation in conservative form,

$$\rho \frac{D}{Dt}\left(E + \frac{1}{2}u^2 + E_R \tau\right) + \nabla \cdot (P_m \bar{u} + \bar{F}_R - E_R \bar{u}) - \dot{q} = 0 \tag{A.51}$$

The total specific energy associated with the material, consisting of material-specific internal and kinetic energies plus the specific radiation energy, changes according to the surface contributions of the work performed by the material pressure, the radiant flux, and the radiation energy convected into the system.

A.3 Numerical Analysis

A.3.1 Difference Equations

This section describes the implicit difference scheme employed for the solution of the non-equilibrium diffusion equations. The form of the difference equations in space and time determines the numerical stability, truncation errors, and ease of solution of the system of equations. Related to the truncation error, or accuracy of solution, are considerations of smoothness of the solution in time and positivity of the radiation energy.

The partial differential equations to be simulated numerically are Eqs. (A.11) and (A.12) including the Compton scattering terms, Eq. (A.23), in the zero moment equation. Rewriting these equations from Section A.2.2, one has

$$\frac{\partial E_\nu}{\partial t} + \frac{1}{r^{\alpha-1}}\frac{\partial\left(r^{\alpha-1}F_\nu\right)}{\partial r} = c\mu_a'\left(\frac{4\pi B_\nu}{c} - E_\nu\right) + T_\nu\,, \tag{A.52}$$

and

$$\frac{1}{c}\frac{\partial F_\nu}{\partial t} + c\left[\frac{\partial(fE_\nu)}{\partial r} + \frac{\alpha-1}{2r}(3f-1)E_\nu\right] = -(\mu_a' + \mu_s)F_\nu\,, \tag{A.53}$$

where the terms are as defined in the previous section. These defining equations are valid in one-dimensional plane or spherical geometry. In Section A.2.5, one noted that to write the energy equation in conservative form the zeroth moment equation was expressed in the Lagrangian frame of reference. For accuracy we likewise will express the zeroth moment equation as follows:

$$\rho\frac{\partial(E_\nu\tau)}{\partial t} + \frac{1}{r^{\alpha-1}}\frac{\partial r^{\alpha-1}(F_\nu - UE_\nu)}{\partial r} = c\mu_a'\left(\frac{4\pi B_\nu}{c} - E_\nu\right) + T_\nu \tag{A.54}$$

In Eqs. (A.52) and (A.53) several terms have been omitted from the moment equations of the radiation transfer equation. The omitted terms arise from the requirement that the absorption coefficient and the source function be evaluated in the reference frame of the moving material rather than the laboratory frame. Except for these omissions, the effect of material velocity has been accounted for in the transformation from Eulerian to

Lagrangian derivative and in the work term associated with the radiation pressure.

It will be ultimately desirable to include the terms which have been omitted. Unfortunately, frequency derivatives for the multigroup formulation are involved which complicates the reduction to difference equations. Moreover, the terms can have an important effect in regions in which the radiation energy is not negligible compared to material energy. In particular, when the equilibrium diffusion approximation is valid, it is clear that the omitted terms contribute to the work term of the radiation pressure; in their absence, the work term has the wrong magnitude and sign. Currently, a correction term has been added to the zeroth moment radiation equation. The added term restores the correct form of the solution in regions where the equilibrium diffusion approximation is valid but is not generally correct otherwise. Fortunately, the terms in question are only significantly large in regions where $|F| \leq uE$. This inequality is not ordinarily satisfied outside of diffusion regions so that the error being made in the above correction should be small.

The correction can be understood in terms of the transformation of radiative quantities from the laboratory reference frame (no superscript) to the center of mass (superscript 0). The pertinent relations between frequency-integrated quantities are given by

$$F = F^0 + uE^0 + u \cdot P^0 \,, \tag{A.55}$$

$$E = E^0 + \frac{2u}{c^2} F^0 \,, \tag{A.56}$$

$$P = P^0 + \frac{u}{c^2} F^0 + F^0 \frac{u}{c^2} \,. \tag{A.57}$$

In a diffusion region the flux is so small that (a) the correction terms may be quite important in the F equation, and (b) $E = E^0$ and $P = P^0$.

When the velocity terms are omitted from Eqs. (A.52) and (A.53) in a diffusion regime, the resulting flux is F^0, the quantity evaluated in the co-moving frame, instead of F. In order to obtain the quantity, $F - uE \simeq F - uE^0 = F^0 + uP^0$, required by the zero moment equation, the zeroth moment equation can be written as

$$\rho \frac{\partial (E_\nu \tau)}{\partial t} + \frac{1}{r^{\alpha-1}} \frac{\partial r^{\alpha-1}(F_\nu + UP_\nu)}{\partial r} = c\mu_a' \left(\frac{4\pi B_\nu}{c} - E_\nu \right) + T_\nu \,, \tag{A.58}$$

where $U = u$, the relative velocity between Lagrangian and Eulerian coordinates. With Eq. (A.58) and the remaining constitutive equations one can now derive the expected conservative equation of total energy from the interaction of radiation and moving material in a diffusion region. The solution of this system of equations for E_ν and F_ν in an implicit manner requires the value of the source term B_ν at the advanced time. To estimate this term, a form of the energy equation is employed. Using the equivalent forms of the radiation terms and including the scattering effects, one can write the internal energy equation as follows:

$$\rho\frac{DE}{Dt} + \rho P_m \frac{D\tau}{Dt} - \dot{q} = \int_0^\infty d\nu \left\{ c\mu'_a \left(\frac{4\pi B_\nu}{c} - E_\nu \right) + T_\nu + O\left(\tfrac{u}{c}\right) \right\}. \tag{A.59}$$

Presently, the terms of $O\left(\frac{u}{c}\right)$ will be neglected except for the convection term. Thus the difference equations to be solved will simulate Eqs. (A.52) and (A.53) with the appropriate form of the zeroth moment equation and Eq. (A.59).

In deriving the difference equations, the first parameter to be quantized will be the frequency interval $(0, \infty)$. The moment equations are integrated over a frequency band (ν_j, ν_{j+1}), the boundaries of which are chosen by the conditions of the problem. The average absorption coefficients are chosen to be Planck average, μ_P, in the zeroth moment equation and the Rosseland average, μ_R or $\lambda_R = \mu_R^{-1}$, in the first moment equation. These choices were made so that the solution to the equations would be properly limited in the optically thin and diffusion cases. Thus, the constitutive equations can be written as

$$\rho\frac{D(E_j\tau)}{Dt} + \frac{1}{r^{\alpha-1}}\frac{\partial r^{\alpha-1}(F_j + UfE_j)}{\partial r} = c\mu_{P_j}\left(a\theta^4 b_j - E_j\right) + T_j\,, \tag{A.60}$$

$$\frac{1}{c}\frac{\partial F_j}{\partial t} + c\left[\frac{\partial(fE_j)}{\partial r} + \frac{(\alpha-1)}{2r}(3f-1)E_j\right] = -\frac{F_j}{\lambda_{R_j}}\,, \tag{A.61}$$

$$\rho\frac{DE}{Dt} = -\rho P_m\frac{D\tau}{Dt} + \dot{q} - \sum_j \left[c\mu_{P_j}\left(a\theta^4 b_j - E_j\right) + T_j\right]. \tag{A.62}$$

where b_j is the integral of the normalized Planck function over (ν_j, ν_{j+1}). Assuming that the internal energy is a function of θ and τ, the material energy equation can be rewritten as

$$\frac{D\phi}{Dt} = \frac{4a\theta^3}{\rho C_V}\left\{-\left(P_m + \frac{\partial E}{\partial \tau}\right)\rho\frac{D\tau}{Dt} + \dot{q} + \sum_j \left[c\mu_{P_j}(E_j - b_j\phi) + T_j\right]\right\}, \tag{A.63}$$

where $\phi = a\theta^4$ and $C_V = (\partial E/\partial\theta)$.

This system of equations is linear in the unknown quantities E_j, F_j and ϕ, when the time dependence is made discrete. The equations are also simultaneous: not only do the unknowns occur together in the three equations, but also the equations are coupled through the space and frequency variables. The spatial coupling enters through the spatial derivatives and the frequency coupling through the j summation in the material energy equation. It is not possible by any simple algorithm to solve the resulting system of simultaneous equations. While iterative and splitting techniques could be applied, a simpler method has been devised. According to this scheme, the coupling between frequency group is removed through the introduction of auxiliary equations replacing the material energy equation. The auxiliary equation for the jth frequency group is

$$\frac{D(b_j\phi)}{Dt} = \frac{4a\theta^3}{\rho C_V} b_j \left[-\left(P_m + \frac{\partial E}{\partial \tau} \right) \rho \frac{D\tau}{Dt} + \dot{q} \right] + \frac{4a\theta^3}{\rho C_V} [c\mu_{P_j}(E_j - b_j\phi) - T_j] . \tag{A.64}$$

This equation is used to evaluate the term $b_j\phi$ which occurs in the zero moment equation for j^{th} group. In essence, the auxiliary equation provides an estimate of the temperature change due to the emission and absorption of radiation in the jth frequency group alone. When this change is taken into account in an implicit way in the jth zero moment equation, the system is rendered unconditionally stable with respect to the emission and absorption within the frequency group. Subsequently, at the end of the time step, the temperature change resulting from the influence of all frequency group is calculated. It is the latter quantity which is used to initialize the temperature at the beginning of the next cycle of the calculation. Even though the coupling between frequency groups is reintroduced in this way, the equations are unconditionally stable with respect to emission and absorption. A weak stability condition from the explicit treatment of the Compton scattering term remains. As is shown below in more detail, the equations which result from the use of the auxiliary equation contain coupling only in the spatial dimension. Having a coefficient matrix of tridiagonal form, they are easily solved by a well known simple algorithm.

The differencing of the system of equations in space and time have been developed with the idea of making the equations compatible with one-dimensional Lagrangian code. The first decision this consideration influences is how E and F will be centered. Since ϕ is a zone-centered quantity, it is clearly desirable that E also be zone-centered in the material energy equation. Figure A.1 will indicate the centering of other variables of interest. It is convenient to use the integral indexing for zone-centered quantities.

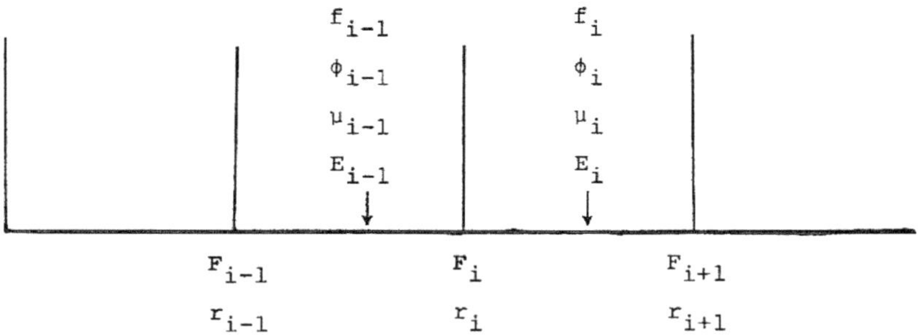

Figure A.1. Centering of variables.

Spatially, the radiation energy equation is differenced in a conservative fashion, i.e. radiation flux out of one zone enters the neighboring zone undiminished. Thus, the equation centered at $i + 1/2$ is

$$\frac{D(V_i E_i)}{Dt} + \tilde{F}_{i+1,j} + A_{i+1}\bar{f}_{i+1}\hat{E}_{i+1,j}U_{i+1} - (\tilde{F}_{ij} + A_i\bar{f}_i\hat{E}_{ij}U_i)$$
$$= cV_i\mu_{P_{ij}}(b_j\phi_j - E_{ij}) + V_iT_i\,, \tag{A.65}$$

where

$$\hat{E}_i = \frac{V_iE_i + V_{i-1}E_{i-1}}{V_i + V_{i-1}}\,, \tag{A.66}$$

$$\tilde{F}_{ij} \equiv \alpha r^{\alpha-1}F_{ij}\,, \tag{A.67}$$

and

$$\bar{f}_i = \left(\frac{f_i\mu_i\Delta r_i + f_{i-1}\mu_{i-1}\Delta r_{i-1}}{\mu_i\Delta r_i + \mu_{i-1}\Delta r_{i-1}}\right), \tag{A.68}$$

$$V_i = \text{volume of zone } i\,,$$

$$A_i = \alpha r_i^{\alpha-1}\,. \tag{A.69}$$

The temporal spacing of the equation is determined by stability and accuracy considerations. To be second order accurate, the terms in the equation should be centered at $n + 1/2$, where n is the time index. The term $\nabla\cdot(\bar{F} + UfE)$ is written in a fully implicit fashion to avoid undamped oscillations. The first term on the right-hand side of the equation is also centered at $n+1$ to guarantee unconditional stability. This term need not be fully implicit for unconditional stability. A discussion and result of variable centering of these terms can be found in Section A.4. The scattering term T_{ij} is calculated at time n to avoid the frequency coupling that would occur if this term were implicit. Thus, one can write the following difference equation

$$\frac{V_i^{n+1}E_{ij}^{n+1} - V_i^n E_{ij}^n}{\Delta t} + \tilde{F}_{i+1}^{n+1} + A_{i+1}^{n+1} f_{i+1} \hat{E}_{i+1,j}^{n+1} U_{i+1}^{n+1} - (\tilde{F}_{ij}^{n+1} + A_i^{n+1} f_i \hat{E}_{ij}^{n+1} U_i^{n+1}) = c\mu_{ij} V_i^n (b_j^{n+1}\phi_i^{n+1} - E_{ij}^{n+1}) + V_i^n T_i^n \,, \tag{A.70}$$

where

$$\mu_{ij} = \mu_{P_{ij}}(\rho^n, \theta^n)\,, \tag{A.71}$$

and

$$\hat{E}_i^{n+1} = \frac{V_i^{n+1}E_i^{n+1} + V_{i-1}^n E_{i-1}^n}{V_i^n + V_{i-1}^n}\,. \tag{A.72}$$

The first moment equation is differenced in an analogous fashion. Rewriting the equation, one has

$$\frac{1}{c}\frac{\partial F_j}{\partial t} + c\left\{\left(\frac{3f-1}{2f}\right)\frac{1}{r^{\alpha-1}}\frac{\partial(r^{\alpha-1}fE_j)}{\partial r} + \left(\frac{1-f}{2f}\right)\frac{\partial(fE_j)}{\partial r}\right\} = -\frac{F_j}{\lambda_{R_j}}\,. \tag{A.73}$$

This differential form is preferred because interpolations of E in space are not required. Spatially, the equation is centered at the ith boundary. Except for the time derivative term, all terms are evaluated at the advanced time, $n+1$. The equation was written in a fully implicit manner to avoid stability criteria and to improve numerical damping. The function f could be evaluated at either zone centers or boundaries. Test calculations that indicate zone centering improve the results. Thus, the difference equation is given by

$$\frac{\tilde{F}_{ij}^{n+1} - \tilde{F}_{ij}^n}{c\Delta t} + \alpha r_i^{\alpha-1} c \left\{\left(\frac{3f-1}{2f}\right)_i^* \frac{1}{r_i^{\alpha-1}} \times \left[\frac{E_{ij}^{n+1} f_i (r^{\alpha-1})_i^* - E_{i-1}^{n+1} f_{i-1} (r^{\alpha-1})_{i-1}^*}{\frac{1}{2}(r_{i+1} - r_{i-1})}\right] + \left(\frac{1-f}{2f}\right)_i^* \times \left[\frac{E_{ij}^{n+1} f_i - E_{i-1}^{n+1} f_{i-1}}{\frac{1}{2}(r_{i+1} - r_{i-1})}\right]\right\} = -\frac{\tilde{F}_{ij}^{n+1}}{\lambda_{ij}}\,, \tag{A.74}$$

where

$$(r^{\alpha-1})_i^* = \left(\frac{r_i + r_{i+1}}{2}\right)^{\alpha-1}, \tag{A.75}$$

$$\left(\frac{3f-1}{2f}\right)_i^* = \frac{3\bar{f}_i - 1}{2\bar{f}_i}, \tag{A.76}$$

and

$$\left(\frac{1-f}{2f}\right)_i^* = \frac{1-\bar{f}_i}{2\bar{f}_i}. \tag{A.77}$$

The terms in the equation without superscript are evaluated at time n. These moment equations are valid for $1 < i < N$ where N is the total number of mesh points. For $i = 1$ and N the boundary conditions for E and F must be employed to reduce the equations to a soluble form. The numerical analysis for the boundary conditions is presented in Section A.3.2. The boundary quantity λ_{ij} on the right-hand side of the equation requires special treatment. The theory and analysis of this term are presented in Section A.3.3.

To complete the system of equations a difference procedure is required for the auxiliary equation. Centering the time derivative term at $n + 1/2$ and centering the radiation coupling term in the same fashion as the zeroth moment equation results in

$$\begin{aligned}\frac{b_j^{n+1}\phi_i^{n+1} - b_j^n \phi_i^n}{\Delta t} &= \frac{4a\theta_i^3}{\rho_i C_{V_i}}\left[c\mu_{ij}\left(E_{ij}^{n+1} - b_j^{n+1}\phi_i^{n+1}\right) - T_{ij}^n\right] \\ &\quad + \frac{4a\theta_i^3 b_j^n}{\rho_i C_{V_i}}\left[-\left(P_m + \frac{\partial E}{\partial \tau}\right)\rho\frac{\partial \tau}{\partial t} + \dot{q}\right]_i .\end{aligned} \tag{A.78}$$

Terms without superscripts are evaluated at time n except the term $\frac{\partial \tau}{\partial t}$, which is evaluated at time $n + 1/2$.

Through Taylor series expansions, one can show that the truncation error of the difference equations is

$$O(\Delta t) + O(\Delta x_i - \Delta x_{i+1}) + O(\Delta x^2). \tag{A.79}$$

The logic in solving the equations is to solve Eq. (A.74) for F_{ij}^{n+1} and Eq. (A.78) for $b_j^{n+1}\phi_i^{n+1}$ in terms of E_{ij}^{n+1}. These quantities are then substituted into Eq. (A.70) eliminating all terms at $n + 1$ except E^{n+1}. The resulting equation is

$$A_i E_{i-1,j}^{n+1} + B_i E_{ij}^{n+1} + C_i E_{i+1,j}^{n+1} + D_i = 0\,, \tag{A.80}$$

$$\begin{aligned} A_i = &-P_i Q_i R_i \left[3f_{i-1} \left(\frac{1-f}{2f} \right)_i^* + 3f_{i-1} (r^{\alpha-1})_{i-1}^* \left(\frac{1}{r_i^{\alpha-1}} \right) \left(\frac{3f-1}{2f} \right)_i^* \right] \\ &- \frac{A_i^{n+1} U_i^{n+1} V_{i-1}^n \bar{f}_i}{V_i^n (V_i^n + V_{i-1}^n)}\,, \end{aligned} \tag{A.81}$$

$$\begin{aligned} C_i = &-P_i Q_{i+1} R_{i+1} \left[3f_{i+1} \left(\frac{1-f}{2f} \right)_{i+1}^* + 3f_{i+1} (r^{\alpha-1})_{i+1}^* \left(\frac{1}{r_{i+1}^{\alpha-1}} \right) \right. \\ &\times \left. \left(\frac{3f-1}{2f} \right)_{i+1}^* \right] - \frac{A_{i+1}^{n+1} U_{i+1}^{n+1} V_{i+1}^n \bar{f}_{i+1}}{V_i^n (V_{i+1}^n + V_i^n)} \end{aligned} \tag{A.82}$$

$$\begin{aligned} B_i = &\, P_i Q_i R_i \left[3f_i \left(\frac{1-f}{2f} \right)_i^* + 3f_i (r^{\alpha-1})_i^* \left(\frac{1}{r_i^{\alpha-1}} \right) \left(\frac{3f-1}{2f} \right)_i^* \right] \\ &+ \frac{A_i^{n+1} U_i^{n+1} \bar{f}_i}{(V_i^n + V_{i-1}^n)} + \frac{V_i^{n+1}}{\Delta t V_i^n} + \frac{c\mu_{ij}}{1+\gamma_i} + P_i Q_{i+1} R_{i+1} \\ &\times \left[3f_i \left(\frac{1-f}{2f} \right)_{i+1}^* + 3f_i \left(r^{\alpha-1} \right)_i^* \left(\frac{1}{r_{i+1}^{\alpha-1}} \right) \left(\frac{3f-1}{2f} \right)_{i+1}^* \right] \\ &+ \frac{A_{i+1}^{n+1} U_{i+1}^{n+1} \bar{f}_{i+1}}{(V_{i+1}^n + V_i^n)}\,, \end{aligned} \tag{A.83}$$

$$\begin{aligned} D_i^* = &-\frac{E_{ij}^n}{\Delta t} + \frac{P_i}{4c\Delta t} (Q_{i+1} \tilde{F}_{i+1,j}^n - Q_i \tilde{F}_{ij}^n) - \frac{c\mu_{ij} b_j^n}{1+\gamma_i} \phi_{ij}^n - \frac{b_j^n \gamma_i}{1+\gamma_i} \\ &\times \left[\dot{q} - \left(P_m + \frac{\partial E}{\partial t} \right) \frac{\partial \tau}{\partial t} \right] - \frac{1}{1+\gamma_i} T_{ij}^n \end{aligned} \tag{A.84}$$

Note: * indicates that the term in square brackets should be multiplied by (volume fraction/cm^3) to yield consistent units
where

$$P_i = \frac{1}{r_{i+1}^\alpha - r_i^\alpha}\,, \tag{A.85}$$

$$Q_i = \left(\frac{1}{4c\Delta t} + \frac{1}{4\lambda_i} \right), \tag{A.86}$$

$$R_i = \frac{c}{6}\frac{\alpha r_i^{\alpha-1}}{r_{i+1} - r_{i-1}}, \tag{A.87}$$

$$\gamma_i = \frac{4a\theta_i^3 \Delta t c \mu_{ij}}{\rho_i C_{V_i}}. \tag{A.88}$$

The technique of solving the tridiagonal set of equations is to assume a solution of the form

$$E_{ij}^{n+1} = (1 + G_i)E_{i+1,j}^{n+1} + H_i. \tag{A.89}$$

By substitution, one finds G_i and H_i to be

$$G_i = -\left(\frac{B_i + A_i + C_i + A_i G_{i-1}}{B_i + A_i(1 + G_{i-1})}\right), \tag{A.90}$$

$$H_i = -\left(\frac{D_i + A_i H_{i-1}}{B_i + A_i(1 + G_{i-1})}\right). \tag{A.91}$$

The radiation energy E_{ij}^{n+1} is then evaluated from Eq. (A.89), and $\tilde{F}_{ij}^{n+1}$ is evaluated from Eq. (A.74). With the radiation energy and fluxes evaluated one can then determine the radiation contribution to the energy equation by either the right- or left-hand side of the zeroth moment equation. Numerically it has been determined that the right-hand side of the zeroth moment equation is preferred because of round-off errors in the difference form.

A.3.2 Boundary Conditions

The boundary conditions for the non-equilibrium diffusions are more difficult to specify than for the transport solution where one must know the inward specific intensity at all points of the boundary. In the VE approach the single quantity corresponding most closely to the inward intensity is the current directed to the surface normal, $\bar{N}$,

$$F_+ = \int_{2\pi} I(\bar{\Omega})\bar{\Omega} \cdot \bar{N} d\Omega. \tag{A.92}$$

In the diffusion approximation, i.e. $f = 1/3$, this quantity can be related to the flux F and radiation energy E, by

$$F_R = -2F_+ + \frac{c}{2}E_R, \tag{A.93}$$

where the subscript R refers to the right boundary. This equation does not limit properly in the case of streaming radiation, i.e. $f = 1$. One can include both the optically thick and the optically thin case in one equation as follows

$$F_R = -2F_+ + cGE_R\,, \tag{A.94}$$

where the factor G is defined by

$$G = \frac{1}{4}\left[1 + \sqrt{1 + 4(3f_B - 1)}\right]. \tag{A.95}$$

The variable Eddington factor f_B (evaluated in this case at the right boundary) is defined in Section A.3.5.

To close the system of VE moment equations, the net interface flux F_{ij} at the boundary must be specified. This can be accomplished by introducing a relation between the dependent variables: the net interface flux, F_{ij}, the zone-centered radiation energy density, E_{ij}, and the VE factor, f. To achieve this functional dependence a number of approximations have been employed. The scheme which is the most satisfactory to date for problems of interest is as follows.

The differential equation to be solved at the boundary for the flux is taken as the steady state first moment equation,

$$F = -\lambda c\frac{\partial(fE)}{\partial r} = -c\frac{\partial(fE)}{\partial\zeta}\,. \tag{A.96}$$

Assuming that the flux is constant in the zone adjacent to the right boundary, one can integrate this equation between (ζ_N, ζ_R) to find

$$f_B E_R = f_N E_N - \frac{F_R}{c}\Delta\zeta\,, \tag{A.97}$$

where

$$\Delta\zeta = \zeta_R - \zeta_N\,. \tag{A.98}$$

The subscript R denotes the boundary interface, while the subscript N denotes the last zone center. Thus $\Delta\zeta$ refers to the outer half zone optical depth.

Equations (A.94) and (A.97) may be solved simultaneously to yield

$$F_R = \frac{f_N G}{(f_B + G\Delta\zeta)}\left[cE_N - \frac{2f_B}{Gf_N}F_+\right]. \tag{A.99}$$

In order to evaluate the coefficients A_i, B_i, C_i, D_i of Section A.3.1, it is convenient to form the quantity $F_R + U_R f_B E_R$ corresponding to the $F + uP$ form in the zeroth moment equation.

To do this, multiply Eq. (A.97) by u_R and add F_R to both sides yielding

$$F_R + u_R f_B E_R = u_R f_N E_N + \left(1 - \frac{u_R \Delta\zeta}{c}\right) \frac{f_N G}{(f_B + G\Delta\zeta)} \left[cE_N - \frac{2f_B}{Gf_N} F_+\right]. \tag{A.100}$$

The left-hand boundary case is analogous. The two conditions imposed are

$$F_L = 2F_+ - cGE_L \,, \tag{A.101}$$

$$f_B E_L = f_1 E_1 - F_L \frac{\Delta\zeta}{c} \,. \tag{A.102}$$

Solving simultaneously gives the boundary flux

$$F_L = \frac{f_1 G}{(f_B + G\Delta\zeta)} \left[\frac{2f_B}{Gf_1} F_+ - cE_1\right], \tag{A.103}$$

and the $F + uP$ form expressed in terms of known quantities is

$$F_L + u_L f_B E_L = u_L f_1 E_1 + \left(1 - \frac{u_L \Delta\zeta}{c}\right) \frac{f_1 G}{(f_B + G\Delta\zeta)} \left[\frac{2f_B}{Gf_1} F_+ - cE_1\right]. \tag{A.104}$$

A.3.3 Opacity Averaging Procedure

The difference approximations to the non-equilibrium diffusion equations were described in Section A.3.1, but the averaging technique required to obtain the absorption coefficient has not been discussed. The numerical approximations required to specify an absorption coefficient at the boundary in the first moment equation [Eq. (A.74)] are now considered. The difficulty in evaluating an average value to be used at a boundary is that the dependence of temperature and density on position between zone centers is unknown. Only the zone-centered quantities are available. One must calculate the optical depth, or the number of mean free paths between zone centers, i.e.

$$\Delta\zeta_i = \int_{\frac{r_i + r_{i-1}}{2}}^{\frac{r_i + r_{i+1}}{2}} (\mu_a' + \mu_s)\, dr \,, \tag{A.105}$$

or in terms of the Rosseland opacity, $\kappa_{R'}$

$$\Delta\zeta_i = \int_{\frac{r_i + r_{i-1}}{2}}^{\frac{r_i + r_{i+1}}{2}} \kappa_R \rho\, dr \,, \tag{A.106}$$

since

$$\mu_a' + \mu_s = \kappa_R \rho \,. \tag{A.107}$$

Two approaches to finding $\Delta\zeta$ will be considered. A comparison of the approaches has been done and Method 2 is recommended.

Method 1: Arithmetic averaging of zone-centered opacities
In this approach, optical depths $\Delta\zeta$ are computed for two adjacent zones,

$$\Delta\zeta(i) = \kappa_R(i)\rho_i(r_{i+1} - r_i)\,, \tag{A.108}$$
$$\Delta\zeta(i+1) = \kappa_R(i+1)\rho_{i+1}(r_{i+2} - r_{i+1})\,, \tag{A.109}$$

based on opacities evaluated with zone temperatures and densities.

If the change in $\Delta\zeta$ from zone (i) to zone $(i+1)$ is large compared to some input parameter, say AC, the minimum $\Delta\zeta$ is used in calculating $\lambda_i^{-1} = \left(\mu_a' + \mu_s\right)_i$ as indicated below. Otherwise, the two $\Delta\zeta$'s are arithmetically averaged. Thus, if

$$\left|\frac{\Delta\zeta(i) - \Delta\zeta(i+1)}{\Delta\zeta(i) + \Delta\zeta(i+1)}\right| < AC\,, \tag{A.110}$$

then

$$\lambda_{i+1}^{-1} = \frac{\Delta\zeta(i) + \Delta\zeta(i+1)}{(r_{i+2} - r_i)}\,. \tag{A.111}$$

Otherwise,

$$\lambda_{i+1}^{-1} = \text{Min}\left\{\frac{\Delta\zeta(i)}{(r_{i+1} - r_i)}, \frac{\Delta\zeta(i+1)}{(r_{i+2} - r_{i+1})}\right\}. \tag{A.112}$$

Method 2. Integration using an interpolated temperature dependence

The second opacity averaging scheme incorporates the temperature in adjacent zones and an interpolated temperature in the development of the average mean free path. The assumption is that for a particular material M the Rosseland opacity is a function of the $n_M th$ power of the temperature in the zone, where n_M is characteristic of the material. Figure A.2 illustrates the calculational technique used.

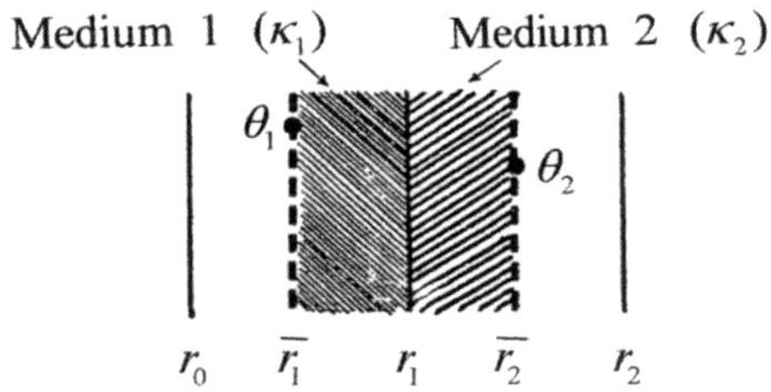

Figure A.2. Mean free path averaging scheme.

It is assumed that

$$\kappa = \kappa_i \left(\frac{\theta_i}{\theta}\right)^{n_i} . \tag{A.113}$$

The total optical depth between $\bar{r}_1$ and $\bar{r}_2$ is given by

$$\nabla\zeta = \nabla\zeta_1 + \nabla\zeta_2 , \tag{A.114}$$

where λ^{-1} at the zone interface is defined by

$$\lambda^{-1} = \frac{\nabla\zeta}{\bar{r}_2 - \bar{r}_1} , \tag{A.115}$$

$$\nabla\zeta_1 = \rho_1\kappa_1 \int_{\bar{r}_1}^{r_1} \left(\frac{\theta_1}{\theta}\right)^{n_1} dx , \tag{A.116}$$

and

$$\nabla\zeta_2 = \rho_2\kappa_2 \int_{r_1}^{\bar{r}_2} \left(\frac{\theta_2}{\theta}\right)^{n_2} dx . \tag{A.117}$$

To determine the temperature distribution, a further approximation is required. Assuming that the equilibrium diffusion approximation is applicable and that the flux F is constant between zone centers, one can derive the following result for the temperature:

$$\theta^{n_i+4}(r) = \theta_i^{n_i+4} - \overline{AA}_i F_i (r - \bar{r}_i) , \tag{A.118}$$

where

$$\overline{AA}_i = \frac{3}{4ac}(n_i + 4)\rho_i\kappa_i\theta_i^{n_i} . \tag{A.119}$$

Employing this result, one finds

$$\nabla\zeta_1 = \frac{ac}{3F_1}\left[\theta_i^4 - (A_1 + B_1 r_1)^{\frac{4}{n_1+4}}\right] , \tag{A.120}$$

where

$$A_1 = \theta_1^{n_1+4} + \bar{r}_1 \overline{AA}_1 F_1 \,, \tag{A.121}$$

and

$$B_1 = -F_1 \overline{AA}_1 \,. \tag{A.122}$$

Likewise,

$$\nabla\zeta_2 = \frac{ac}{3F_1}\left[(A_2 + B_2 r_1)^{\frac{4}{n_2+4}} - \theta_2^4\right] , \tag{A.123}$$

where

$$A_2 = \theta_2^{n_2+4} + \bar{r}_2 \overline{AA}_2 F_1 \,, \tag{A.124}$$

$$B_2 = -F_1 \overline{AA}_2 \,. \tag{A.125}$$

Defining $TEMP(10)$ and $TEMP(11)$ as

$$TEMP(10) = \frac{\overline{AA}_1 F_1 (r_1 - \bar{r}_1)}{\theta_1^{n_1+4}} \,, \tag{A.126}$$

and

$$TEMP(11) = \frac{\overline{AA}_2 F_1 (\bar{r}_2 - r_1)}{\theta_2^{n_2+4}} \,. \tag{A.127}$$

Rewriting the equation for the $\Delta\zeta$'s, if

$$0.01 < TEMP(10) < 1.0 \,, \tag{A.128}$$

$$\Delta\zeta_1 = \frac{ac}{3F_1}\theta_1^4\left[1 - (1 - TEMP(10))^{\frac{4}{n_1+4}}\right] , \tag{A.129}$$

and if

$$-0.01 > TEMP(11) > -1.0 \,, \tag{A.130}$$

$$\Delta\zeta_2 = \frac{ac}{3F_1}\theta_2^4\left[(1 + TEMP(11))^{\frac{4}{n_2+4}} - 1\right] . \tag{A.131}$$

if $|TEMP(10)| < 0.01$ a binomial expansion of the term is performed yielding

$$\Delta\zeta_1 = (r_1 - \bar{r}_1)\rho_1 \kappa_1 \,. \tag{A.132}$$

If $TEMP(10) > 1.0$ we set $TEMP(10) = 1.0$ so that

$$\Delta\zeta_1 = \frac{ac}{3F_1}\theta_1^4 \,. \tag{A.133}$$

At $TEMP(10) = 1.0$, then one obtains

$$\Delta\zeta_1 = \frac{n_1 + 4}{4}(r_1 - \bar{r}_1)\rho_1\kappa_1\,. \tag{A.134}$$

Similarly, if $|\,TEMP(11)| < 0.01$, we expand and obtain

$$\Delta\zeta_2 = (\bar{r}_2 - r_1)\rho_2\kappa_2\,. \tag{A.135}$$

In the case $TEMP(11) < -1.0$, we set $TEMP(11) = -1.0$, so that

$$\Delta\zeta_2 = \frac{ac}{3F_1}\theta_2^4\,, \tag{A.136}$$

and when $TEMP(11) = -1.0$

$$\Delta\zeta_2 = \frac{n_2 + 4}{4}(\bar{r}_2 - r_1)\rho_2\kappa_2\,. \tag{A.137}$$

The terms $\Delta\zeta_1, \Delta\zeta_2$ are summed to provide λ_i^{-1}. The variable $R4LAMB$ is defined as

$$R4LAMB = \frac{1}{4}\left(\frac{\Delta\zeta_1 + \Delta\zeta_2}{\bar{r}_2 - \bar{r}_1}\right)\,. \tag{A.138}$$

A.3.4 Variable Eddington Factor

This section considers the VE approximation and the Variable Eddington factor (f) in more detail. Since the accuracy of the VE approach depends on the method of evaluating $f(= P/E)$, which is a function of space, time, and frequency, it is important that f be evaluated as accurately as possible. Although the VE factor (f) could be evaluated through a detailed transport theory calculation of $P$and E, this would be very uneconomical, especially if f can be satisfactorily evaluated from geometrical considerations.

In order to develop a useful scheme for the determination of $f(\cdot)$, the mathematical models of typical problems will be considered and formulations for $f(\cdot)$ will then be developed based on these mathematical models. Whether such a treatment is applicable and effective depends on the ability to incorporate retardation effects through the retarded values of the radiation energy density (E) and the radiative flux (F).

This treatment is consistent with the basic assumption that $f(\cdot)$ is a slowly varying function in time, and it is satisfactory to make the calculation of $f(\cdot)$ explicit in time. The effects of scattering on $f(\cdot)$ are also included in the sense that the values of E and F used to calculate $f(\cdot)$ reflect the effects of any scattering which might be presented.

The diffusion and streaming of radiation through a medium (region) are examples of cases for which f is known. The value of f approaches the diffusion limit of $1/3$ whenever the radiation intensity is isotropic or linearly anisotropic, for example, in the interior of an optically thick medium. The streaming case can be exemplified by the situation in which a localized source streams through a surrounding transparent medium such that the radiation intensity becomes increasingly concentrated in the $\mu = 1$ direction with increasing distance from the source. Thus, P approaches E, and f approaches the value of 1.0. These cases are incorporated as the limiting values of $f(\cdot)$ for the technique considered.

A number of options for determining $f(\cdot)$ have been developed and tested. One or more of these options are applicable to most problems of interest. As experience with the VE approach is acquired it is likely that better methods of calculating f will be developed.

In the following, two idealized problems will be considered, and formulations for $f(\cdot)$ will be developed based on these models. Therefore, each option will be discussed, the assumptions inherent in the derivation of the basic formula will be reviewed, and the advantages and limitations of the formulation presented.

(a) Option 0 — the diffusion value, $f = 0.3333$.
(b) Option 1 — based on the idealized problem (Problem 1) of a uniform temperature body radiating into a vacuum, Option 1 utilizes the expression

$$f = \frac{1}{3}(1 - \varepsilon + \varepsilon^2)\,, \tag{A.139}$$

where

$$\varepsilon = \left|\frac{2F}{cE}\right|\,. \tag{A.140}$$

This expression is derived under the assumptions:

(1) the flux is one-sided,
(2) the medium is a vacuum,
(3) the source is a constant in time, and
(4) $\left|\frac{2F}{cE}\right|$ is determined either by geometry or retardation considerations.

Option 1 inherently includes retardation effects through E and F. The derivation of Option 1 is as follows:

Problem 1: Radiation into a vacuum from a uniform temperature body with a rectangular pulse or a step function temperature history.

Spillman [A.2] has developed expressions for f by considering a uniform temperature body whose surface is symmetrical about the r axis, radiating into a vacuum region. The surface may be of any shape, but for the purpose of illustrating a one-dimensional problem, the spherical body shown in Fig. A.3 is assumed.

Assuming that the radiation from the body is isotropic, radiation energy density, E, flux, F, and r-r component of the pressure tensor, P, are given by

$$E = \frac{2\pi B}{c}(1 - \mu_1)\,, \tag{A.141}$$

$$F = \pi B(1 - \mu_1^2)\,, \tag{A.142}$$

$$P = \frac{2\pi B}{3c}(1 - \mu_1^3)\,, \tag{A.143}$$

since the radiation intensity is constant in the angular interval $\mu_1 \leq \mu \leq 1$. The μ_1 may be established by retardation or by geometry since it corresponds to the cosine of the largest angle with respect to the radius vector that the sphere illuminates at (r, t). For instance, if the radiating body is a sphere of radius $r_i < r$, with the source function given by

$$\begin{aligned} \text{source function} &= 0, \quad \text{when } t < t_0\,, \\ &= B_0, \quad \text{when } t \geq t_0\,, \end{aligned}$$

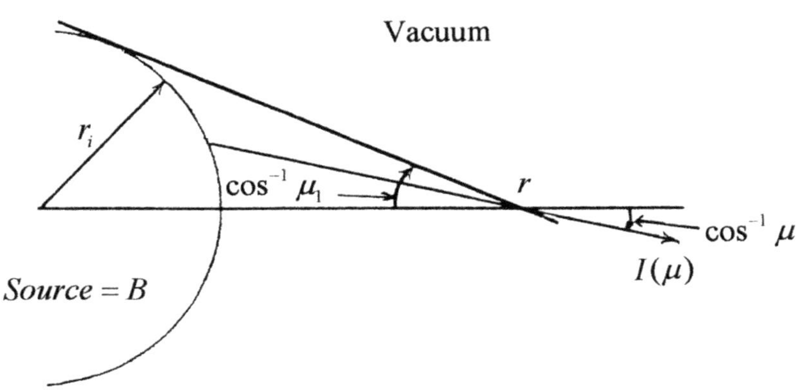

Figure A.3. Isothermal sphere with μ_1 geometry limited.

then

$$\mu_1 = \max\left\{\frac{\left[r - \frac{1}{2r}\left(r^2 + r_i^2 - c^2t^2\right)\right]}{ct}, \sqrt{1 - \left(\frac{r_i}{r}\right)^2}\right\}, \qquad \text{(A.144)}$$

the first form being used if t is less than the transit time of photons to the point r along a tangent ray from the sphere, i.e. if $t < t_0 + \frac{r}{\mu_1 c}$.

Equations (A.141)–(A.143) inherently assume steady state with respect to the source function B. Another configuration yielding the same equations would be in a plane surface where μ_1 is determined by retardation.

From Eqs. (A.141) and (A.143),

$$f \equiv \frac{P}{E} = \frac{1}{3}\frac{1-\mu_1^3}{1-\mu} = \frac{1}{3}\left(1 + \mu_1 + \mu_1^2\right). \qquad \text{(A.145)}$$

Using Eqs. (A.141) and (A.142) to eliminate μ_1 in favor of E and F yields

$$f = \frac{1}{3}\left(1 + \left[\frac{2F}{cE}\right]\mu_1\right), \qquad \text{(A.146)}$$

or, since $\frac{2F}{cE} = 1 + \mu_1$,

$$f = \frac{1}{3}\left(1 - \frac{2F}{cE} + \left[\frac{2F}{cE}\right]^2\right). \qquad \text{(A.147)}$$

Inherent in the derivation of Eq. (A.147) is the assumption that the radiation intensity is one-sided because the configuration in Fig. A.3 is assumed. In the one-sided case, $\frac{2F}{cE}$ is restricted,

$$1.0 \le \frac{2F}{cE} \le 2.0, \qquad \text{(A.148)}$$

being 1.0 when the intensity is isotropic for $1.0 > \mu \ge 0$ and zero otherwise, and being 2.0 in the streaming case from a point source about $\mu = 1$. For this reason, we set $f = 1/3$ whenever Eq. (A.146) is being used and (A.148) is not satisfied, where the absolute value of $\frac{2F}{cE}$ is used in (A.148). Later in this section, it will be shown that Eq. (A.146) is valid for cases when the flux is not one-sided, so the restriction (A.148) is not imposed on this formulation.

(c) Option 2 — There is no Option 2 at this time.
(d) Option 3, 4, and 5 — Based on the idealized problem (problem 2) of a vacuum region between two radiating bodies. All three of these options use the expression

$$f = \frac{1}{3}(1 + \varepsilon\mu)\,, \tag{A.149}$$

where

$$\varepsilon = \left|\frac{2F}{cE}\right|\,, \tag{A.150}$$

and where μ is calculated differently for each option.

Problem 2: The determination of f in a vacuum region between two radiating uniform-temperature bodies.

Here we consider a vacuum region between two radiating bodies schematically shown in Figure A.4. For purposes of illustration, the annular region between two concentric spheres will be utilized. In the case shown in the figure, μ_i is determined by geometry, while μ_0 is established by radiation. These bodies may be two slabs, two spheres, or two concentric shells, etc. The body near the origin is denoted by the subscript i (inner), while 0 is used for the subscript of the outer body. For this case, the expressions E, F,

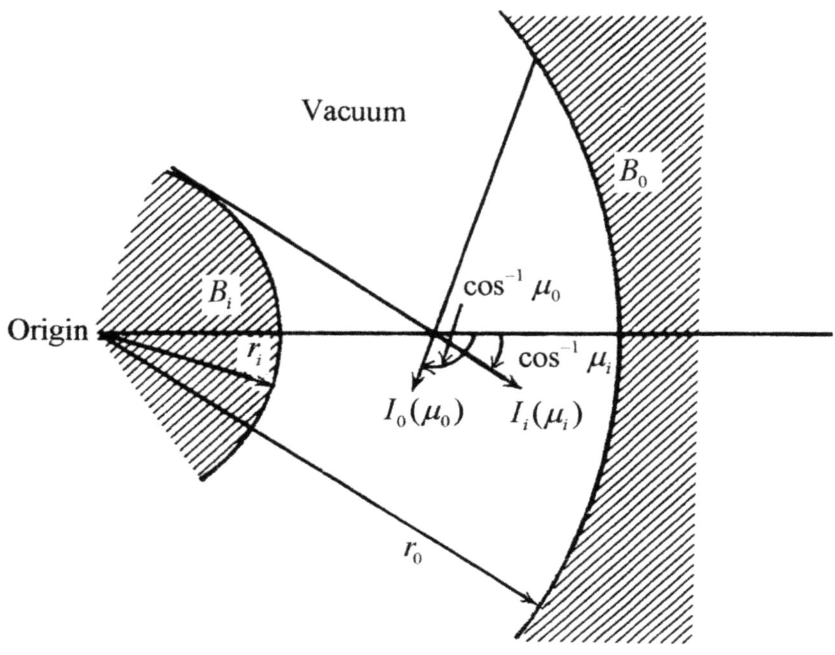

Figure A.4. Vacuum annular region between two concentric spheres.

and P become

$$E = \frac{2\pi}{c}[B_i(1-\mu_i) + B_0(1+\mu_0)]\,, \quad \text{when } \mu_0 < 0 \tag{A.151}$$

$$F = \pi\left[B_i\left(1-\mu_i^2\right) - B_0\left(1-\mu_0^2\right)\right], \tag{A.152}$$

$$P = \frac{2\pi}{3c}\left[B_i\left(1-\mu_i^3\right) + B_0\left(1+\mu_0^3\right)\right], \tag{A.153}$$

so that

$$f = \frac{P}{E} = \frac{1}{3}\left[\frac{B_i\left(1-\mu_i^3\right) + B_0\left(1+\mu_0^3\right)}{B_i(1-\mu_i) + B_0(1+\mu_0)}\right], \tag{A.154}$$

which can be written as

$$f = \frac{1}{3}\left\{1 + \frac{2P}{cE}(\mu_i + \mu_0) - \frac{2\pi\mu_i\mu_0}{cE}\left[\frac{B_i}{\mu_i}\left(1-\mu_i^2\right) - \frac{B_0}{\mu_0}\left(1-\mu_0^2\right)\right]\right\}. \tag{A.155}$$

As in Problem 1, μ_i and μ_0 may be determined by either geometry or retardation. If the problem is geometry limited and $\mu_0 = \mu_i$, the last term of Eq. (A.155) becomes $-\frac{2F}{cE}\mu_0$, and the result is

$$f = \frac{1}{3}\left(1 + \frac{2F}{cE}\mu_i\right), \tag{A.156}$$

which is identical to Eq. (A.146). In most real problems, the definition of μ_i is not unique. There is usually no well defined material-vacuum interface, and the density and opacity usually vary considerably more slowly as a function of radius.

Option 3 utilizes the concept of the Milne-Eddington model in which the source function at the surface of a radiating body is assumed to be a linear function of the optical depth measured from the surface. For a sphere,

$$B_\tau = B_0(1 + 1.5\tau)\,. \tag{A.157}$$

Thus, for the f of Option 3 (for sphere) at a point r, the μ of Eq. (A.156) is determined by the tangent drawn from r to a sphere, centered at the origin, and whose surface passes through a point two-thirds of a mean free path back along the radius ($\overline{\Delta r_1} = 2/3$) from the point of interest, r.

$$\mu = \sqrt{1 - \sin^2\theta} = \sqrt{\frac{\overline{\Delta r_1}}{r}\left(2 - \frac{\overline{\Delta r_1}}{r}\right)} \tag{A.158}$$

since

$$\sin\theta = \frac{r - \overline{\Delta r_1}}{r} \,. \tag{A.159}$$

Options 4 and 5 reflect the recognition that the Option 3 choice of $\overline{\Delta r_1}$ is not always appropriate. For instance, near the interior surface of a thick spherical shell region having an adjacent interior thin region, the μ of Option 3 would be determined by a sphere with a radius much smaller than that of the current thick region. To eliminate this difficulty, two empirical representations for $\overline{\Delta r_1}$ were incorporated as Options 4 and 5. Options 4 and 5 involve calculations of $\overline{\Delta r_1}$ by the integral

$$\overline{\Delta r_1} = a\int_{-\infty}^{\zeta_1} (r_1 - r)e^{a(\zeta-\zeta_1)}d\zeta \,, \tag{A.160}$$

where ζ is a function of μ_a (the absorption coefficient) and r.

For Option 4,

$$\zeta = \int_{-\infty}^{r} \mu_a(r')dr' \,, \tag{A.161}$$

so that ζ corresponds to the optical depth, while for Option 5,

$$\zeta = \int_{-\infty}^{r} \mu_a^2(r')r'dr' \,. \tag{A.162}$$

Thus, for Option 4, ζ is the optical depth. Of Options 4 and 5, Option 5 has the more correct behavior near the surface of an isotheral sphere.

Determination of f during the computational process for either Option 1 or 3, i.e. Eqs. (A.147) or (A.156), the optical depth, $\Delta\zeta$, of each zone is required. Option 1 uses the following formula for calculating the zone-centered Variable Eddington factor.

$$f_{i+1/2} = \frac{1}{3}(1 - |\eta| + \eta^2) \,, \tag{A.163}$$

where

$$\eta = \frac{2F_{i+1/2}}{cE_{i+1/2}} \,. \tag{A.164}$$

The values of E and F from the preceding cycle are used in the above equations. The $F_{i+1/2}$ is a zone-centered quantity, being the arithmetic mean between the fluxes at the boundaries of the zone. One advantage of this method is that it accounts for retardation effects to the extent that such effects are contained in the flux and radiation energy density. On the other hand, since this f is completely dependent on E and F, faulty values of the latter two variables will certainly produce poor values of f.

For the third option f is assumed to be a geometry-limited formulation and is therefore applicable to spherical problems.

$$f_{i+1/2} = \frac{1}{3}\left(1 + \frac{2F_{i+1/2}}{cE_{i+1/2}}\mu\right) . \tag{A.165}$$

The μ in the above equation is the cosine of the angle subtended by a sphere, centered at the origin which lies 2/3 of a mean free path away from the local point in question (toward the center). Since μ is a simple geometric quantity, not subject to large fluctuations in time or space, this formulation produces generally well behaved values for f. Values for f at the zone interfaces are required by the differencing equations. These are formed by taking a weighted average of the $f_{i+1/2}$ in optical depth space.

$$f_i = \frac{f_{i-1/2}\Delta\zeta_{i-1} + f_{i+1/2}\Delta\zeta_i}{\Delta\zeta_{i-1} + \Delta\zeta_i} . \tag{A.166}$$

The values of f at the right and left boundaries, f_B, are evaluated by considering several possible options. In the spherical case f_B on the left-side boundary is defined to be 1/3. The f_B factor for the right-side is evaluated as follows:

A geometric and an extrapolated boundary value f_μ and f_x respectively, are calculated by

$$f_\mu = \frac{1}{3}(1 + \mu + \mu^2) \tag{A.167}$$

where μ is the cosine of the angle subtended by a sphere lying a distance 2/3 away in optical depth space and

$$f_x = f_N + \frac{(f_N - f_{N-1})\Delta\zeta_N}{\Delta\zeta_N + \Delta_{N-1}} , \tag{A.168}$$

where f_N is the quantity associated with the last zone. Next, f_B is maximized according to

$$f'_B = \max(0.40476, f_\mu, f_x, f_N) , \tag{A.169}$$

and finally limited by

$$f_B = \min(1.0, f'_B) . \tag{A.170}$$

The constant 0.40476 is the value of f determined at the surface of a grey Milne-Eddington slab.

In plane problems both the right- and left-side boundary f factors are determined similarly; however, the quantity f_μ is not applicable in this case.

A.3.5 Scattering Terms

This section of the appendix develops the difference equations used to evaluate the Compton scattering term T^n_{ij} in Section A.3.1 (i refers to the space index and j the frequency index). From Eq. (A.23) the monochromatic scattering contribution to the zeroth moment equation is

$$T = -\frac{c\mu_s}{mc^2}\left\{eE - \frac{\partial e^2 E}{\partial e} - \theta\left[4E - 6\frac{\partial eE}{\partial e} + \frac{\partial^2 e^2 E}{\partial e^2}\right] + G\left(\frac{E^2}{e^2} - \frac{\partial E^2/e}{\partial e}\right)\right\}, \tag{A.171}$$

where e is the photon energy and

$$G = \frac{3c^3h^3}{32\pi}\left[\left(\frac{3}{2} - f + \frac{3}{2}f^2\right) + \frac{F^2}{c^2E^2}(-1.7 + 0.6f - 0.9f^2)\right]. \tag{A.172}$$

The terms in the equation are as defined previously. This equation can also be written as (ignoring derivatives of G)

$$T = \frac{c\mu_s e}{mc^2}\frac{\partial}{\partial e}\left[eE + \theta\left(6\frac{\partial eE}{\partial e} - 4E\right) + G\frac{E^2}{e^2}\right]. \tag{A.173}$$

Although the expressions of Eqs. (A.171) and (A.173) for T are identical, they will serve as guides for obtaining different difference approximations having different truncation errors associated with them. The form of Eq. (A.171) is called the energy conservation form, while Eq. (A.173) is called the photon conservation form. The reasons for the name can be seen by integrating over the photon energy variable resulting in

$$\int_0^\infty T de = -\frac{c\mu_s}{mc^2}\left[\int_0^\infty \left(eE - 4\theta E + G\frac{E^2}{e^2}\right) de\right], \tag{A.174}$$

and

$$\int_0^\infty \frac{T}{e} de = \frac{c\mu_s}{mc^2}\left[eE + \theta\left(6\frac{\partial eE}{\partial e} - 4E\right) + G\frac{E^2}{e^2}\right]\Bigg|_0^\infty = 0. \tag{A.175}$$

The first equation provides the correct energy exchange rate between the material and radiation due to Compton interactions. The second equation demonstrates that the rate of change of photon number density is zero. The latter is required by the scattering process itself: photons are neither created nor destroyed in a Compton scattering event. A priori, it is not clear which form of the equations should be used. Initially, numerical approximations to Eq. (A.171) were developed and tested. Due to truncation errors, a slow departure from the desired equilibrium solution resulted. Subsequently, Eq. (A.173) is recommended. While both methods are capable of describing Compton scattering energy exchange satisfactorily when large numbers of frequency groups are used, it appears that the imposition of the number conservation requirement inherent in Eq. (A.173) appreciably improves the accuracy.

The numerical analysis for the scattering contribution T will be presented only for the scheme based on Eq. (A.173). However, many of the approximations used are applicable to the first form as well. The entire radiation spectrum is partitioned into radiation groups. To evaluate T_{ij}^n an integration of T is performed over the interval (e_j, e_{j+1}). The subscript denotes that the quantity is to be evaluated at the photon energy constituting the lower boundary of the jth photon energy group. Thus

$$T_{ij} = \frac{c\mu_s}{mc^2}\bar{e}_j \left[eE\Big|_j^{j+1} + \theta_j \left(6\frac{\partial eE}{\partial e} - 4E \right)\Bigg|_j^{j+1} + G\frac{E^2}{e^2}\Bigg|_j^{j+1} \right] = 0\,, \tag{A.176}$$

where

$$\bar{e}_j = \frac{\int_{e_j}^{e_{j+1}} eE(e)de}{\int_{e_j}^{e_{j+1}} E(e)de}\,. \tag{A.177}$$

The scheme of Eq. (A.164) is to be implemented explicitly, i.e. quantities are evaluated in terms of the quantities (denoted by cycle n) available at the beginning of the cycle. In order to obtain values of E_j and $(\partial E/\partial e)_j$, an interpolation scheme using the known radiation group energies

$$E_{ij} = \int_{e_j}^{e_{j+1}} E(e)de\,, \tag{A.178}$$

is established. The functional dependence assumed in the interpolation scheme is considered first. Different empirical fits for $E(e)$ are used in three cases:

(1) the infinite group, (e_K, ∞).
(2) interior group $(e_1, e_2), \ldots, (e_{K-1}, e_K)$.
(3) the lowest frequency group, $(0, e_1)$.

The number and values of the frequency boundaries e_i are functions of the particular problem under consideration.

Case 1: The infinite group, (e_K, ∞).

The scattering contribution from the infinite group is developed using the assumption that $E(e)$ follows an exponential distribution on the interval (e_K, ∞), i.e.

$$E(e) = C_1 \exp(-e/C_2)\,. \tag{A.179}$$

Now

$$E_{ij} = \int_{e_j}^{e_{j+1}} E(e)de\,; \tag{A.180}$$

thus

$$E_{iK} = \int_{e_K}^{\infty} C_1 \exp(-e/C_2)de = C_1 C_2 \exp(-e_K/C_2)\,, \tag{A.181}$$

and

$$E_{iK-1} = \int_{e_{K-1}}^{e_K} C_1 \exp(-e/C_2)de = C_1 C_2[\exp(-e_{K-1}/C_2) - \exp(-e_K/C_2)]\,. \tag{A.182}$$

Letting

$$\alpha = \frac{E_{iK}}{E_{iK-1}}\,, \tag{A.183}$$

we see that

$$\alpha = \frac{-1}{1 - \exp[(e_K - e_{K-1})/C_2]}\,; \tag{A.184}$$

then

$$C_2 = \frac{(e_K - e_{K-1})}{\ell n\left(\frac{1+\alpha}{\alpha}\right)}\,, \tag{A.185}$$

and

$$C_1 = \frac{E_{iK}}{C_2} \exp(e_K/C_2)\,. \tag{A.186}$$

In this case the contributions to the scattering term T_{ij} are summarized in Table A.3.1.

Table A.3.1 Numerical approximations in scattering in Case 1.

Term in Scattering Equation		Exponential Assumption Numerical Approximation
$E(e)\|_{e_K}$	$=$	$E(e) = C_1 \exp(-e/C_2) = E_{iK}/C_2$
$eE(e)\|_{e_K}$	$=$	$e_K E_{iK}/C_2$
$\frac{\partial}{\partial e}[eE(e)]\|_{e_K}$	$=$	$E(e_K)\left(1 + \frac{e_K}{C_2}\right)$
$\left.\left(\frac{E(e)}{e}\right)^2\right\|_{e_K}$	$=$	$\left(\frac{E(e_K)}{e_K}\right)^2$
$\int_{e_K}^{\infty} E(e)de$	$=$	E_{iK}
$\int_{e_K}^{\infty} eE(e)de$	$=$	$E(e_K)(e_K + C_2)$

Case 2: Interior group.

The interior groups are evaluated assuming that $E(e)$ satisfies a power law, $E(e) = d_1 e^{d_2}$.

Figure A.5 shows the E_{ij} group, which is to be evaluated. The approach used is:

(1) Some terms depend on values at the boundary e_{j+1}. The evaluation of these terms is carried out using a power law fit from the interval (e_j, e_{j+2}) (Fit I).
(2) These terms must also be evaluated at the boundary e_j. Fit I does not depend on conditions in the group (e_{j-1}, e_j) and hence is not appropriate for this evaluation. Therefore, a second power law fit is made on the interval (e_{j-1}, e_{j+1}) and Fit II is used for the e_j boundary.

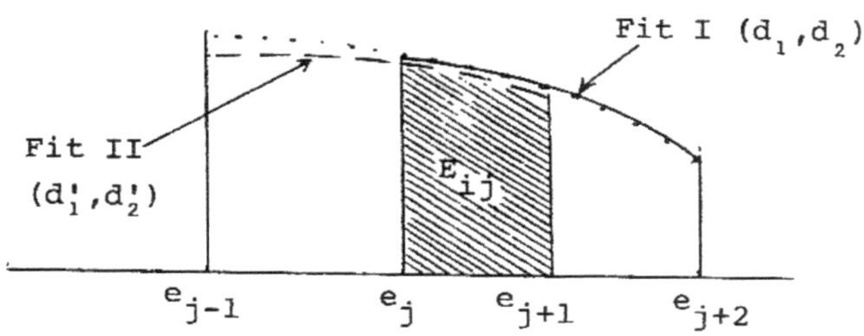

Figure A.5. Structure of scheme for evaluating interior groups.

(3) The term $\int_{e_j}^{e_{j+1}} eE(e)de$ is computed using the average parameters; $\hat{d}_2 = 0.5(d_2 + d_2')$, $\hat{d}_1 = 0.5(d_1 + d_1')$, since as an integral it represents an average over the group (e_j, e_{j+1}). $\hat{d}_1$ and $\hat{d}_2$ are used in Table A.3.2.

The quantities E_{ij} and $E_{i,j-1}$ are known. They are used to develop the exponent d_2 for the interval (e_{j-1}, e_{j+1}) in the following fashion:

$$E_{ij} = \int_{e_j}^{e_{j+1}} d_1 e^{d_2} de = \frac{d_1}{d_2+1}\left(e_{j+1}^{d_2+1} - e_j^{d_2+1}\right), \tag{A.187}$$

$$E_{i,j-1} = \int_{e_{j-1}}^{e_j} d_1 e^{d_2} de = \frac{d_1}{d_2+1}\left(e_j^{d_2+1} - e_{j-1}^{d_2+1}\right). \tag{A.188}$$

Next define

$$\alpha = \frac{E_{ij}}{E_{i,j-1}}, \tag{A.189}$$

$$\gamma_P = \frac{e_{j+1}}{e_j}, \tag{A.190}$$

$$\gamma_M = \frac{e_{j-1}}{e_j}, \tag{A.191}$$

so that

$$\alpha = \frac{e_{j+1}^{d_2+1} - e_j^{d_2+1}}{e_j^{d_2+1} - e_{j-1}^{d_2+1}} = \frac{\gamma_P^{d_2+1} - 1}{1 - \gamma_M^{d_2+1}}. \tag{A.192}$$

The solution of this equation for d_2 satisfies

$$h(d_2) \equiv (1+\alpha) - \alpha\gamma_M^{d_2+1} - \gamma_P^{d_2+1} = 0. \tag{A.193}$$

This equation has the following properties:

(1) $\alpha = \dfrac{E_{ij}}{E_{i,j-1}} > 0,$ (A.194)

(2) $0 < \gamma_M < 1 < \gamma_P,$ (A.195)

(3) $\ell n(\gamma_P) > 0,$ (A.196)

(4) $\ell n(\gamma_M) < 0,$ (A.197)

(5) $h'(d_2) = -\alpha\gamma_M^{d_2+1}\ell n(\gamma_M) - \gamma_P^{d_2+1}\ell n(\gamma_P),$ (A.198)

(6) $h(-1) = 0,$ (A.199)

(7) $h''(d_2) = -\alpha\gamma_M^{d_2+1}\ell n^2(\gamma_M) - \gamma_P^{d_2+1}\ell n^2(\gamma_P)\,.$ (A.200)

Since $h''(d_2) < 0$ for all d_2, the curve $h(d_2)$ is concave downward. Further,

$$\lim_{d_2\to\infty}[h(d_2)] = -\infty\,, \tag{A.201}$$

$$\lim_{d_2\to-\infty}[h(d_2)] = -\infty\,. \tag{A.202}$$

Figure A.6 shows the possible sketches of $h(d_2)$, where $(d_2)_{\max}$ is the value of d_2 where $h(d_2)$ takes its maximum value.

One can ensure that a Newton-Raphson iteration of Eq. (A.165) for d_2 will converge (with one exceptional case, $d_2 = -1$) as long as the initial estimate, d_2^0, satisfies:

(1) $d_2^0 < (d_2)_{\max}$ if $(d_2)_{\max} < -1$,
(2) $d_2^0 > (d_2)_{\max}$ if $(d_2)_{\max} > -1$.

In general, $d_2^0 = 2(d_2)_{\max} - 1$ for the initial iteration.

If $-\alpha\ell n(\gamma_M) \approx \ell n(\gamma_P)$, there is a solution $d_2 \approx -1$. Since Eq. (A.165) has introduced an extraneous root $d_2 = -1$ (property 6), it is necessary to avoid the iterative procedure in the neighborhood of $d_2 = -1$. The difference $\ell n(\gamma_P) - \alpha\ell n(\gamma_M)$ is therefore compared to zero, and d_2 is set equal to -1 if the comparison is close enough. In this case, the Newton-Raphson iteration is bypassed.

Evaluation of the group (e_{K-1}, e_K) where K is the total number of groups requires special handling because the two-interval span involved is (e_{K-1}, ∞).

In this case it is convenient to choose a point e' in the interval (e_K, ∞) for developing the power law fit. Arbitrarily, e' is chosen such that the interval (e_K, e') contains one-half the energy in the infinite group.

$$\int_{e_K}^{e'} C_1 \exp(-e/C_2)de = 0.5E_{iK}\,, \tag{A.203}$$

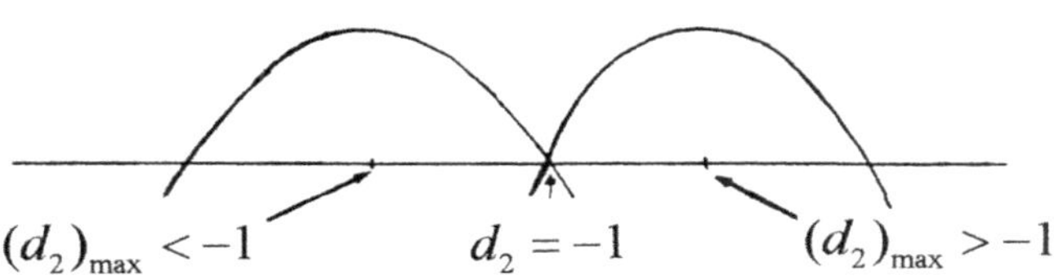

Figure A.6. Sketches of function $h(d_2)$.

Table A.3.2 Numerical approximations to scattering in Case 2.

Term in Scattering Equation		Power Law Assumption Numerical Approximation
$eE(e)\big\vert_{e_j}$	=	$\dfrac{E_{ij}(d_2+1)}{\gamma_P^{d_2+1}-1}$
$E(e)\big\vert_{e_j}$	=	$\dfrac{E_{ij}(d_2+1)}{(\gamma_P^{d_2+1}-1)e_j}$
$\dfrac{\partial}{\partial e}eE(e)\big\vert_{e_j}$	=	$(d_2+1)E(e_j)$
$\left[\dfrac{E(e)}{e}\right]^2\bigg\vert_{e_j}$	=	$\left[\dfrac{E_{ij}(d_2+1)}{(\gamma_P^{d_2+1}-1)e_j}\right]^2 e_j^{-2}$
Terms evaluated at e_{j+1}		Saved from computations in the previous frequency group
$\displaystyle\int_{e_j}^{e_{j+1}} eE(e)de$	=	$\dfrac{\hat{d}_1}{\hat{d}_2+2}e_j^{\hat{d}_2+2}(\gamma_P^{\hat{d}_2+1}-1)$

$$e' = C_2\ell n(2) + e_K\,. \tag{A.204}$$

Then, replacing e_{K+2} with e' and E_{ij} with $0.5E_{ij}$, the scheme can be carried out. The numerical approximations to scattering for Case 2 are summarized in Table A.3.2.

Case 3: The lowest frequency group $(0, e_1)$.

The exponent d_2 computed from the interval $(0, e_2)$ is used for the lowest frequency group. The contributions are:

Term in Scattering Equation		Numerical Approximation Lowest Frequency Group
All terms evaluated at zero		0
All terms evaluated at e_1		Saved from previous frequency group
$\displaystyle\int_0^{e_1} eE(e)de$	=	$\dfrac{e_1(E_i)_1(d_2+1)}{d_2+2}$

A.3.6 Difference Equation for the Conservation of Momentum and Energy

The difference formulations of the momentum and energy equations are described below.

Conservation of Momentum

From Section 2, Eqs. (A.41), we have the momentum equation employing an equivalent radiation form

$$\rho\frac{D\bar{u}}{Dt} + \nabla P_m = \frac{1}{c}\int_0^\infty (\mu_a' + \mu_s)F_\nu d\nu\,. \tag{A.205}$$

The finite difference is given by

$$\begin{aligned} u_i^{n+1/2} &= u_i^{n-1/2} - \Delta t_1\big[2\big(P_i^n - P_{i-1}^n\big) \\ &\qquad + \big(r_i^n - r_{i-1}^n\big)M_i^n\big]\frac{A_i^n}{g_i + g_{i-1}}\,, \end{aligned} \tag{A.206}$$

$$\Delta r_i^{n+1} = \Delta r_i^n + \Delta t_2\big(u_{i+1}^{n+1/2} - u_i^{n+1/2}\big)\,, \tag{A.207}$$

$$r_{i+1}^{n+1} = r_i^{n+1} + \Delta r_i^{n+1}\,, \tag{A.208}$$

where

$$M_i^n = -\frac{1}{c}\sum_j \frac{F_{ij}^n}{\lambda_{ij}}\,, \tag{A.208.a}$$

$$\lambda_{ij}^{-1} = \big(\mu_a' + \mu_s\big)_{ij}\,, \tag{A.208.b}$$

$$\Delta t_1 = \frac{t^{n+1} - t^{n-1}}{2}\,, \tag{A.208.c}$$

$$\Delta t_2 = t^{n+1} - t^n\,, \tag{A.208.d}$$

$$g_i = \alpha\int_{r_i}^{r_{i+1}} \rho r^{\alpha-1} dr\,, \tag{A.208.e}$$

$$A_i^n = \alpha(r_i^n)^{\alpha-1}\,. \tag{A.208.f}$$

The radius of the left-hand boundary is assumed stationary for spherical problems. For planar problems the relation

$$r_1^{n+1} = r_1^n + u_1^{n+1/2}\Delta t\,, \tag{A.209}$$

is used.

The zone rate of expansion term for planes is

$$\dot{v}_i^{n+1/2} = u_{i+1}^{n+1/2} - u_i^{n+1/2} . \tag{A.210}$$

The zone expansion rate for spheres is given by

$$\begin{aligned}\dot{v}_i^{n+1/2} &= u_{i+1}^{n+1/2}\left[r_{i+1}^{n+1}\left(r_{i+1}^{n+1} + r_{i+1}^{n}\right) + \left(r_{i+1}^{n}\right)^2\right] \\ &\quad - u_i^{n+1/2}\left[r_i^{n+1}\left(r_i^{n+1} + r_i^{n}\right) + \left(r_i^{n}\right)^2\right] .\end{aligned} \tag{A.211}$$

The term P_i^n includes both the material pressure P_i and the artificial viscosity $P2$. Following the method of von Neumann and Richtmyer [A.3] for each zone, the artificial or dissipative pressure $P2$ is defined as

$$P2_i^n = \frac{c_0}{\tau_i^n}\left(\frac{\delta(Au)_i^n(\delta r)_i^n}{g_i\tau_i^n}\right) \quad \text{if } \delta(Au)_i^n < 0 , \tag{A.212}$$

$$P2_i^n = 0 \quad \text{if } \delta(Au)_i^n \geq 0 , \tag{A.213}$$

where

$$\delta(Au)_i^n = A_{i+1}^n u_{i+1}^{n-1/2} - A_i^n u_i^{n-1/2} , \tag{A.214}$$

and

$$c_0 = 1.6 .$$

The centering of the momentum equation is at the ith boundary at time n. Notice that the radiation contribution M_i^n is a boundary quantity known at integral times. Thus, the truncation error is unaffected. However, one notes that by Taylor series expansions, the truncation, error is of order

$$O(\Delta t^2) + O(g_i - g_{i-1}) + O(g_i^n) . \tag{A.215}$$

Since the momentum equation is explicit, the Courant stability condition must be established.

Conservation of Energy

From Section A.2, Eq. (A.42), we have the internal energy equation with the equivalent radiation forms

$$\rho\frac{DE}{Dt} + \rho P_m\frac{D\tau}{Dt} - \dot{q} + \int_0^\infty \left[c\mu_P\left(\frac{4\pi B_\nu}{c} - E_\nu\right) + T + \frac{u}{c}\frac{F_\nu}{\lambda_\nu}\right] d\nu = 0 , \tag{A.216}$$

where the terms are as defined in Section A.2. The centering of Eq. (A.216) is at $i+1/2, n+1/2$. Thus one has the formula for the radiative heating rate ER_i,

$$ER_i^{n+1/2} = -\sum_j c\mu_{ij}(b_j^{n+1}\phi_i^{n+1} - E_{ij}^{n+1}) + \frac{1}{8}\big[\big(u_i^{n+1/2} - u_{i+1}^{n+1/2}\big) \times \big(M_i^n + M_{i+1}^n + M_i^{n+1} + M_{i+1}^{n+1}\big)\big] \tag{A.217}$$

The terms involving M_i^n have been centered correctly in time to minimize the truncation errors of these terms. The equation for the updated material energy E_i^{n+1} is given by

$$E_i^{n+1} = E_i^n + \big(-P_i^n\dot{\tau}_i^{n+1/2} + \dot{Q}_i^{n+1/2}\big)\Delta t_i^n\,, \tag{A.218}$$

where

$$\dot{Q}_i^{n+1/2} = \frac{1}{g_i}\big(ER_i^{n+1/2} + \dot{q}_i^{n+1/2}\big)\,, \tag{A.218.a}$$

$$\dot{\tau}_i \equiv \frac{D\tau}{Dt}\,. \tag{A.218.b}$$

Using an equation of state of the form $E(\theta, \tau)$, θ can be evaluated iteratively. For an initial value the temperatures are evaluated from

$$\theta_i^{n+1} = \theta_i^n + \left[-\left(P + \frac{\partial E}{\partial \tau}\right)_i^n \dot{\tau}_i^{n+1/2} + \dot{Q}_i^{n+1/2}\right]\frac{\Delta t_i^n}{C_{V_i}^n} \tag{A.219}$$

The truncation error in the energy equation is

$$O(\Delta t) + O(g_i^2)\,. \tag{A.220}$$

A.4 Formulae for the First Moment of the Radiative Transfer Equation in 1D Geometry

The zeroth and first moments of the radiative transfer equation can be written as

$$\frac{\partial E}{\partial t} + \nabla\cdot\bar{F} = c\mu_a'(4\pi B - E)\,, \tag{A.221}$$

$$\frac{1}{c}\frac{\partial \bar{F}}{\partial t} + c\nabla\cdot\bar{\bar{P}} = -(\mu_a' + \mu_s)\bar{F}\,, \tag{A.222}$$

where

$$E = \frac{1}{c}\int I d\Omega \quad \text{is the (scalar) radiation energy density}\,, \tag{A.223}$$

$$\bar{F} = \frac{1}{c}\int I\bar{\Omega} d\Omega \quad \text{is the (vector) flux}\,, \tag{A.224}$$

$$\bar{\bar{P}} = \frac{1}{c}\int I\bar{\Omega}\bar{\Omega} d\Omega \quad \text{is the (tensor) radiation pressure}\,. \tag{A.225}$$

The LTE approximation for the source leads to the introduction of the Planck function

$$B = \frac{2h\nu^3}{c^2(e^{h\nu/\theta}-1)}\,, \tag{A.226}$$

where θ is the material temperature. The absorption coefficient corrected for induced emission is μ'_a, and scattering has been taken into account in the Thomson approximation with the scattering coefficient μ_s.

It is desirable to reduce these equations to a more explicit form for the cases of one-dimensional flow, i.e. planar, cylindrical or spherical geometries, in which all quantities are constant on planar, infinite cylindrical or spherical surfaces. Consequently, it is required to obtain expressions for $\nabla \cdot \bar{F}$ and $\nabla \cdot \bar{\bar{P}}$ in orthogonal curvilinear coordinates and to determine the forms of the components of the vectors and tensors involved.

The orthogonal coordinates are denoted by (x_1, x_2, x_3), the non-zero components of the metric tensor by $g_{11} = h_1^2, g_{22} = h_2^2, g_{33} = h_3^2$, and $\sqrt{g} = h_1 h_2 h_3$. The divergence of the flux is given by

$$F^i_{,i} = \frac{1}{\sqrt{g}}\frac{\partial}{\partial x^i}\left(\sqrt{g}F^i\right), \tag{A.227}$$

where the summation convention is used and F^i is the i-component of the flux given by Eq. (A.224) and the Ω^i component of the unit direction vector. The corresponding formula for the i-component of the divergence of the pressure tensor is

$$P^{ij}_{,j} = \frac{1}{\sqrt{g}}\frac{\partial}{\partial x^j}\left(\sqrt{g}P^{ij}\right) + P^{mn}\left\{ {i \atop m\ \ n} \right\}. \tag{A.228}$$

In Eq. (A.228) $\left\{ {i \atop m\ \ n} \right\}$ denotes the Christoffel symbol, having the values

$$\left\{ {k \atop i\ \ j} \right\} = 0\,, \tag{A.229}$$

$$\left\{ {j \atop i \quad i} \right\} = -\frac{h_i}{h_j^2}\frac{\partial h_i}{\partial x^j}, \tag{A.230}$$

$$\left\{ {i \atop i \quad j} \right\} = \frac{1}{h_i}\frac{\partial h_i}{\partial x^j}, \tag{A.231}$$

$$\left\{ {i \atop i \quad i} \right\} = \frac{1}{h_i}\frac{\partial h_i}{\partial x^i}, \tag{A.232}$$

in which the summation convention is not applicable.

The above formulae are first applied in Cartesian, spherical and cylindrical coordinates without restriction to 1D symmetry. This latter restriction will be imposed by determining the special form which the components take in that case.

Cartesian Coordinates

$$x_1 = x, \quad x_2 = y, \quad x_3 = z, \quad h_1 = h_2 = h_3 = 1, \quad \sqrt{g} = 1\,. \tag{A.233}$$

$$F^i_{,i} = \frac{\partial F^x}{\partial x} + \frac{\partial F^y}{\partial y} + \frac{\partial F^z}{\partial z}, \tag{A.234}$$

$$P^{ij}_{,j} = \frac{\partial P^{ix}}{\partial x} + \frac{\partial P^{iy}}{\partial y} + \frac{\partial P^{iz}}{\partial z}\,. \tag{A.235}$$

Cylindrical Coordinates

$$x_1 = r, \quad x_2 = \theta, \quad x_3 = z, \quad h_1 = 1, \quad h_2 = r,$$

$$h_3 = 1, \quad \sqrt{g} = r\,, \tag{A.236}$$

$$F^i_{,i} = \frac{1}{r}\frac{\partial rF^r}{\partial r} + \frac{\partial F^\theta}{\partial \theta} + \frac{\partial F^z}{\partial z}, \tag{A.237}$$

$$P^{rj}_{,j} = \frac{1}{r}\frac{\partial (rP^{rr})}{\partial r} + \frac{\partial P^{r\theta}}{\partial \theta} + \frac{\partial P^{rz}}{\partial z} - rP^{\theta\theta}, \tag{A.238}$$

$$P^{\theta j}_{,j} = \frac{1}{r}\frac{\partial (rP^{\theta r})}{\partial r} + \frac{\partial P^{\theta\theta}}{\partial \theta} + \frac{\partial P^{\theta z}}{\partial z} + \frac{1}{r}(P^{r\theta} + P^{\theta r})\,, \tag{A.239}$$

$$P^{zj}_{,j} = \frac{1}{r}\frac{\partial (rP^{zr})}{\partial r} + \frac{\partial P^{z\theta}}{\partial \theta} + \frac{\partial P^{zz}}{\partial z}\,. \tag{A.240}$$

Spherical Coordinates

$$x_1 = r, \quad x_2 = \theta, \quad x_3 = \phi, \quad h_1 = 1,$$

$$h_2 = r, \quad h_3 = r\sin\theta, \quad \sqrt{g} = r\sin\theta, \tag{A.241}$$

$$F^i_{,i} = \frac{1}{r^2}\frac{\partial(r^2 F^r)}{\partial r} + \frac{1}{\sin\theta}\frac{\partial(\sin\theta F^\theta)}{\partial\theta} + \frac{\partial F^\phi}{\partial\phi}, \tag{A.242}$$

$$P^{rj}_{,j} = \frac{1}{r^2}\frac{\partial(r^2 P^{rr})}{\partial r} + \frac{1}{\sin\theta}\frac{\partial(\sin\theta P^{r\theta})}{\partial\theta} + \frac{\partial P^{r\phi}}{\partial\phi}$$
$$- rP^{\theta\theta} - rP^{\phi\phi}\sin^2\theta, \tag{A.243}$$

$$P^{\theta j}_{,j} = \frac{1}{r^2}\frac{\partial(r^2 P^{\theta r})}{\partial r} + \frac{1}{\sin\theta}\frac{\partial(\sin\theta P^{\theta\theta})}{\partial\theta} + \frac{\partial P^{\theta\phi}}{\partial\phi} + \frac{P^{\theta r} + P^{r\theta}}{r}$$
$$- P^{zz}\sin\theta\cos\theta, \tag{A.244}$$

$$P^{\phi j}_{,j} = \frac{1}{r^2}\frac{\partial(r^2 P^{\phi r})}{\partial r} + \frac{1}{\sin\theta}\frac{\partial(\sin\theta P^{\phi\theta})}{\partial\theta} + \frac{\partial P^{\phi\phi}}{\partial\phi} + \frac{P^{\phi r} + P^{r\phi}}{r}$$
$$+ (P^{\phi\theta} + P^{\theta\phi})\cot\theta. \tag{A.245}$$

In order to evaluate the flux and pressure in terms of the radiation intensity and describe the direction of photon travel, a polar coordinate system with polar axis parallel to (1) the z-axis for planes, (2) the r-axis for cylinders, and (3) the r-axis for spheres is introduced. In this system,

$$\bar{\Omega} = \bar{i}\cos\theta + \sin\theta(\bar{j}\sin\phi + \bar{k}\cos\phi), \tag{A.246}$$

and

$$d\Omega = d(\cos\theta)d\phi, \tag{A.247}$$

where θ is the polar angle, ϕ the angle of azimuth, $\bar{i}$ is a unit vector along the polar direction, and $\bar{j}$ and $\bar{k}$ are mutually perpendicular directions lying parallel to the remaining (curvilinear) coordinate directions. In this system, the flux becomes

$$\bar{F} = \begin{pmatrix} \int I\mu d\mu d\phi \\ \int I\sqrt{1-\mu^2}\sin\phi d\phi \\ \int I\sqrt{1-\mu^2}\cos\phi d\phi \end{pmatrix}, \tag{A.248}$$

where $\mu = \cos\theta$ and intensity $I = I(\mu, \phi)$. The dependence on position has been suppressed. The pressure tensor is

$$cP = \begin{pmatrix} \int I\mu^2 d\mu d\phi & \int I\mu\sqrt{1-\mu^2} \sin\phi d\mu d\phi & \int I\mu\sqrt{1-\mu^2} \cos\phi d\mu d\phi \\ \int I\mu\sqrt{1-\mu^2} \sin\phi d\mu d\phi & \int I(1-\mu^2) \sin^2\phi d\mu d\phi & \int I(1-\mu^2) \sin\phi\cos\phi d\mu d\phi \\ \int I\mu\sqrt{1-\mu^2} \cos\phi d\mu d\phi & \int I(1-\mu^2) \sin\phi\cos\phi d\mu d\phi & \int I(1-\mu^2) \cos^2\phi d\mu d\phi \end{pmatrix} \tag{A.249}$$

For the 1D geometries there are considerable simplifications arising from two sources: (1) the components depend only on a single spatial variable, and (2) some of the components either vanish or are reduced due to symmetry arguments. In the case of planes, all quantities are taken to be independent of x and y. In addition, the intensity is independent of ϕ. Consequently, as given by Eqs. (A.234) and (A.235), the divergence of the flux and the z-component of the divergence of the pressure tensor (all other components vanish) are

Planes:

$$F^i_{,i} = \frac{\partial F^z}{\partial z}, \tag{A.250}$$

$$P^{zj}_{,j} = \frac{\partial P^{zz}}{\partial z}, \tag{A.251}$$

where

$$F^z = 2\pi \int I\mu d\mu, \tag{A.252}$$

$$P^{zz} = \frac{2\pi}{c} \int I\mu^2 d\mu. \tag{A.253}$$

For 1D spheres all derivatives of F and P with respect to θ and ϕ vanish and, again by symmetry, the intensity is independent of ϕ. The results are

Spheres:

$$F^i_{,i} = \frac{1}{r^2}\frac{\partial(r^2 F^r)}{\partial r}, \tag{A.254}$$

$$P^{rj}_{,j} = \frac{1}{r^2}\frac{\partial(r^2 P^{rr})}{\partial r} - rP^{\theta\theta} - rP^{\phi\phi}\sin^2\theta\,, \tag{A.255}$$

where

$$F^r = 2\pi \int I\mu d\mu\,, \tag{A.256}$$

$$P^{rr} = \frac{2\pi}{c}\int I\mu^2 d\mu\,, \tag{A.257}$$

$$rP^{\theta\theta} = \frac{\pi}{c}\int I(1-\mu^2)d\mu = \frac{1}{2}(E - P^{rr})\,, \tag{A.258}$$

$$rP^{\phi\phi}\sin^2\theta = \frac{\pi}{c}\int I(1-\mu^2)d\mu = \frac{1}{2}(E - P^{rr})\,. \tag{A.259}$$

For 1D cylinders the derivatives of F and P with θ and z vanish. Differing from the two preceding cases, the intensity is not independent of ϕ. Consequently,

Cylinders:

$$F^i_{,i} = \frac{1}{r}\frac{\partial(rF^r)}{\partial r}\,, \tag{A.260}$$

$$P^{rj}_{,j} = \frac{1}{r}\frac{\partial(rP^{rr})}{\partial r} - rP^{\theta\theta}\,, \tag{A.261}$$

where

$$F^r = \int I\mu d\mu d\phi\,, \tag{A.262}$$

$$P^{rr} = \frac{1}{c}\int I\mu^2 d\mu d\phi\,, \tag{A.263}$$

$$rP^{\theta\theta} = \frac{1}{c}\int I(1-\mu^2)\sin^2\phi d\mu d\phi\,. \tag{A.264}$$

This case differs also in that it is not possible to evaluate $P^{\theta\theta}$ in terms of previously introduced quantities. The first two cases are contained in the following expression:

$$F^i_{,i} = \frac{1}{r^{\alpha-1}}\frac{\partial(r^{\alpha-1}F^r)}{\partial r}\,, \tag{A.265}$$

$$P^{rj}_{,j} = \frac{\partial(P^{rr})}{\partial r} + \frac{\alpha-1}{2r}(3P^{rr} - E)\,, \tag{A.266}$$

in which $\alpha = 1$ for planes, $\alpha = 3$ for spheres, and where r stands for the z coordinate when $\alpha = 1$. For cylinders with$\alpha = 2$, the flux divergence expression is also correct in Eqs. (A.265) and (A.266). Equations (A.260) and (A.261) can be rewritten to place the pressure term in a form closer to Eqs. (A.265) and (A.266):

$$P^{rj}_{,j} = \frac{\partial(P_{rr})}{\partial r} + \frac{1}{r}(P^{rr} - r^2 P^{\theta\theta}) \,. \tag{A.267}$$

If in this expression it is possible to consider the intensity to be independent of ϕ or if $\overline{\sin^2\phi} = 1/2$, then $r^2 P^{\theta\theta} = 0.5(E - P^{rr})$ and Eq. (A.267) is given by the $\alpha = 2$ case of Eq. (A.266). Such is the case for the diffusion approximation which appears to be a useful expression for the variable Eddington approximation.

A.5 Properties of an Implicit Difference Approximation to the Wave Equation

The equations of radiative transfer in the Eddington approximation contain terms accounting for the propagation of electro-magnetic waves. These terms are emphasized in regions where absorption and scattering are so small that the radiation may stream freely. In such an event, the Eddington approximation equations become

$$\frac{\partial E}{\partial t} + \nabla \cdot \bar{F} = 0 \,, \tag{A.268}$$

$$\frac{\partial \bar{F}}{\partial t} + \frac{c^2}{3}\nabla E = 0 \,, \tag{A.269}$$

where E is the spectral radiation energy density, $\bar{F}$ is the (vector) spectral radiation flux, and c is the speed of light. In one-dimensional plane geometry, which is considered since the equations have constant coefficients, the equations become

$$\frac{\partial E}{\partial t} + \frac{\partial F}{\partial x} = 0 \,, \tag{A.270}$$

$$\frac{\partial F}{\partial t} + \frac{c^2}{3}\frac{\partial E}{\partial x} = 0 \,, \tag{A.271}$$

where F is now the x-component of the flux, all others being assumed to vanish.

Difference Equations

The difference expression in which the Eddington approximation equations are evaluated, is an implicit one in which the quantities in the spatial terms are evaluated at the advanced time. The times on cycles n and $n+1$ are denoted by t^n and t^{n+1}, such that $\Delta t = t^{n+1} - t^n$, and positions corresponding to spatial indices i and $i+1$ are denoted by x_i and x_{i+1}, such that $\Delta x = x_{i+1} - x_i$. Implicit difference equations are

$$\frac{E_i^{n+1} - E_i^n}{\Delta t} + \frac{\theta}{\Delta x}(F_{i+1}^{n+1} - F_i^{n+1}) + \frac{(1-\theta)}{\Delta x}(F_{i+1}^n - F_i^n) = 0\,, \tag{A.272}$$

$$\frac{F_{i+1}^{n+1} - F_{i+1}^n}{\Delta t} + \frac{c^2\theta}{3\Delta x}(E_{i+1}^{n+1} - E_i^{n+1}) + \frac{c^2(1-\theta)}{3\Delta x}(E_{i+1}^n - E_i^n) = 0\,. \tag{A.273}$$

In Eqs. (A.272) and (A.273) the fact that the equations are centered in position is masked by the notation in which integral subscripts are employed. The quantity E_i in fact refers to a position one-half zone farther to the right than the position of F_i. (The replacement of i by $i+1/2$ in the E-subscripts would clarify this centering at the expense of more cumbersome notation.) The quantity θ is a dimensionless parameter which determines the degree of implicitness of the formulation.

Accuracy

To determine the accuracy of the difference approximation, Eqs. (A.272) and (A.273), relative to the differential equations, Eqs. (A.270) and (A.271), an investigation is made into the rate of convergence of the difference equation to the differential equation as Δx and Δt approach zero. In order to do this, the exact equation is substituted into the Eqs. (A.272) and (A.273) and subsequently developed in Taylor's series in such a way as to recover the differential equation. The lowest order remainder terms dominate the truncation error and determine the order of accuracy of the method in question. In performing the series expansion it is naturally a necessary requirement that the solution be sufficiently well behaved to contain the indicated partial derivatives.

Carrying out the expansions about the time $t = t^n + \frac{\Delta t}{2}$ and position $x = x_i + \frac{\Delta x}{2}$ (the position associated with E_i) the first equation becomes

$$\begin{aligned}&\frac{E_i^{n+1} - E_i^n}{\Delta t} + \frac{\theta}{\Delta x}(F_{i+1}^{n+1} - F_i^{n+1}) + \frac{(1-\theta)}{\Delta x}(F_{i+1}^n - F_i^n)\\ &= \frac{\Delta t}{2}(2\theta - 1)\frac{\partial^2 F}{\partial x \partial t} + O(\Delta x^2)\,,\end{aligned} \tag{A.274}$$

where the coefficient of Δt contains the derivative of the exact solution evaluated at the above indicated position, and the remaining terms are a shorthand notation for additional derivative terms having quadratic coefficients in Δt and Δx. The second equation, centered about the time $t = t^n + \Delta t$ and position $x = x_{i+1}$, is

$$\frac{F_{i+1}^{n+1} - F_{i+1}^{n}}{\Delta t} + \frac{c^2\theta}{3\Delta x}(E_{i+1}^{n+1} - E_i^{n+1}) + \frac{c^2(1-\theta)}{3\Delta x}(E_{i+1}^{n} - E_i^{n})$$
$$= \frac{c^2(2\theta - 1)}{6}\frac{\partial^2 E}{\partial x \partial t}\Delta t + O(\Delta t^2) + O(\Delta x^2)\,. \tag{A.275}$$

Equations (A.274) and (A.275) show that the difference equations have truncation errors of order Δt and Δx^2 unless $\theta = 1/2$.

Stability

The equations describing light wave propagation in Eqs. (A.268) and (A.269) are identical to those for sound wave propagation when a suitable identification of constants is made. Consequently, the discussion of the stability of approximations to the sound wave equations by Richtmyer and Morton [A.4] is relevant. In particular, the analysis for an implicit difference approximation, dealing with special cases of Eqs. (A.272) and (A.273), indicated unconditional stability. A similar analysis of Eq. (A.272) is carried out below. The stability conditions applicable to two-level schemes with constant coefficients derived by Richtmyer and Morton [A.4] and based on Fourier transformation are applicable.

The equation for the amplitude of the kth Fourier component

$$v(k) = \begin{pmatrix} e(k) \\ f(k) \end{pmatrix}, \tag{A.276}$$

is obtained by substitution in Eqs. (A.272) and (A.273). It is

$$Av^{n+1} = Bv^n\,, \tag{A.277}$$

where

$$A = \begin{pmatrix} 1 & ia\theta \\ \frac{iac^2}{3}\theta & 1 \end{pmatrix}, \tag{A.278}$$

$$B = \begin{pmatrix} 1 & -ia(1-\theta) \\ \dfrac{-iac^2}{3}(1-\theta) & 1 \end{pmatrix}, \tag{A.279}$$

and $a = \Delta x \frac{2\Delta t}{\Delta x} \sin \frac{k}{2}$. The amplification matrix of Eq. (A.277) is

$$G(\Delta t, k) = \begin{pmatrix} \dfrac{1 - a^2c^2\theta(1-\theta)/3}{1 + a^2c^2\theta^2/3} & \dfrac{-ia}{1 + a^2c^2\theta^2/3} \\ \dfrac{-iac^2/3}{1 + a^2c^2\theta^2/3} & \dfrac{1 - a^2c^2\theta(1-\theta)/3}{1 + a^2c^2\theta^2/3} \end{pmatrix}, \tag{A.280}$$

having eigenvalues

$$\lambda = \frac{1 - a^2c^2\theta(1-\theta)/3}{1 + a^2c^2\theta^2/3} \pm \frac{iac/\sqrt{3}}{1 + a^2c^2\theta^2/3}. \tag{A.281}$$

The spectral radius of G is given by

$$\lambda = \sqrt{1 + \frac{a^2c^2(1-2\theta)/3}{1 + a^2c^2/3}}, \tag{A.282}$$

which requires $\theta \geq 0.5$ for stability. Both of the eigenvalues of G have the same absolute value which is

$$|\lambda| = \frac{1}{\sqrt{1 + a^2c^2/3}}, \quad \text{for } \theta = 1. \tag{A.283}$$

It may be noted that there exists no damping at all for the centered equations with $\theta = 1/2$.

References

A.1 Campbell, PM, and Nelsen, RG, Numerical methods for non-linear radiation transport calculations, UCRL-7838 (1964).
A.2 Spillman, GR, Formulation of the Eddington factor for use in one-dimensional non-equilibrium diffusion calculations, Private communication.
A.3 von Neumann, J, and Richtmyer, RD, A method for the numerical calculations of hydrodynamic shocks, J. Appl. Phys., 21, p. 232 (1950).
A.4 Richtmyer, RD, and Morton, KW, Difference methods for initial value problems, Interscience Publisher, Inc, New York (1957).

APPENDIX B

THERMONUCLEAR BURN OF DEUTERIUM-TRITIUM SPHERE

Notations

A	coefficient of energy transfer between fields
$\bar{A}$	atomic mass (g)
c	speed of light (3×10^{10} cm/s)
C_v	specific heat at constant volume
e	constant used in Eqs. (B.7), (B.9), (B.12), (B.13), (B.14), (B.16) and (B.17), ($e = \sqrt{\hbar c}$)
h	Planck constant
$\hbar$	reduced Planck constant ($\hbar = h/2\pi$)
I	specific internal energy per unit mass (KJ/μg)
k	Boltzmann constant
K	thermal conductivity
m	mass (g)
N_A	Avogadro's number
N_D	ion number density for deuterium
N_{Li}	ion number density for lithium
N_T	ion number density for tritium
n	ion number density (#/cm^3)
n_0^1	neutron particle
p	pressure (KJ/μcm^3)
p_1^1	proton particle or H_1^1
q	artificial viscosity (KJ/μcm^3)
r_0	electron radius
$\dot{S}$	specific energy source rate
S_n	discrete ordinate method
t	time (ps)
T	temperature (keV)
u	velocity in x direction (cm/ps)
V	specific volume (cm^3/g)
x	horizontal coordinate (cm)

y vertical coordinate (cm)
$\bar{Z}$ atomic number

Greek Letters

α alpha particle, i.e. He_2^4
ϕ radiation intensity (energy/time-area-frequency)
λ mean-free-path
λ_a absorption mean-free-path
λ_R Rosseland mean-free-path
λ_s scattering mean-free-path
Λ harmonic mean of the classical and quantum results
μ $\mu = \cos\theta, \theta$ is the angle (see Eq. (B.47) and the description followed)
ν frequency
ρ density (g/cm^3)
σ Stephan-Boltzmann constant
$\overline{\sigma\nu}$ Maxwell averaged cross-section
Ω unit vector in the direction of the photon transport

Subscripts

0 initial value
DD D-D reaction
DT D-T reaction
e electron
er between electron and radiation
i ion
ie between ion and electron
nLi neutron-lithium reaction
r radiation
t derivative with respect to time
x derivative with respect to x coordinate

B.1 Introduction

The dynamic behavior of the thermonuclear burn for deuterium-tritium fuel is very complicated, and the computer simulation of the burning physics involves solving the hydrodynamic, radiation transport, charge particle transport, nuclear fusion reaction, and neutron transport equations. The modeling of the hydrodynamic problems is described in Chapters 2 and 3.

The neutron transport equation can be solved by using diffusion or Monte Carlo methods as described in Refs. [B.1] and [B.2].

In this Appendix, the couplings between the photon (radiation) and the electron and between the electron and the ion (material) will be discussed assuming homogeneous (same velocity) but non-equilibrium conditions (different temperatures). The nuclear fusion reaction kinetics and the reaction rate will be determined by the Maxwellian-average reaction rate parameter $\langle \sigma \nu \rangle$, which depends on temperature. Most of the material in this Appendix is obtained from Ref. [B.3].

B.2 The Governing Equations

In the 1D Lagrangian code, the momentum equation is given by

$$\rho \frac{\partial u}{\partial t} = -(p+q)\frac{\partial u}{\partial x} \tag{B.1}$$

where

$$p = p_r + p_i + p_e \,, \tag{B.2}$$

and q is the usual artificial viscosity as described in Chapter 5. Pressure p_r, p_i and p_e can be obtained from the EOS (equation of state). The energy equations for ions, electrons, and photons are

$$\begin{aligned}\frac{\partial T_i}{\partial t} = \frac{1}{C_{vi}} \Bigg[\dot{S}_i - \left(p_i + q + \frac{\partial I_i}{\partial V} \right) \frac{\partial V}{\partial t} - A_{ie}(T_i - T_e) \\ + V \left(\frac{\partial}{\partial x} K_i \frac{\partial T_i}{\partial x} + \frac{\partial}{\partial y} K_i \frac{\partial T_i}{\partial y} \right) \Bigg] \,,\end{aligned} \tag{B.3}$$

$$\begin{aligned}\frac{\partial T_e}{\partial t} = \frac{1}{C_{ve}} \Bigg[\dot{S}_e - \left(p_e + \frac{\partial I_e}{\partial V} \right) \frac{\partial V}{\partial t} - A_{er}(T_e - T_r) + A_{ie}(T_i - T_e) \\ + V \left(\frac{\partial}{\partial x} K_e \frac{\partial T_e}{\partial x} + \frac{\partial}{\partial y} K_e \frac{\partial T_e}{\partial y} \right) \Bigg] \,,\end{aligned} \tag{B.4}$$

and

$$\begin{aligned}\frac{\partial T_r}{\partial t} = \frac{1}{C_{vr}} \Bigg[- \left(p_r + \frac{\partial I_r}{\partial V} \right) \frac{\partial V}{\partial t} + A_{er}(T_e - T_r) \\ + V \left(\frac{\partial}{\partial x} K_r \frac{\partial T_r}{\partial x} + \frac{\partial}{\partial y} K_r \frac{\partial T_r}{\partial y} \right) \Bigg] \,.\end{aligned} \tag{B.5}$$

The coupling terms A_{ei} and A_{er} and the thermal conductivities K_i, K_e and K_r are

$$A_{ei} = C_{vi}\nu_{eq} \tag{B.6}$$

where

$$\nu_{eq} = \frac{8(2\pi)^{1/2} m_e^{1/2} e^4 N_A^2}{3}\left(\frac{Z^2}{A^2}\right)(\ln \Lambda_{ei})\frac{\rho}{(kT_e)^{3/2}}\left[1 - \frac{3}{2}\frac{T_i}{T_e}\frac{m_e}{m_i}\right], \tag{B.7}$$

and

$$A_{er} = C_{ve}(\nu_b + \nu_c)\,, \tag{B.8}$$

where

$$\nu_b = \frac{32}{3}\left(\frac{2}{\pi m_e}\right)^{1/2}\frac{e^4 N_A^2}{\hbar c}k\left(\frac{Z^2}{A^2}\right)\frac{\rho Z}{(kT_e)^{1/2}}G(T_r/T_e)\,, \tag{B.9}$$

with

$$G(\gamma) = \int_0^\infty \frac{f(\xi)\left[1 - e^{-\xi\left(\frac{1}{\gamma}-1\right)}\right]}{(\gamma - 1)\left(1 - e^{-\xi/\gamma}\right)}d\xi\,, \tag{B.10}$$

containing

$$f(\xi) = \int_0^\infty \ln\left(\sqrt{x} + \sqrt{x+1}\right)e^{-\xi x}dx\,, \tag{B.11}$$

and

$$\nu_c = \frac{128}{3}\frac{\pi e^2\sigma}{(m_e c^2)^2}N_A\left(\frac{Z}{A}\right)T_r^4 k\,, \tag{B.12}$$

$$K_i = 20\left(\frac{2}{\pi}\right)^{3/2}\frac{(kT_e)^{5/2}k}{m_i^{1/2}e^4 Z^4 \ln \Lambda_i}\,, \tag{B.13}$$

$$K_e = 20\left(\frac{2}{\pi}\right)^{3/2}\frac{(kT_e)^{5/2}k\varepsilon\delta_T}{m_e^{1/2}e^4 Z^4 \ln \Lambda_{ei}}\,, \tag{B.14}$$

from Spitzer [B.4], in which

$$\varepsilon\delta_T = \frac{0.43Z}{[3.44 + Z + 0.26\ln(Z)]}\,, \tag{B.15}$$

and

$$\ln \Lambda_{ei} = \max \left\{ 1, \ln \left[\frac{3}{2e^3} \left(\frac{A}{Z} \frac{k^3 T_e^3}{\pi \rho N_A} \right) \frac{1}{\left(Z + \frac{1}{2\alpha c} \sqrt{\frac{3kT_e}{m_e}} \right)} \right] \right\}, \tag{B.16}$$

$$\ln \Lambda_{i} = \max \left\{ 1, \ln \left[\frac{3}{2e^3 Z^2} \left(\frac{A}{Z} \frac{k^3 T_i^3}{\pi \rho N_A} \right) \frac{1}{\left(Z + \frac{T_i}{T_e} \right)} \right] \right\}. \tag{B.17}$$

Once the T_i, T_e and T_r are calculated, the internal energy I_i, I_e and I_r can be obtained from

$$I_i = \frac{3kT_i}{2m_i}\,, \tag{B.18}$$

$$I_e = \frac{3kT_i}{2m_i}\,, \tag{B.19}$$

and

$$I_r = \frac{4\sigma T_r^4}{\rho c}\,. \tag{B.20}$$

Pressure, p_i, p_e and p_r, and the specific heat, C_{vi}, C_{ve} and C_{vr} are

$$p_i = \rho I_i\,, \tag{B.21}$$

$$p_e = \rho I_e\,, \tag{B.22}$$

$$p_r = \rho I_r\,, \tag{B.23}$$

$$C_{vi} = \frac{3}{2} \frac{k}{m_i}\,, \tag{B.24}$$

$$C_{ve} = \frac{3}{2} \frac{k}{m_i}\,, \tag{B.25}$$

$$C_{vr} = \frac{4I_r}{T_r}\,, \tag{B.26}$$

where

$$\alpha = \frac{e^2}{hc} = cons.\,, \tag{B.27}$$

and

$$e^2 = \hbar c\,, \tag{B.28}$$

with

$c =$ speed of light $= 3 \times 10^{10}$ cm/s

$\sigma =$ Stephan-Boltzmann constant,

$k =$ Boltzmann constant.

B.3 The Thermonuclear Reaction Rate

The thermonuclear reaction rates of deuterium, tritium, and lithium can be calculated from

$$\frac{\partial N_D}{\partial t} = -N_D^2 \left(\overline{\sigma v}\right)_{DD} - N_D N_T \left(\overline{\sigma v}\right)_{DT}, \tag{B.29}$$

$$\frac{\partial N_T}{\partial t} = -N_D N_T \left(\overline{\sigma v}\right)_{DT} + 0.25 N_D^2 \left(\overline{\sigma v}\right)_{DD}, \tag{B.30}$$

$$\frac{\partial N_{Li_3^6}}{\partial t} = -N_{n_0^1} N_{Li_3^6} \left(\overline{\sigma v}\right)_{nLi}, \tag{B.31}$$

and

$$\frac{\partial N_{Li_3^7}}{\partial t} = -N_{n_0^1} N_{Li_3^7} \left(\overline{\sigma v}\right)_{nLi}, \tag{B.32}$$

where the ion number density of deuterium, tritium, lithium-6, and lithium-7 is N_D, N_T, $N_{Li_3^6}$, and $N_{Li_3^7}$ respectively. σ is the cross section in barn, v is the ion velocity, $(\overline{\sigma v})$ is the average of the cross section multiplied by velocity over each ion. The analytical fits of the Maxwell average $(\overline{\sigma v})$ is provided by Tuck [B.5]. The *D-D*, *D-T* and *n-Li* reactions are indexed by *DD*, *DT* and *nLi* respectively. The 0.25 value in Eq. (B.30) is representative of the fact that 1/4 of *D-D* reactions will produce tritium. The highest rate at low temperatures is provided by the *D-T* reaction:

$$D_1^2 + T_1^3 \rightarrow \alpha(3.5\,\text{MeV}) + n(14.1\,\text{MeV}), \tag{B.33}$$

where the 17.6 MeV($= 2.82 \times 10^{-12}$ joule) reaction energy (the Q value) is partitioned between a 3.5 MeV alpha particle $\left(He_2^4\right)$ and a 14.1 MeV neutron. The *D-D* fusion reaction is characterized by two reactions of approximately equal probability:

$$D_1^2 + D_1^2 \rightarrow He_2^3(0.82\,\text{MeV}) + n_0^1(2.45\,\text{MeV}), \tag{B.34}$$

$$D_1^2 + D_1^2 \rightarrow T_1^3(1.01\,\text{MeV}) + p_1^1(3.02\,\text{MeV}). \tag{B.35}$$

The reaction products, He_2^3 and T_1^3, have a high probability of reaction with deuterium as they slow down:

$$(T_1^3)^* + D_1^2 \to He_2^4(3.5\,\text{MeV}) + n_0^1(14.1\,\text{MeV})\,, \tag{B.36}$$

$$(He_2^3)^* + D_1^2 \to He_2^4(3.6\,\text{MeV}) + p_1^1(14.7\,\text{MeV})\,, \tag{B.37}$$

where the asterisk denotes a non-thermal particle. In fact, the largest fraction of the overall Q-value is a result of such secondary reactions. The total reaction

$$6D_1^2 \to 2He_2^4 + 2p_1^1 + 2n_0^1\,, \tag{B.38}$$

has a Q-value of 43.25 MeV. This results in a specific yield of 345 MJ/mg for the D-D reaction compared with 339 MJ/mg for D-T.

Since tritium is a radioactive nuclide (with a half life of 12.3 years), it is not naturally occurring but rather must be produced artificially to fuel a fusion system. Fortunately, tritium can be produced quite easily using neutron capture reactions in lithium:

$$n_0^1 + Li_3^6 \to T_1^3 + He_2^4 + 4.8\,\text{MeV}\,, \tag{B.39}$$

$$n_0^1 + Li_3^7 \to T_1^3 + He_2^4 + n_0^1 - 2.4\,\text{MeV}\,. \tag{B.40}$$

The first process is a capture reaction involving neutrons of any energy. The second process is effectively an inelastic scattering reaction requiring a neutron energy in excess of the 2.4 MeV threshold. Since each D-T reaction produces one neutron, it is possible to breed tritium by surrounding the fusion reaction region with a lithium blanket.

At the very beginning of a time step, one can use Eq. (B.29) to calculate the decreased amount of N_D, assuming that all of the decreased deuterium ions participate in the reaction as described by Eqs. (B.33) and (B.38). Therefore, one will know the total amount of energy released from these reactions. Then, from Eq. (B.30), one can determine the decreased amount of tritium ions which will be used by Eq. (B.29) to determine the amount of the deuterium ions consumed by the reaction. Equations (B.31) and (B.39) illustrate the amount of the energy released from the Li_3^6 reaction. If Li_3^7 appears in the system, then Eqs. (B.32) and (B.40) can be used to determine the energy released from the Li_3^7 reaction. For a system starting from time $t = 0.0$ sec, the reaction given by Eq. (B.33) will bring the system to a high temperature and subsequently tritium will be produced from the reactions given by Eqs. (B.39) or (B.40). By adding Eqs. (B.33) and (B.39), the result is

$$D_1^2 + Li_3^6 \rightarrow 2He_2^4 + 22.4\,\text{MeV}\,. \tag{B.41}$$

If a system contains $Li_3^6D_1^2$, then one can assume 1/2 of the Li_3^6 ions reacted with neutrons as described by Eq. (B.39) and the other 1/2 reacted with deuterium as shown by Eq. (B.41). All of the tritium produced by Eq. (B.39) will be used by Eq. (B.33). The remaining deuterium ions will be used by Eq. (B.38). Therefore, the energy released from $Li_3^6D_1^2$ can be calculated by using Eqs. (B.33), (B.38), (B.39) and (B.41). If the system contains Li_3^7, then one will have to add Eq. (B.40).

The α particles and neutrons are transported in accordance with the Boltzmann equation and calculated by S_n method as described in Ref. [B.2] with a typical value of $n = 2$ for α particles and $n = 4$ for neutrons. The energy from neutron scattering is distributed to background ions. For the n-Li reaction, the deposited energy will be distributed to ions and electrons.

The artificial viscosity used in Eqs. (B.1) and (B.3) is obtained from

$$\begin{aligned} q &= 2\rho_{j+1/2}(u_{j+1} - u_j)^2, && \text{if } \dot{V} < 0, \text{ i.e. when it is in compression.} \\ &= 0, && \text{if } \dot{V} > 0, \text{ i.e. when it is in expansion.} \end{aligned}$$

B.4 Calculation Procedure

If one chooses the 2D Lagrangian method as described in Chapter 2 as the basic code for the thermonuclear burn calculation, then, the calculation procedures provided in Section 2.5 can be incorporated with the present non-equilibrium ion, electron, and photon computations. Based on the known dependent variable at time $t = t^n$, the following steps can be used:

Step 1. Calculate U_R, U_Z, V_R and V_Z.

Step 2. If the sliding interface treatment is required, then update $U^{n+1/2}, V^{n+1/2}, R^{n+1/2}$, and $Z^{n+1/2}$ due to the effects of sliding.

Step 3. Calculate ρ^{n+1} and ρ_t from R^n, Z^n, and M.

Step 4. Calculate G and Y from Chapter 8.

Step 5. Calculate $(S^{ij})^n$ from Eqs. (2.38) through (2.41) for pure elastic deformation.

Step 6. Check the von Mises yield condition.

Step 7. Calculate $(S^{ij})^n$ with Prandtl-Reuss plastic flow term, i.e. Eq. (2.42).

Step 8. Calculate equivalent plastic strain.

Step 9. Add the rigid body rotation correction terms to the deviatoric stresses.

Step 10. Calculate the principle stresses to check the fracture conditions.

Step 11. Calculate velocities $U^{n+1/2}$ and $V^{n+1/2}$ from Eqs. (2.46) and (2.47).

Step 12. Calculate the artificial viscosity ${}_1Q^n_{k,\ell}$, ${}_2Q^n_{k,\ell}$, ${}_3Q^n_{k,\ell}$ and ${}_4Q^n_{k,\ell}$.

Step 12.1. If the non-equilibrium ion, electron and photon calculations are required, then, one can use Eqs. (B.29)–(B.31) to obtain the thermonuclear reaction rates for deuterium, tritium and lithium. From the reaction rate, one can calculate the energy released from each reaction and therefore the energy sources will be known.

Step 12.2. Calculate the ion, electron and photon transports and depositions.

Note: For 2D problems, one should skip Steps 12.3–12.5 since they are used for solving 1D problems only.

Step 12.3. With the specific source rate $\dot{S}_{i,e}$ available, one can use Eqs. (B.3)–(B.5) to calculate the temperatures T_i, T_e, and T_r, i.e. $T_i = I_i/C_{vi}$, $T_e = I_e/C_{ve}$, and $T_r = I_r/C_{vr}$.

Step 12.4. Calculate the new pressure p^n using $p^n = p^n_e + p^n_i + p^n_r$ where $p^n_e = \rho I^n_e$, $p^n_i = \rho I^n_i$, and $p^n_r = \rho I^n_r$.

Step 12.5. For the non-equilibrium T_i, T_e, and T_r calculations, one can skip Steps 13 and 14.

Step 13. Calculate energy ε^{n+1} from Eq. (2.48).

Step 14. Calculate pressure P^n from the equation of state, i.e. $P = f(\rho, \varepsilon)$. In our case, it is a table lookup.

Steps 1–14 completes one-time-step calculations, and we return to Step 1 for the next time increment computations.

B.5 Radiation Transport Calculation

At Step 12.2, the transport of radiation, neutrons, and fast reaction products for 2D problems is accomplished by Monte Carlo simulation. The standard Monte Carlo scheme to radiation transport is highly inefficient in high opacity regions where photons have a high probability of absorption. The following method is used to improve the efficiency for 1D problems. The radiation transport equation is

$$\left(\frac{1}{c}\frac{\partial}{\partial t} + \vec{\Omega}\cdot\vec{\nabla} + \frac{1}{\lambda_a} + \frac{1}{\lambda_s}\right)\phi(\nu,\Omega) - \int A(\nu',\Omega' \to \nu,\Omega)\phi(\nu',\Omega')d\nu' d\Omega'$$
$$\equiv L(\phi) = S(\nu) = \frac{1}{\lambda_a}B(\nu)\,, \tag{B.42}$$

in which c is the speed of light, $\vec{\Omega}$ is a unit vector in the direction of the photon transported, $d\Omega$ is the differential solid angle around this direction, and ϕ is the radiation intensity (energy/time-area-frequency). The function $S(\nu)$ is the frequency ν dependent emission source. In fully ionized regimes, this is bremsstrahlung, which has the non-relativistic form

$$S(\nu) = \frac{4}{3\pi}\left(\frac{2}{\pi m_e k T_e}\right)^{1/2}\frac{\bar{Z}n_n^2\left(\bar{Z}^2\right)e^4}{\hbar c}(K_0 h)\left(\frac{h\nu}{2kT_e}\right)e^{-\frac{h\nu}{kT_e}}\,. \tag{B.43}$$

The unit of $S(\nu)$ is energy/time-volume-freq-steradian, and $B(\nu)$ is the Planckian distribution

$$B(\nu) = \frac{2h\nu^3}{c^2}\frac{1}{\left(e^{\frac{h\nu}{kT_e}} - 1\right)}\,, \tag{B.44}$$

which together define the absorption mean-free-path, corrected for stimulated emission, λ_a. Scattering is represented by the differential inverse Compton scattering length $A(\nu,\Omega \to \nu',\Omega')$, which is given analytically in Ref. [B.6]. A first order relativistic approximation to the total Compton scattering length is

$$\frac{1}{\lambda_a} \equiv \int A(\nu,\Omega \to \nu',\Omega')d\nu' d\Omega'$$
$$= \frac{8\pi}{3}\bar{Z}n_i r_0^2\left\{1 - \frac{2h\nu}{mc^2}\left[\frac{K_1(\xi)}{K_2(\xi)} + \frac{4}{\xi}\right]\right\}, \tag{B.45}$$

containing r_0, the classical electron radius, and $\xi = mc^2/kT_e$. The K's are Bessel function of the indicated order. The effects of the material motion in Eq. (B.42) are neglected, so the source and cross-sections are isotropic.

To establish a "reduced source", we note that the left hand side of Eq. (B.42) is a linear operator acting on ϕ. Thus, if ϕ is an approximation to the true solution ϕ, then the deviation $\delta = \phi - \phi_e$ from the estimate obey

$$L(\delta) = L(\phi) - L(\phi_e) = S - L(\phi_e) \equiv S_R\,, \tag{B.46}$$

which is Eq. (B.42), with $S(\nu)$ replaced by the reduced source S_R. Our procedure is to find a suitable estimated intensity ϕ_e by straightforward means and then to solve for the intensity deviation δ by Monte Carlo simulation. Since the number of Monte Carlo "particles" required for an accurate simulation is proportional to the square of the source intensity, considerable computation time can be saved by the transport of δ, in lieu of ϕ, if ϕ_e is good enough so $\delta \ll \phi_e = \phi$ and $S_R \ll S$.

The estimate ϕ_e is found by replacing the kernel in Eq. (B.42) by an isotropic kernel that produces no frequency change (i.e. coherent scattering): $A(\nu, \Omega \to \nu', \Omega') \to \delta(\nu - \nu')/4\pi\lambda_s$. Thus, for each frequency

$$\begin{aligned}\left[\frac{1}{c}\frac{\partial}{\partial t} + \mu\frac{\partial}{\partial r} + \frac{(1-\mu^2)}{r}\frac{\partial}{\partial \mu} + \frac{1}{\lambda_a} + \frac{1}{\lambda_s}\right]\phi_e - \frac{1}{2\lambda_s}\int_{-1}^{+1}\phi_e(\mu')d\mu'&\\ = S(\nu)&\\ = \frac{1}{\lambda_a}B(\nu)\,,&\end{aligned} \tag{B.47}$$

which is reduced to spherical coordinates. The variable $\mu = \cos\theta$, where θ is the angle between the radius vector (from the center of the pellet to a photon) and the velocity vector of the photon. This equation is solved by the "spherical harmonic" method [B.7], with truncation after the first Legendre Polynomial, i.e. with a "P_1 approximation". That is, the angular dependence is approximated by

$$\phi_e = \phi_0 + \phi_1\mu\,, \tag{B.48}$$

Equations for ϕ_0 and ϕ_1 are obtained by taking the zeroth and first moments of Eqs. (B.47)–(B.48) over μ; thus

$$\frac{1}{c}\frac{\partial\phi_0}{\partial t} + \frac{1}{3}\frac{\partial\phi_1}{\partial r} + \frac{2}{3}\frac{\phi_1}{r} + \frac{1}{\lambda_a}\phi_0 = \frac{1}{\lambda_a}B(\nu) \tag{B.49}$$

and

$$\frac{1}{c}\frac{\partial\phi_1}{\partial t} + \frac{\partial\phi_0}{\partial r} + \left(\frac{1}{\lambda_s} + \frac{1}{\lambda_a}\right)\phi_1 = 0\,. \tag{B.50}$$

These equations are solved, semi-analytically, for a set of frequency groups (typically thirty). For each new advancement in time, the time derivatives are replaced by their difference form, e.g.

$$\frac{\partial \phi_0}{\partial t} \to \frac{\phi_0^{n+1} - \phi_0^n}{\Delta t_n}, \tag{B.51}$$

ϕ_0^n is known, and the remaining terms are associated with the $n+1$ time level. The exact values for ϕ_0^{n+1} and ϕ_1^{n+1} are found analytically across each zone, assuming constants λ_s, λ_a, and $B(\nu)$ are within a zone. This relates the $\phi's$ at the zone boundaries, leading to exact solutions at these boundaries by the inversion of a quasi-diagonal matrix. With ϕ_e thus determined, the reduced source $S_R(\nu)$ for each frequency group is simply

$$\begin{aligned} S_R(\nu) = (1-\mu^2)\left\{\frac{1}{\lambda_a}[B(\nu)-\phi_0] - \frac{\phi_1}{r} - \frac{1}{c}\frac{\partial \phi_0}{\partial t}\right\} - \frac{\phi_0}{\lambda_s} \\ + \int \phi_0(\nu')a_0(\nu',\nu)d\nu' + \frac{2\mu}{3}\int \phi_1(\nu')a_1(\nu',\nu)d\nu' \end{aligned} \tag{B.52}$$

where

$$a_0(\nu',\nu) = \int A(\nu',\Omega' \to \nu,\Omega)d\Omega' \tag{B.53}$$

and

$$a_1(\nu',\nu) = \int A(\nu',\Omega' \to \nu,\Omega)\cos(\vec{\Omega}\cdot\vec{\Omega}')d\Omega'. \tag{B.54}$$

For Thomson scattering, $a_1 \to 0$ and $a_0 \to \delta(\nu-\nu')/\lambda_s$, so the 2 integral terms in Eq. (B.52) cancel. In high opacity regions, where $\ell/\lambda_a \ll 1$ (with ℓ the characteristic length of flow gradients), absorption dominates scattering, the radiation is nearly isotropic ($\phi_1 \ll \phi_0$), and slowly varying ($\partial\phi_0/\partial t \ll c\phi_0/\lambda_a$), so $S_R(\nu) \to (1-\mu^2)[B(\nu)-\phi_0]/\lambda_a$. Furthermore, by Eqs. (B.49) and (B.50), $\phi_0 \to B(\nu)$, so, indeed $S_R \ll S$, and $\delta \ll \phi_e$. In the $\rho = 3000\,\mathrm{g/cm^3}$, $T = 3\,\mathrm{keV}$ DT burn calculation, for example, $S_R/S \approx 0.1$, allowing 100 fewer Monte Carlo particles for an accurate simulation.

Given that the solution $\phi(\equiv \phi_e + \delta)$ to Eq. (B.47) is obtained by the above methods, there is energy exchange with the electron field such that in Eq. (B.4), the replacement

$$A_{er}(T_r - T_e) \to \frac{4\pi}{\rho}\int\left[\frac{\phi - B(\nu)}{\lambda_a} + C(\nu)\phi\right]d\nu \tag{B.55}$$

should be made. Here

$$C(\nu) = \int_0^\infty \frac{a_0(\nu',\nu)(\nu-\nu')}{\nu'}d\nu'. \tag{B.56}$$

The first term in Eq. (B.55) is the bremsstrahlung exchange, and the second is the Compton energy transfer. At each time step $t = t^{n+1}$, Eqs. (B.3), (B.4), (B.49), (B.50), and (B.55) are solved together implicitly, but with $\delta = \delta^n$ assumed for T_e^{n+1}, T_i^{n+1}, and ϕ_e^{n+1}. δ^{n+1} is then determined explicitly by the Monte Carlo scheme.

Regarding the transport of the fast products from nuclear reactions, we note that scatterings are treated stochastically. The original fast particle sources are the *D-D* and *D-T* thermonuclear reactions. Additional fast particles are created by nuclear scatterings and "in-flight" reactions with field ions.

B.6 Some Important Fission Reactions

Neutrons can be produced by the reaction of alpha particles on light elements such as beryllium, boron, or lithium. Alpha particles can be easily obtained from the radium or polonium emitter. The reaction of alpha particles with beryllium is

$$Be_4^9 + He_2^4 \rightarrow C_6^{12} + n_0^1 \tag{B.57}$$

Some of the reactions between neutrons and matter will emit a photon γ

$$H_1^1 + n_0^1 \rightarrow H_1^2 + \gamma \tag{B.58}$$

$$Cd_{48}^{113} + n_0^1 \rightarrow Cd_{48}^{114} + \gamma \tag{B.59}$$

$$In_{49}^{115} + n_0^1 \rightarrow In_{49}^{116} + \gamma \tag{B.60}$$

$$U_{92}^{238} + n_0^1 \rightarrow U_{92}^{239} + \gamma \tag{B.61}$$

The important fission reactions between neutrons and matter are

$$U_{92}^{235} + n_0^1 \rightarrow Ba_{56}^{137} + Kr_{36}^{97} + 2n_0^1 \tag{B.62}$$

which releases 193 MeV per fission. Approximately 97% of the released energy is deposited directly into the fuel material. The other important fission reaction is with plutonium-239 and neutrons, i.e.

$$Pu_{94}^{239} + n_0^1 \rightarrow Ba_{56}^{137} + Sr_{38}^{100} + 3n_0^1 \,. \tag{B.63}$$

Again, about 198.5 MeV energy is released for each fission.

References

B.1 Carlson, BG, Neutron diffusion theory-the transport approximation, Los Alamos Scientific Laboratory Report, LA-1061 (1950).

B.2 Duderstadt, JJ and Hamilton, LJ, Nuclear reactor analysis, Wiley, New York (1976).

B.3 Fraley, GS, Linnebur, EJ, Mason, RJ and Morse, RL, Thermonuclear burn characteristics of compressed deuterium-tritium microspheres, Los Alamos Scientific Laboratory Report, LA-5403-MS (1973).

B.4 Spitzer, L, Physics of fully ionized gases, Chapter V, Interscience, New York (1969).

B.5 Tuck, JL, Nuclear fusion, **1**, p. 202 (1961).

B.6 Fraley, GS, Los Alamos Scientific Laboratory Report, LA-4592 (1971).

B.7 Richtmeyer, RD and Morton, KW, Difference methods for initial value problems, Wiley, New York, Chapters 8 and 9 (1967).

APPENDIX C

HUGONIOT DATA

Hugoniot data fitted by the equation, $u_s = c_0 + su_p + qu_p^2$

Material	ρ_0 (g/cm^3)	c_0 (km/s)	s	q (s/km)	γ_0	Comments
Element						
Antimony	6.700	1.983	1.652		0.60	
Barium	3.705	0.700	1.600		0.55	Above $p = 115$ and $u_s = 2.54$
Beryllium	1.851	7.998	1.124		1.16	
Bismuth	9.836	1.826	1.473		1.10	
Cadmium	8.639	2.434	1.684		2.27	
Calcium	1.547	3.602	0.948		1.20	
Cesium	1.826	1.048	1.043	0.051	1.62	
Chromium	7.117	5.173	1.473		1.19	
Cobalt	8.820	4.752	1.315		1.97	
Copper	8.930	3.940	1.489		1.99	
Germanium	5.328	1.750	1.750		0.56	Above $p = 300$ and $u_s = 4.20$
Gold	19.240	3.056	1.572		2.97	
Hafnium	12.885	2.954	1.121		0.98	Below $p = 400$ and $u_s = 3.86$
Hafnium	12.885	2.453	1.353		0.98	Above transition
Indium	7.279	2.419	1.536		1.80	
Iridium	22.484	3.916	1.457		1.97	
Iron	7.850	3.574	1.920	−0.068	1.69	Above $u_s = 5.0$
Lead	11.350	2.051	1.460		2.77	

Material	ρ_0 (g/cm^3)	c_0 (km/s)	s	q (s/km)	γ_0	Comments
Lithium	0.530	4.645	1.133		0.81	
Magnesium	1.740	4.492	1.263		1.42	
Mercury	13.540	1.490	2.047		1.96	
Molybdenum	10.206	5.124	1.233		1.52	
Nickel	8.874	4.602	1.437		1.93	
Niobium	8.586	4.438	1.207		1.47	
Palladium	11.991	3.948	1.588		2.26	
Platinum	21.419	3.598	1.544		2.40	
Potassium	0.860	1.974	1.179		1.23	
Rhenium	21.021	4.184	1.367		2.44	
Rhodium	12.428	4.807	1.376		1.88	
Rubidium	1.530	1.134	1.272		1.06	
Silver	10.490	3.229	1.596		2.38	
Sodium	0.968	2.629	1.223		1.17	
Strontium	2.628	1.700	1.230		0.41	Above $p = 150$ and $u_s = 3.63$
Sulfur	2.020	3.223	0.959			
Tantalum	16.654	3.414	1.201		1.60	
Thallium	11.840	1.862	1.523		2.25	
Elements						
Thorium	11.680	2.133	1.263		1.26	
Tin	7.287	2.608	1.486		2.11	
Titanium	4.528	5.220	0.767		1.09	Below $p = 175$ and $u_s = 5.74$
Titanium	4.528	4.877	1.049		1.09	Above transition
Tungsten	19.224	4.029	1.237		1.54	
Uranium	18.950	2.487	2.200		1.56	
Vanadium	6.100	5.077	1.201		1.29	
Zinc	7.138	3.005	1.581		1.96	
Zirconium	6.505	3.757	1.018		1.09	Below $p = 260$ and $u_s = 4.63$
Zirconium	6.505	3.296	1.271		1.09	Above transition

Material	ρ_0 (g/cm^3)	c_0 (km/s)	s	q (s/km)	γ_0	Comments
Alloys						
Brass	8.450	3.726	1.434		2.04	
2024 Aluminum	2.785	5.328	1.338		2.00	
921-T Aluminum	2.833	5.041	1.420		2.10	
Lithium-Magnesium Alloy	1.403	4.247	1.284		1.45	
Magnesium Alloy	1.775	4.516	1.256		1.43	
Stainless Steel	7.896	4.569	1.490		2.17	
U-3 wt%Mo	18.450	2.565	2.20		2.03	
Synthetics						
Adiprene	0.927	2.332	1.536		1.48	
Epoxy resin	1.186	2.730	1.493		1.13	Below $p = 240$ and $u_s = 7.0$
Epoxy resin	1.186	3.234	1.255		1.13	Above transition
Lucite	1.181	2.260	1.816		0.75	
Neoprene	1.439	2.785	1.419		1.39	
Nylon	1.140	2.570	1.849	0.081	1.07	
Paraffin	0.918	2.908	1.560		1.18	
Phenoxy	1.178	2.266	1.698		0.55	
Plexiglas	1.186	2.598	1.516		0.97	
Polyethylene	0.915	2.901	1.481		1.64	
Polyrubber	1.010	0.852	1.865		1.50	
Polystyrene	1.044	2.746	1.319		1.18	
Polyurethane	1.265	2.486	1.577		1.55	Below $p = 220$ and $u_s = 6.5$
Silastic (RTV-521)	1.372	0.218	2.694	−0.208	1.40	
Teflon	2.153	1.841	1.707		0.59	
Compounds						
Periclase (MgO)	3.585	6.597	1.369		1.32	Above $p = 220$ and $u_s = 7.45$
Quartz	2.204	0.794	1.695		0.90	Stishovite Above $P = 400$

Material	ρ_0 (g/cm^3)	c_0 (km/s)	s	q (s/km)	γ_0	Comments
Sodium chloride	2.165	3.528	1.343		1.60	Transition ignored
Water	0.998	1.647	1.921	0.096		

Notations

u_s the shock velocity (km/s)
u_p particle velocity (km/s)
c_0 material dependent constant (km/s)

AUTHOR INDEX

www.ingramcontent.com/pod-product-compliance
Ingram Content Group UK Ltd.
Pitfield, Milton Keynes, MK11 3LW, UK
UKHW020401150926
8624IPUK00004B/277

9 789812 566232